Application of Data-Driven Method for HVAC System

Application of Data-Driven Method for HVAC System

Editors

Yabin Guo
Zhanwei Wang
Yunpeng Hu
João M. M. Gomes

Basel • Beijing • Wuhan • Barcelona • Belgrade • Novi Sad • Cluj • Manchester

Editors
Yabin Guo
School of Civil Engineering
Zhengzhou University
Zhengzhou
China

Zhanwei Wang
Institute of Building Energy
and Thermal Science
Henan University of Science
and Technology
Luoyang
China

Yunpeng Hu
Department of Building
Environment and Energy
Engineering
Wuhan Business University
Wuhan
China

João M. M. Gomes
Electrical Engineering
University of Algarve
Faro
Portugal

Editorial Office
MDPI
St. Alban-Anlage 66
4052 Basel, Switzerland

This is a reprint of articles from the Special Issue published online in the open access journal *Processes* (ISSN 2227-9717) (available at: www.mdpi.com/journal/processes/special_issues/hvac).

For citation purposes, cite each article independently as indicated on the article page online and as indicated below:

Lastname, A.A.; Lastname, B.B. Article Title. *Journal Name* **Year**, *Volume Number*, Page Range.

ISBN 978-3-0365-9691-4 (Hbk)
ISBN 978-3-0365-9690-7 (PDF)
doi.org/10.3390/books978-3-0365-9690-7

Contents

About the Editors . . . . **vii**

Preface . . . . **ix**

Yabin Guo, Yaxin Liu, Zhanwei Wang and Yunpeng Hu
Application of Data-Driven Methods for Heating Ventilation and Air Conditioning Systems
Reprinted from: *Processes* **2023**, *11*, 3133, doi:10.3390/pr11113133 . . . . **1**

Xi Chen, Yahui Gao, Liu Yang, Yang Liu, Miaomiao Qin and Jialing Xia et al.
An Experimental Study on Temperature, Relative Humidity, and Concentrations of CO and
CO_2 during Different Cooking Procedures
Reprinted from: *Processes* **2023**, *11*, 2648, doi:10.3390/pr11092648 . . . . **5**

Zhenlan Dou, Lu Jin, Yinhui Chen and Zishuo Huang
Optimization of Cost–Carbon Reduction–Technology Solution for Existing Office Parks Based
on Genetic Algorithm
Reprinted from: *Processes* **2023**, *11*, 2452, doi:10.3390/pr11082452 . . . . **19**

Junjie Jin, Peiyao Duan, Yu Liu, Honglin Chen and Tingting Yu
Experimental Study on Convection and Heat Conduction Heating of an Air-Conditioned Bed
System under Winter Lunch Break Mode
Reprinted from: *Processes* **2023**, *11*, 2391, doi:10.3390/pr11082391 . . . . **41**

Zhanwei Wang, Boyang Liang, Jingjing Guo, Lin Wang, Yingying Tan and Xiuzhen Li et al.
Fault Diagnosis Based on Fusion of Residuals and Data for Chillers
Reprinted from: *Processes* **2023**, *11*, 2323, doi:10.3390/pr11082323 . . . . **61**

Shuizhong Zhao, Jiangfeng Si, Gang Chen, Hong Shi, Yusong Lei and Zhaoyang Xu et al.
Research on Passive Design Strategies for Low-Carbon Substations in Different Climate Zones
Reprinted from: *Processes* **2023**, *11*, 1814, doi:10.3390/pr11061814 . . . . **82**

Yabin Guo, Jiangyan Liu, Changhai Liu, Jiayin Zhu, Jifu Lu and Yuduo Li
Operation Pattern Recognition of the Refrigeration, Heating and Hot Water Combined
Air-Conditioning System in Building Based on Clustering Method
Reprinted from: *Processes* **2023**, *11*, 812, doi:10.3390/pr11030812 . . . . **105**

Weilin Li, Yonghui Liang, Jianli Wang, Zhenhe Lin, Rufei Li and Yu Tang
Peak Load Shifting Control for a Rural Home Hotel Cluster Based on Power Load Characteristic
Analysis
Reprinted from: *Processes* **2023**, *11*, 682, doi:10.3390/pr11030682 . . . . **124**

Lanbo Lai, Xiaolin Wang, Gholamreza Kefayati and Eric Hu
Cooling and Water Production in a Hybrid Desiccant M-Cycle Evaporative Cooling System with
HDH Desalination: A Comparison of Operational Modes
Reprinted from: *Processes* **2023**, *11*, 611, doi:10.3390/pr11020611 . . . . **144**

Xi Zhao, Cheng Li, Jiayin Zhu, Yu Chen and Jifu Lu
Data Analysis and Optimization of Thermal Environment in Underground Commercial
Building in Zhengzhou, China
Reprinted from: *Processes* **2022**, *10*, 2584, doi:10.3390/pr10122584 . . . . **169**

Ruixin Li, Gaoyi Liu, Yuanli Xia, Olga L. Bantserova, Weilin Li and Jiayin Zhu
Pollution Dispersion and Predicting Infection Risks in Mobile Public Toilets Based on
Measurement and Simulation Data of Indoor Environment
Reprinted from: *Processes* **2022**, *10*, 2466, doi:10.3390/pr10112466 . . . . **194**

Yan Wang, Changnv Zeng and Chaoxin Hu
Thermal Performance and Energy Conservation Effect of Grain Bin Walls Incorporating PCM
in Different Ecological Areas of China
Reprinted from: *Processes* **2022**, *10*, 2360, doi:10.3390/pr10112360 . . . . **213**

Martim L. Aguiar, Pedro D. Gaspar, Pedro D. Silva, Luísa C. Domingues and David M. Silva
Real-Time Temperature and Humidity Measurements during the Short-Range Distribution of
Perishable Food Products as a Tool for Supply-Chain Energy Improvements
Reprinted from: *Processes* **2022**, *10*, 2286, doi:10.3390/pr10112286 . . . . **230**

Guangtao Fan, Haoran Chang, Chenkai Sang, Yibo Chen, Baisong Ning and Changhai Liu
Evaluating Indoor Carbon Dioxide Concentration and Ventilation Rate of Research Student
Offices in Chinese Universities: A Case Study
Reprinted from: *Processes* **2022**, *10*, 1434, doi:10.3390/pr10081434 . . . . **245**

About the Editors

Yabin Guo

Guo Yabin, PhD, graduated from Huazhong University of Science and Technology. He is a senior member of the Chinese Association of Refrigeration. He hosted and participated in multiple national and horizontal projects, and he has published over 30 academic papers both domestically and internationally. Among them, he has published 12 SCI indexed papers, 1 EI indexed paper, and 3 international conference papers. The first inventor authorizes one patent. He participated in the compilation of the book "Refrigeration and Air Conditioning Encounters Big Data - Industry Revolution". The main research directions include: fault detection and diagnosis of air conditioning systems based on artificial intelligence methods; optimization of control for heat pump air conditioning systems; data mining and energy-saving optimization of building air conditioning systems.

Zhanwei Wang

Wang Zhanwei graduated with a PhD from Xi'an University of Architecture and Technology. The main research direction is the development and application of energy-saving and intelligent fault diagnosis technology in building energy systems. I have led one National Science Foundation project, one sub project, and four provincial and ministerial level projects. I have also published 17 scientific research papers as the first (corresponding) author, including 8 papers indexed by SCI; applied for 6 national invention patents; and granted 3 authorizations.

Yunpeng Hu

Hu Yunpeng graduated from Huazhong University of Science and Technology. He hosted one project funded by the Hubei Provincial Natural Science Foundation and one project funded by the Hubei Provincial Excellent Young and Middle-aged Science and Technology Innovation Team in Higher Education Institutions. More than 20 of his academic papers have been published. The main research directions are fault detection of air conditioning systems, optimization of building energy efficiency, etc.

João M. M. Gomes

João Gomes graduated with a PhD from University of Algarve. He has published, as co-author, 9 scientific research articles in indexed journals and more than 40 articles in conference proceedings and book chapters in the fields of building energy conservation, HVAC control systems, among others. The current research directions include building energy savings, fault detection for electrical energy systems, HVAC control systems and renewable energies.

Preface

Artificial intelligence has become a hot topic in current research. Various industries are using data-driven methods to solve the problems they face. The HVAC industry is an important and long-standing industry with considerable relevance to everyday life and industrial production. Moreover, as the operation of HVAC systems consumes a large amount of energy, it is of great significance to utilize data-driven methods to improve the operational reliability and energy-saving effects of HVAC systems.

The data-driven approach is currently mainly applied in various aspects of HVAC systems, such as fault detection and diagnosis, energy consumption prediction, predictive control, and data mining. The application of data-driven methods in the field of HVAC is gradually shifting from simple models to complex models and from shallow models to deep models. Gradually, the study of the transferability, interpretability, and other aspects of the model are being studied. From current research results, it can be seen that data-driven methods have shown great potential in HVAC applications.

It is a great honor to be invited by MDPI Publishing, which has received support from dozens of peer authors and published multiple high-quality and very interesting research articles, to organize a Special Issue on "Application of Data-Driven Method for HVAC System". Once again, we would like to express our gratitude to MDPI.

Yabin Guo, Zhanwei Wang, Yunpeng Hu, and João M. M. Gomes
Editors

Editorial

Application of Data-Driven Methods for Heating Ventilation and Air Conditioning Systems

Yabin Guo [1,*], **Yaxin Liu** [1], **Zhanwei Wang** [2,*] **and Yunpeng Hu** [3,*]

[1] School of Civil Engineering, Zhengzhou University, Zhengzhou 450001, China; yaxin_liu@gs.zzu.edu.cn
[2] Institute of Building Energy and Thermal Science, Henan University of Science and Technology, Luoyang 471000, China
[3] Department of Building Environment and Energy Engineering, Wuhan Business University, Wuhan 430010, China
* Correspondence: guoyb_h@zzu.edu.cn (Y.G.); wzhanwei@haust.edu.cn (Z.W.); yunpenghu@wbu.edu.cn (Y.H.)

1. Introduction

At present, with the continuous global energy crisis, buildings, as a significant factor in energy consumption, have significant importance in achieving the energy-saving operation of buildings [1]. In buildings, the main energy-consuming pieces of equipment are heating ventilation and air conditioning (HVAC) systems. However, some of the limitations that HVAC systems currently face, such as control optimization [2], pattern recognition [3], and fault diagnosis [4], are difficult to solve effectively using traditional approaches. There is an urgent need for new technologies and methods to provide valid solutions. Data-driven methods utilize machine learning and artificial intelligence technology, from a data perspective, to construct a new solution with the ability to overcome the limitations of physical models [5,6]. Data-driven methods have been widely applied in fields such as image recognition [7] and speech recognition [8], and have already become a research hotspot in the field of HVAC systems [9,10]. The application of data-driven methods in the field of HVAC systems is gradually diversifying, and the depth of research is also gradually increasing.

This Special Issue, "Application of Data-Driven Methods for HVAC Systems", is a collection of 13 interesting articles mainly covering three aspects: building environment data analysis, HVAC system control, and the application of data-driven methods. We are deeply honored to serve as guest editors and extend our heartfelt gratitude to all the authors who have contributed to this Special Issue. The papers authored by our colleagues will make substantial contributions to the application of data-driven methods within the field of HVAC systems.

2. Data Analysis in Experimental Research in Built Environments

Some studies have obtained data through conducting experiments related to HVAC systems. Thereby, some valuable conclusions have been drawn through various data analysis methods. Real-time measurements and simulations of the indoor thermal environment have been conducted. Indoor environmental data have thereby been obtained. The CFD method has been used to analyze internal pollutant diffusion patterns and concentration changes. Finally, regression analysis was used to study the relationship between the indoor thermal environment and aerosol diffusion paths. Moreover, an experiment has been conducted on an air-conditioned bed system. Experimental data on two heating forms, heat conduction-dominated and heat convection-dominated heating, were collected. After analyzing the experimental data, the results indicate that a higher temperature range is required when using convective heating. At the same time, convective heating temperature increases rapidly. However, it is easy to develop a dry atmosphere in the long term. Finally,

Citation: Guo, Y.; Liu, Y.; Wang, Z.; Hu, Y. Application of Data-Driven Methods for Heating Ventilation and Air Conditioning Systems. *Processes* **2023**, *11*, 3133. https://doi.org/10.3390/pr11113133

Received: 8 October 2023
Revised: 27 October 2023
Accepted: 31 October 2023
Published: 2 November 2023

it was concluded that heat conduction heating can bring better thermal comfort and higher energy efficiency.

A series of experiments have been conducted on the indoor environmental quality in kitchens. The indoor temperature, relative humidity, and CO and CO_2 concentrations in kitchens were measured under conditions of different cooking temperature. After an analysis of the experimental data, the results indicate that the heat and gas consumed during the cooking process are closely related to the temperature and CO and CO_2 concentration. In addition, natural wind has a significant impact on various indoor parameters. Experiments on the internal environment of granaries have been conducted. The latent heat and phase change temperature of the prefabricated phase change material (PCM) warehouse wall were optimized through numerical simulation. By analyzing experimental data, the thermal regulation performance of prefabricated panels on the grain warehouse wall was optimized. The results indicate that the application of PCM in granaries has advantages. Aguiar et al. [11] conducted real-time measurement and collection of temperature and relative humidity parameters during the transportation of corrosion-prone products. Optimizing parameters will help decision making in logistics and route management, as well as in the diagnosis and timely prevention of food losses. A total of 18 parameters regarding temperature and humidity were collected during the experiment, and the results indicate that cargo monitoring is of great significance for quality control and energy efficiency optimization in the supply chain.

3. Research on the Control of HVAC Systems

One of the major applications for data analysis of HVAC systems is system control and optimization. Basic information and electricity load data from a typical tourist village were collected. The power load characteristics of heating, cooling, and transition seasons have been studied. A cluster control conversion system using phase change energy storage was proposed through data analysis. The system's control logic has been determined and established. Finally, the collected power load data were introduced into the model for practical case analysis, and its feasibility and effectiveness were verified. DEST software was used to study the effects of building envelope thermal parameters, the window-to-wall ratio, and the shape factor on the total energy consumption of buildings in different climate zones. A sensitivity analysis was conducted on different parameters to determine a passive design scheme suitable for substations in different climate zones. The analysis results also indicate that among the thermal parameters associated with the building envelope, the thickness of the roof insulation has the greatest impact on the energy consumption of substation buildings. The conclusions drawn from this study can offer architects valuable strategies and suggestions for energy saving in substations in different climate zones, and provide a reference for building energy-saving designs and selecting appropriate air-conditioning and heating equipment.

A mixed desiccant M-circulating cooling system with an HDH unit has been proposed for the simultaneous cooling and production of water. An analysis was conducted on the refrigeration and water production performance of the system for three typical operating modes. The indoor air temperature, relative humidity, and CO_2 concentration data were measured over a 4-week period. At the same time, the number of indoor residents, room occupancy time, and window opening were recorded. The results indicate that opening windows can effectively reduce the indoor carbon dioxide concentration, thus improving indoor air quality. However, at the same time, it is also necessary to pay attention to the impact of outdoor pollutants on the indoor environment.

4. Applications Based on Data-Driven Methods

Data driven methods are widely used in fault diagnosis, pattern recognition, scheme optimization, and data analysis research for HVAC systems [12,13]. Wang et al. [14] proposed a fault diagnosis model based on Bayesian networks. The proposed approach achieves the fusion of feature residuals and feature data in a single fault diagnosis model.

The proposed hybrid method improves the fault diagnosis performance of chillers, with significant improvements in the diagnosis of refrigerant leakage faults and lubricating oil faults. An unsupervised clustering method for studying the running mode recognition of refrigeration, heating, and combined hot water–air conditioning systems is proposed. K-means, Gaussian mixed model clustering, and spectral clustering are three data-driven methods used to establish pattern recognition models for air-conditioning systems. A correlation analysis was used for the reduction of dimensionality in characteristic variables. The results indicate that the clustering model can identify defrosting patterns of air conditioning. The accuracy rate of pattern recognition reached 98.99%.

The indoor thermal environment and air quality of underground commercial buildings have been studied from a functional perspective. An optimal control strategy for airflow organization was proposed. This study provides a theoretical basis for the creation of a thermal environment and the organization and control of airflow in underground commercial buildings. The relationship between office park costs, energy consumption, and carbon emissions based on genetic algorithms was analyzed. A mathematical optimization model for the carbon reduction transformation of existing office parks has been established with the goal of reducing carbon throughout the entire life cycle of office parks. The results indicate that the established model can provide a comprehensive and optimized energy allocation plan to minimize carbon emissions at various investment costs.

5. Conclusions

The articles in this Special Issue include a series of studies from the perspective of data application. A new research approach has been introduced to HVAC systems, and some new strategies and models have been proposed, yielding valuable insights for improving performance and achieving energy-saving operations in HVAC systems. Furthermore, these results will also play a significant role in promoting the application of data-driven methods in field of HVAC systems.

Author Contributions: Y.G.: Methodology, Writing—Review and Editing. Y.L.: Writing—Review and Editing. Z.W.: Supervision. Y.H.: Supervision. All authors have read and agreed to the published version of the manuscript.

Funding: This research received no external funding.

Conflicts of Interest: The authors declare no conflict of interest.

References

1. Zhang, F.; Saeed, N.; Sadeghian, P. Deep learning in fault detection and diagnosis of building HVAC systems: A systematic review with meta analysis. *Energy AI* **2023**, *12*, 100235. [CrossRef]
2. Gunay, B.; Hobson, B.W.; Darwazeh, D.; Bursill, J. Estimating energy savings from HVAC controls fault correction through inverse greybox model-based virtual metering. *Energy Build.* **2023**, *282*, 112806. [CrossRef]
3. Fan, C.; He, W.; Liu, Y.; Xue, P.; Zhao, Y. A novel image-based transfer learning framework for cross-domain HVAC fault diagnosis: From multi-source data integration to knowledge sharing strategies. *Energy Build.* **2022**, *262*, 111995. [CrossRef]
4. Chen, J.; Zhang, L.; Li, Y.; Shi, Y.; Gao, X.; Hu, Y. A review of computing-based automated fault detection and diagnosis of heating, ventilation and air conditioning systems. *Renew. Sustain. Energy Rev.* **2022**, *161*, 112395. [CrossRef]
5. Yan, Y.; Cai, J.; Tang, Y.; Chen, L. Fault diagnosis of HVAC AHUs based on a BP-MTN classifier. *Build. Environ.* **2023**, *227*, 109779. [CrossRef]
6. Li, G.; Wang, L.; Shen, L.; Chen, L.; Cheng, H.; Xu, C.; Li, F. Interpretation of convolutional neural network-based building HVAC fault diagnosis model using improved layer-wise relevance propagation. *Energy Build.* **2023**, *286*, 112949. [CrossRef]
7. Bensaoud, A.; Kalita, J. Deep multi-task learning for malware image classification. *J. Inf. Secur. Appl.* **2022**, *64*, 103057. [CrossRef]
8. Martinez, A.M.C.; Mallidi, S.H.; Meyer, B.T. On the relevance of auditory-based Gabor features for deep learning in robust speech recognition. *Comput. Speech Lang.* **2017**, *45*, 21–38. [CrossRef]
9. Pinto, G.; Wang, Z.; Roy, A.; Hong, T.; Capozzoli, A. Transfer learning for smart buildings: A critical review of algorithms, applications, and future perspectives. *Adv. Appl. Energy* **2022**, *5*, 100084. [CrossRef]
10. Aguilera, J.J.; Meesenburg, W.; Ommen, T.; Markussen, W.B.; Poulsen, J.L.; Zühlsdorf, B.; Elmegaard, B. A review of common faults in large-scale heat pumps. *Renew. Sustain. Energy Rev.* **2022**, *168*, 112826. [CrossRef]

11. Aguiar, M.L.; Gaspar, P.D.; Silva, P.D.; Domingues, L.C.; Silva, D.M. Real-Time Temperature and Humidity Measurements during the Short-Range Distribution of Perishable Food Products as a Tool for Supply-Chain Energy Improvements. *Processes* **2022**, *10*, 2286. [CrossRef]
12. Singh, V.; Mathur, J.; Bhatia, A. A comprehensive review: Fault detection, diagnostics, prognostics, and fault modeling in HVAC systems. *Int. J. Refrig.* **2022**, *144*, 283–295. [CrossRef]
13. Bellanco, I.; Fuentes, E.; Vallès, M.; Salom, J. A review of the fault behavior of heat pumps and measurements, detection and diagnosis methods including virtual sensors. *J. Build. Eng.* **2021**, *39*, 102254. [CrossRef]
14. Wang, Z.; Liang, B.; Guo, J.; Wang, L.; Tan, Y.; Li, X.; Zhou, S. Fault Diagnosis Based on Fusion of Residuals and Data for Chillers. *Processes* **2023**, *11*, 2323. [CrossRef]

Article

An Experimental Study on Temperature, Relative Humidity, and Concentrations of CO and CO_2 during Different Cooking Procedures

Xi Chen *, Yahui Gao, Liu Yang, Yang Liu, Miaomiao Qin, Jialing Xia and Peng Wang

School of Civil Engineering and Architecture, Henan University of Technology, Zhengzhou 450001, China
* Correspondence: chenxi2014@haut.edu.cn; Tel.: +86-18623719927

Abstract: In order to explore the indoor air quality during different cooking procedures, a very common kitchen in China is selected for experimental research. An indoor air quality meter is used to measure the temperature, relative humidity, and CO and CO_2 concentrations of the indoor air above the stove when people cook four different dishes under different ventilation patterns in the kitchen. The results indicate that the heat and gas consumed during cooking are closely related to the temperature and concentrations of CO and CO_2. Some cooking procedures such as boiling water are related to the indoor air temperature and relative humidity in the kitchen. In addition, in kitchens without mechanical ventilation, natural ventilation shows a more significant positive effect on controlling temperature, relative humidity, and concentrations of CO and CO_2 during cooking procedures.

Keywords: temperature; relative humidity; CO concentration; CO_2 concentration; cooking procedure

check for **updates**

Citation: Chen, X.; Gao, Y.; Yang, L.; Liu, Y.; Qin, M.; Xia, J.; Wang, P. An Experimental Study on Temperature, Relative Humidity, and Concentrations of CO and CO_2 during Different Cooking Procedures. *Processes* **2023**, *11*, 2648. https://doi.org/10.3390/pr11092648

Academic Editors: Yabin Guo, João M. M. Gomes, Zhanwei Wang and Yunpeng Hu

Received: 8 August 2023
Revised: 2 September 2023
Accepted: 3 September 2023
Published: 4 September 2023

1. Introduction

In recent years, indoor air quality (IAQ) has received increasing attention as people recognize the importance of air quality to health and comfort. Great interest has been directed towards indoor obnoxious gases such as CO and CO_2. Kitchens are environments wherein multiple pollutants and airflow combine with the emissions from cooking operations to create considerable indoor air pollution. Cooking plays an important role in people's lives, especially in China with a population of 1.4 billion. The statistics show that Chinese housewives spend an average of 3.4–4 h in kitchens every day [1]. However, the cooking process produces a large amount of particles, which are inhaled by the human body and deposited in the lungs, thereby affecting human health [2]. Therefore, indoor air pollution from the cooking process deserves more attention.

Many investigators have explored the air quality of kitchen environments during cooking. For example, Chiang et al. [3] found that the accumulation location of air contaminants was highly correlated with the location of gas fires. Furthermore, Ryhl Svendsen et al. [4] found through experimental measurements that people in rural areas of Denmark were exposed to a certain level of indoor air pollution. In addition, Singh et al. [5] found that the concentration of pollutants in the kitchen after cooking was higher than the recommended guidelines. Following this work, another study found that kitchen type, fuel type, and kitchen location were also important factors affecting the concentration of pollutants in the kitchen [6]. Torkmahalleh et al. [7] studied the neurological responses caused by human exposure to pollutants from indoor gas ovens by recording electroencephalograms before, during (14 min), and 30 min after cooking. The study suggested that Alzheimer's disease might develop in people chronically exposed to high levels of cooking pollutants. Zhao et al. [8] reported the air temperature, air relative humidity, and generations of CO and CO_2 during the cooking of Eight Cuisines of China as a case study. They pointed out that cooking techniques were not responsible for the serious pollution. The air temperature and humidity were the main factors affecting thermal comfort [9]. Zhou et al. [10] found that the area between the gas stove and the exhaust fan observed the highest temperature when cooking. The results of other researchers indicated that the indoor environment of Chinese-style residential kitchens was too hot in summer [11,12]. Moreover, the questionnaire survey on the Tianjin campus showed that 83.33% of chefs felt hot in summer, but most chefs had neutral wet sensations in four seasons [13]. Liu et al. [14] found that heat diffused significantly after the oil was heated for 30 s. The air temperature around the cook increased by about 10.0 °C throughout the cooking process. Moreover, during the cooking process, high levels of indoor air pollution including CO and CO_2 often accumulate in the kitchen. Exposure to high levels of CO could affect the cardiovascular system, lungs, blood, and central nervous system as well as lead to various health problems [15,16]. Increased levels of CO_2 in the atmosphere stimulated respiration and caused all sorts of systemic confusion [17]. Although many articles have studied the generation of pollutants in the cooking process and the impact of related pollutants on the human body, there are few studies on the temperature, relative humidity, and CO and CO_2 concentrations in the cooking process of different cooking procedures.

Ventilation systems are commonly used in kitchens to remove secure contaminants and provide a comfortable and healthy environment. Related studies have shown that the closer the exhaust outlet of the kitchen range hood is to the pollution source, the higher the efficiency of pollutant removal [18]. Li et al. [19] evaluated the impact of typical ventilation systems of four commercial kitchens in China on their indoor thermal environment and found that the ventilation systems of typical commercial kitchens in China were unable to effectively remove waste heat and impurities. The impact of increased air exchange rates in the test kitchen was discussed by Grabow et al. [20]. Their results indicated that opening the door and window in the test kitchen lowered particulate matter and CO. Afterwards, another research showed that the push–pull ventilation system could provide scientific support for reducing the indoor pollutant concentration in the kitchen [21]. Nejat et al. [22] improved the performance of a two-sided windcatcher under low wind speed conditions by installing anti-short-circuit devices on them. Recently, Lu et al. [23] explored a new way of obtaining make-up air based on a push–pull ventilation system in the kitchen. The experimental results showed that it could be used to create more efficient air distribution and a good environment in the kitchen. Although a lot of work has been conducted on kitchen ventilation, few researchers have studied the air quality during different cooking processes in the kitchen with different ventilation patterns.

This study reports the results of an experimental investigation of the air quality during cooking in a Chinese residential kitchen. Four cooking procedures (boiling, steaming, sautéing, and frying) and different ventilation patterns generate 24 different case scenarios. A TSI 7545 IAQ-Calc Indoor Air Quality Meter is used to monitor the air temperature, relative humidity, and concentrations of CO and CO_2 above the stove.

2. Materials and Methods

2.1. Site Description

The characteristics of Chinese cooking result in the production of large amounts of pollutants. Most Chinese people are exposed to the environment with CO and CO_2 when they are cooking in residential kitchens [24]. This experiment was carried out in May, 2023 in a typical Chinese residential kitchen located on the sixth floor of a 6-floor building, which is far from the streets and in a community of Zhengzhou city. The kitchen is part of a 90 m^2 family apartment with a length of 4440 mm, a width of 1700 mm, and a height of 2600 mm, as shown in Figure 1. The kitchen has an exterior window on the north wall, and the dimension of the window is width × height = 1700 mm × 1550 mm. Opposite the exterior window, there is an interior door on the south wall, and the size of the door is 650 mm × 2000 mm (width × height). In addition to this, the kitchen also has a two-head natural gas stove, which is placed on the hearth. Above the stove, there is an exhaust hood used in the present study. The locations as well as parameters of the stove and the exhaust hood are presented in Figure 2 and Table 1, respectively. In this research, only the left head of the stove is adopted for cooking.

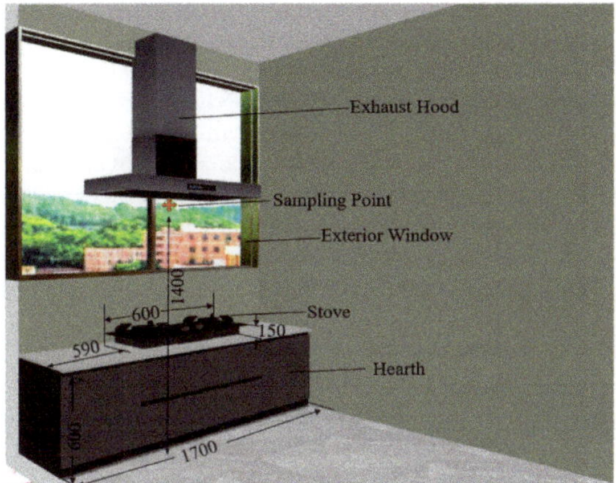

Figure 1. Schematic of the kitchen.

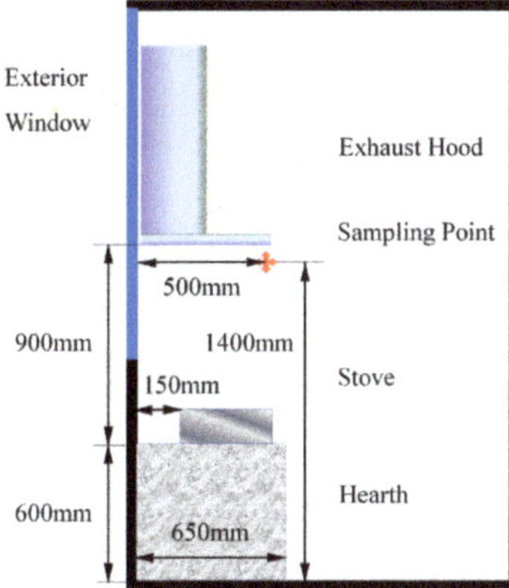

Figure 2. Schematic of the Sampling Point.

Table 1. Parameters of the stove and the exhaust hood.

Name	Model	Size (mm)	Parameters
Two-head natural gas stove	JZT-Q30A (12T)	$710 \times 400 \times 150$	Rated pressure: 2000 Pa
			Weight: 8.1 kg
			Rated heat load: 4.1 kW $\times$ 2
Exhaust hood	CXW-219-JT30 (SN)	$900 \times 520 \times 880$	Maximum pressure: $\geq$300 Pa
			Weight: 30 kg
			Volume flow rate: 0.25 m^3/s

2.2. Studied Cases

The apartment residents involved in this study suffered from variations in the temperature, humidity, and concentrations of CO and CO_2 during the cooking process, which made the residents uncomfortable. At the same time, the apartment residents were affected differently by different cooking processes. A significant contribution to indoor air pollution is made by different styles of cooking operations. Therefore, boiling, steaming, sautéing, and frying, which are very common cooking procedures and have different characteristics, are considered in this work to detect the air quality during different cooking procedures. Four dishes, namely boiled eggs, steamed perch, sautéed mungbean sprouts, and fried chicken middle wings, are chosen to represent the four cooking procedures. The details of these four dishes are listed in Table 2. Exhaust hoods are often used in residential kitchens to ensure the provision of a healthy environment. Opening windows and doors in kitchens is another choice for some people to make improvements to indoor environments during cooking. For a better understanding of the effects of ventilation on the indoor quality in kitchens, various on–off patterns of the interior door, exterior window, and exhaust hood are set in different cases. This study is performed under 24 cases with different cooking procedures as well as ventilation patterns, and the cases are listed in Table 3.

Table 2. The information on the four dishes.

Names	Pictures	Average Consumed Gas (m^3)	Cooking Procedures
Boiled eggs		0.020	Preparation: Put 2 eggs into 425 g water and soaked for 10 min.
			Step 1: Turned on the stove to medium-high and brought to the water boil.
			Step 2: Turned the heat to low and boiled the eggs for 6 min.
			Step 3: Turned off the stove and dished out the eggs.
Steamed perch		0.096	Preparation: Rinsed the perch (450 g) and marinated with salt, scallion, ginger, and light soy sauce for 30 min.
			Step 1: Turned on the stove to high and brought 1200 g of water to a boil.
			Step 2: Put the perch above the water and steamed for 8 min.
			Step 3: Turned off the stove and left the perch in the steamer for a period of time.
Sautéed mungbean sprout		0.023	Step 1: Turned on the stove to medium-high and made the wok hot.
			Step 2: Added 10 g cooking oil, 2 chili peppers, and some scallion into the wok.
			Step 3: When the oil smoked, added 375 g of mungbean sprout. Turned the heat up to high and performed a few quick stirs.
			Step 4: Added some salt as well as light soy sauce and kept on stirring.
			Step 5: Turned off the stove. Added some vinegar and dished out the mungbean sprout.

Table 2. *Cont.*

Names	Pictures	Average Consumed Gas (m^3)	Cooking Procedures
Fried chicken middle wings		0.060	Preparation: Rinsed 15 chicken middle wings (550 g) and marinated with salt, scallion, ginger, garlic, white pepper, and light soy sauce for 30 min.
			Step 1: Turned on the stove to medium-high and made the wok hot.
			Step 2: Added some cooking oil, Sichuan pepper, and star aniseed.
			Step 3: When the oil was heat, turned the heat to low and put in 4 chicken middle wings. Fried the chicken middle wings until golden brown and dished out. (Repeated 3 times.)
			Step 4: Put in 3 chicken middle wings. Fried the chicken middle wings until golden brown and dished out.
			Step 5: Turned off the stove. Added some cumin powder and coriander to the chicken middle wings.

Table 3. Parameters in different cases.

Cases	Dishes	Interior Door	Exterior Window	Exhaust Hood
1	Boiled eggs	off *	off	off
2	Boiled eggs	off	off	on ***
3	Boiled eggs	off	on **	off
4	Boiled eggs	on	on	off
5	Boiled eggs	on	off	on
6	Boiled eggs	off	on	on
7	Steamed perch	off	off	off
8	Steamed perch	off	off	on
9	Steamed perch	off	on	off
10	Steamed perch	on	on	off
11	Steamed perch	on	off	on
12	Steamed perch	off	on	on
13	Sautéed mungbean sprout	off	off	off
14	Sautéed mungbean sprout	off	off	on
15	Sautéed mungbean sprout	off	on	off
16	Sautéed mungbean sprout	on	on	off
17	Sautéed mungbean sprout	on	off	on
18	Sautéed mungbean sprout	off	on	on
19	Fried chicken middle wings	off	off	off
20	Fried chicken middle wings	off	off	on
21	Fried chicken middle wings	off	on	off
22	Fried chicken middle wings	on	on	off
23	Fried chicken middle wings	on	off	on
24	Fried chicken middle wings	off	on	on

* The interior door is closed completely, and there is a crack with a height of 6mm under the door. ** The exterior window is opened to the maximum. *** The exhaust hood is turned on to the high-speed level when the stove is switched on. And 5 min after the stove is turned off, the exhaust hood is switched off.

2.3. Apparatus and Parameters Measured

In the present study, a TSI 7545 IAQ-Calc Indoor Air Quality Meter is selected for monitoring the air temperature, relative humidity, and concentrations of CO and CO_2 at the sampling point above the stove during the cooking procedures. The TSI 7545 IAQ-Calc Indoor Air Quality Meter is widely accepted and has been used in the field of indoor measurements by several authors [25,26]. More information about the TSI 7545 IAQ-Calc Indoor Air Quality Meter can be seen in Table 4 [27]. The sampling point is set in the breathing zone, which is 1400 mm above the ground, in the middle of the two heads of the stove, and 500 mm away from the exterior window, as shown in Figure 2.

Table 4. Specifications of the Indoor Air Quality Meter.

Parameters	Range	Accuracy	Resolution
Temperature	0–60 °C	±0.6 °C	0.1 °C
Relative humidity	5–95% RH	±3% RH	0.1% RH
CO concentration	0–500 ppm	±3% of reading or ±3 ppm, whichever is greater	0.1 ppm
CO_2 concentration	0–5000 ppm	±3% of reading or ±50 ppm, whichever is greater	1 ppm

2.4. Procedure

In this work, the interior door and the exterior window are opened at least 2 h before each measurement to keep the air fresh in the kitchen [28]. The doors of other rooms are closed during the experiment to minimize the impact of other rooms on the kitchen's air quality. When preparations are completed, the TSI 7545 IAQ-Calc Indoor Air Quality Meter starts running, and the operator leaves the kitchen. The Indoor Air Quality Meter tests the initial background levels of the air temperature, relative humidity, and concentrations of CO and CO_2 for 10 min before the stove is turned on. Then, these parameters at the sampling point are continuously measured for 50 min. During the measurements, the data are automatically recorded every 1 s. There are no other people except the cook in the kitchen during the cooking. The cook should stay as far away from the sampling point as possible. The distance of the cook from the sampling point is about 0.5–1 m in different cases. Meanwhile, the cook is asked to wear a mask. In order to reduce errors, the experiments are repeated twice for each case. The purpose of the second test is to verify the trend and correctness of the data in the first test. If the law of the data in the two tests is close, the values presented in this paper use the data obtained from the first test; otherwise, the tests are carried out for the third time or even more.

3. Results and Discussion

3.1. Temperature

In order to reduce the effect of the airflow on the results, the cases with a closed door and window and an idle exhaust hood are chosen to compare the data during the cooking with different procedures. Figure 3 depicts the air temperatures versus time at the sampling point in Cases 1, 7, 13, and 19 with a closed interior door, exterior window, and exhaust hood. The steps in these cases are also marked in Figure 3. We can observe that the air temperatures increase initially and then decrease in all cases. During cooking, the temperatures are obviously above the comfort level without ventilation measures, and productivity will drop according to the research of Wyon [29]. The maximum temperatures in Cases 1, 7, 13, and 19 all appear at 600–1500 s. However, the specific moments and steps with the maximum temperatures are different. The case with the maximum peak temperature is not the one with the highest initial background temperature, which mostly has to do with the cooking process. And the gaps between the maximum and the initial background temperatures increase with an increase in the average consumed gas. It follows that the temperature increment is in connection with the consumed gas during the cooking. It can also be demonstrated that the temperatures dramatically increase after the stove is turned on and decrease as soon as the stove is turned off. This agrees with the result of a previous study in which cooking makes the air temperature increase in commercial kitchens [19]. Thus, the increase in temperatures during the cooking is mainly due to the burning fire. At 60 min (3600 s), the air temperatures are still greater than the initial background values, which is mainly related to the absence of ventilation measures. In addition, it can be seen the temperature is turned to be low after reaching the peak value of 39.7 °C in Case 1. Also, the temperature decreases obviously after Step 2 is carried out. The air temperature in Case 7 continuously rises until the stove is turned off. The comparison of Cases 1 and 7 shows that reducing heat can lower the temperature in the kitchen. In Case 13, a sharp increase in the temperature appears after the stove is turned on. Then, the temperature keeps rising until the stove is turned off. The slope of the temperature curve between Step 1 and Step 3 is observed to be slightly larger than that between Step 3 and Step 5. A possible reason is that the procedures of making the wok hot and oil smoked contribute to the increase in the temperature. The temperature in Case 19 sharply increases between Step 1 and Step 3. When the heat is turned to be low, the temperature initially decreases and then increases slightly till the stove is turned off. This phenomenon indicates that the change in temperature is directly affected by the heat of the stove. A comparison of the data in different cases shows that the temperatures in Cases 1 and 7 increase slowly before the values reach the maximum, whereas the temperatures in Cases 13 and 19 rise sharply. The difference between the cooking procedures (such as boiling water and making the wok hot and oil smoked) is possibly sufficient to trigger this result.

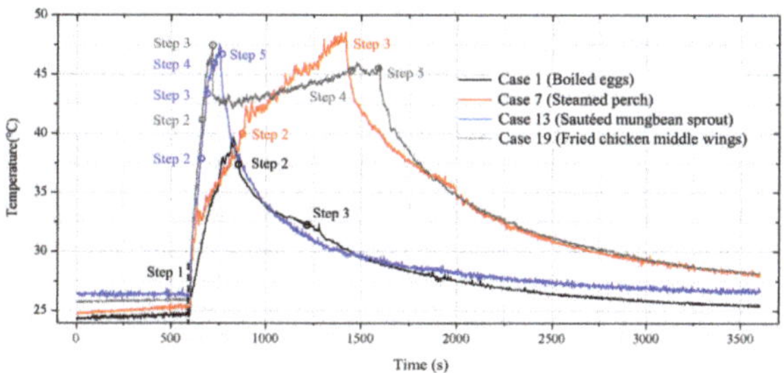

Figure 3. Variations in air temperatures in Cases 1, 7, 13, and 19.

In Figure 4, the temperatures at the sampling point against time in cases with the dish of sautéed mungbean sprout are presented. As it is seen, the air temperatures all increase as soon as the fire is lighted and decline when the stove is turned off. The temperatures come up to the maximum value before Step 5 is performed. This may be caused by a habit of putting the lid on the wok before the stove is turned off. Putting the lid on the wok and turning off the stove happen almost simultaneously. The gaps between the peak and the initial background temperatures are ordered from high to low, starting with Case 15 then Case 13 and Case 16. There is little difference between Cases 13 and 15. The reason may be that only opening the exterior window has a lesser impact on the airflow and temperature control. The interior door and exterior window opened simultaneously, which could promote the airflow in the kitchen and may have lowered the peak temperature to some degree. The maximum temperatures in cases with an idle exhaust hood all exceed 42 °C, while the peak values of temperatures in cases with the working exhaust hood fall below 33 °C. This implies that the working exhaust hood can effectively reduce the temperature in the kitchen during cooking. The gaps between the maximum and the initial background temperatures in cases with the working exhaust hood ordered from high to low are Case 17, then Case 18, and Case 14. This means that the open door or window has little effect on the temperature drop when the exhaust hood keeps working. And the open window shows a better effect on the cooling of heated air than the open door. The larger distance between the interior door and the stove as well as the greater air velocity outdoors possibly leads to this result. It can also be concluded that the temperatures in cases with an idle exhaust hood increase firstly, reach the peak values, and then decrease when the time is increased. However, temperatures in cases with the working exhaust hood present a different trend and increase a bit after the exhaust hood is switched off. One of the possible causes is that the heated air above the stove can be taken away effectively by the exhaust hood.

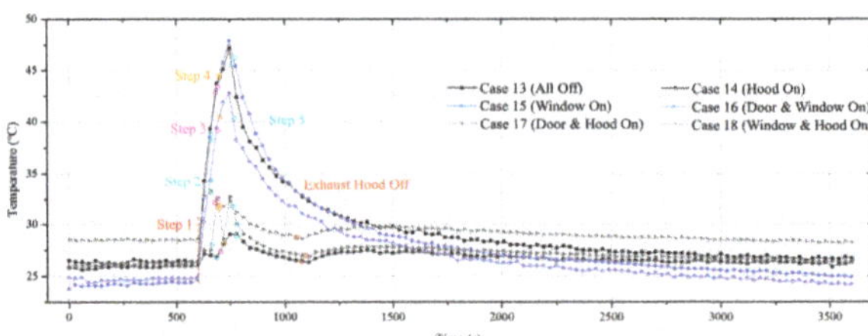

Figure 4. Variations in air temperatures while cooking sautéed mungbean sprout.

3.2. Relative Humidity

The variation in the relative humidity in Cases 1, 7, 13, and 19 with a closed interior door, exterior window, and exhaust hood is plotted in Figure 5. As soon as the stove is turned on, the relative humidity rises in all of these cases. A similar result was obtained in the study by Giwa et al. [30]. The relative humidity has a similar trend to the temperature. We also find that the relative humidity is comparatively volatile during the cooking procedures. In Case 1, after turning on the stove, the relative humidity increases first and then decreases. The relative humidity comes up to a small peak value before the heat is turned to low, which is related to the large amount of steam from the boiling water. While the eggs are being boiled, the relative humidity stays at a higher level. After the stove is turned off, the relative humidity gradually decreases and then tends to be stable. It proves that the relative humidity above the stove is related to the water status in the saucepan. The relative humidity in Case 7 has a similar tendency to that in Case 1. One reason for this could be that Cases 1 and 7 both have the process of boiling water and keep some water in the vessels throughout the cooking time. The difference is that the ascensional range of relative humidity in Case 7 is higher than that in Case 1 after the stove is turned off, which may be due to the lag of dishing out the perch. For Case 13, the relative humidity rises initially and then declines with fluctuations after the fire is lighted. After the stove is turned off, the relative humidity rises a bit and keeps steady later. The relative humidity in Case 19 fluctuates regularly from Step 3 to Step 5, and a reduction in relative humidity appears after every time the chicken middle wings are dished out. It is also shown that the gaps between the maximum and the initial relative humidity in Cases 1 and 7 are greater than those in Cases 13 and 19. One of the possible causes is that Cases 1 and 7 have the same procedure of boiling water, while Cases 13 and 19 do not have this step. By comparing the values in Figure 5 and Table 2, it can be concluded that the average consumed gas has little influence on the relative humidity.

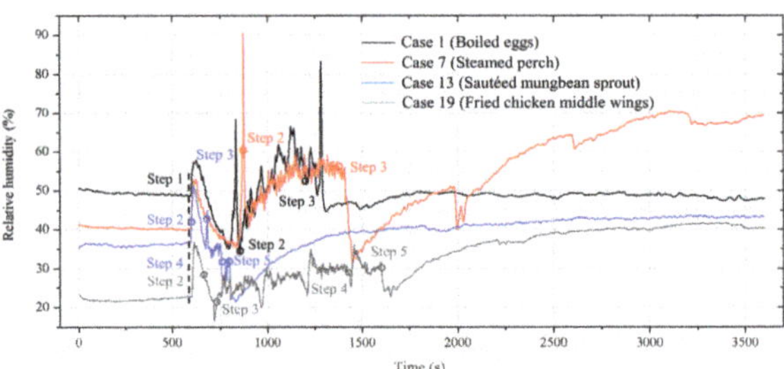

Figure 5. Variations in the relative humidity in Cases 1, 7, 13, and 19.

Figure 6 illustrates the relative humidity versus time at the sampling point in cases with the dish of boiled eggs. The relative humidity in these cases with the maximum peak value of 65.6% increases rarely. After the 1500 s, the relative humidity tends to be stable and is close to the initial background level. In cases with an idle exhaust hood, the relative humidity increases as soon as the stove is switched on and goes down after the stove is turned off. The relative humidity in cases with the working exhaust hood declines when the stove is turned on and has no obvious changes as soon as the stove is turned off. Also, Li et al. [19] found that the relative humidity was reduced due to the use of exhaust hoods during the cooking process in commercial kitchens. The relative humidity in Cases 1, 3, and 4 exhibits a larger fluctuation than that in Cases 2, 5, and 6 during cooking. It seems that the differences in the relative humidity curves between the cases with an idle and working exhaust hood are caused by the vapor, which can be exhausted in a timely manner by the hood. The comparison in Figure 6 also shows that the gaps between the maximum and the initial background relative humidity in Case 1 are the highest, followed by Case 3, with Case 4 the lowest among the cases with an idle exhaust hood. This result indicates that appropriate natural ventilation is beneficial to control the increase in relative humidity during cooking. For the cases with a working exhaust hood, the minimum values of the relative humidity during cooking are lower than the corresponding initial background relative humidity. The gaps between the initial background and the minimum relative humidity among cases with the working exhaust hood appear from large to small in the following order: Case 5, Case 6, and Case 2. This order implies that an open door or window can make the relative humidity change more dramatically during cooking when the exhaust hood keeps working.

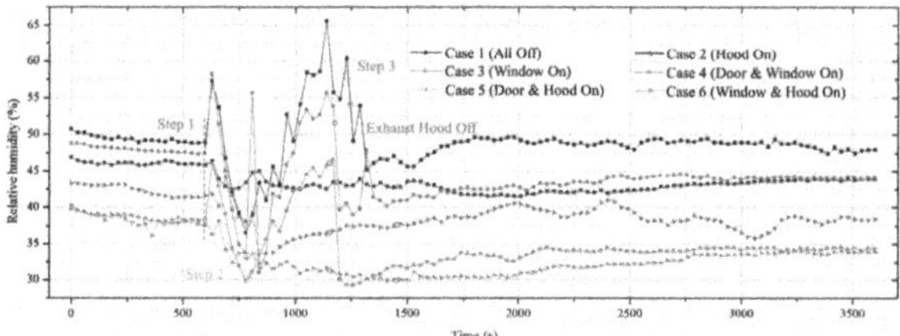

Figure 6. Variations in the relative humidity while cooking boiled eggs.

3.3. Concentration of Carbon Monoxide

Figure 7 displays the variation in the concentrations of CO in Cases 1, 7, 13, and 19 with a closed interior door, exterior window, and exhaust hood. According to the results, the concentrations of CO in all cases exhibit a sharp increase after the fire is lit and stay at a high level during the cooking, which is in consonance with some previous research [31]. The CO concentrations decrease with a delay when the stove is turned off. The acceptable value of the Indoor Air Quality Standard in China is 10 ppm for the CO concentration, and the concentrations of CO in all of these four cases are always below this value [32]. In Case 1, the CO concentration increases quickly between Step 1 and Step 2 and fluctuates smoothly after the heat is turned to low. When the stove is switched off, the concentration of CO shows a rapid decrease with a lag. An obvious increasing trend of CO concentration in Case 7 is observed after the stove is turned on. The value of the CO concentration achieves a peak value of 8.4 ppm and then decreases. Though the process of boiling water exists both in Cases 1 and 7, the continuous increase in the CO concentration shown in Case 7 is absent in Case 1. A possible cause is that there is a step to lowering the heat in Case 1. With a similar trend of CO concentration in Case 7, the concentration of CO in Case 13 shows a shorter rising process and a smaller maximum value. These results can be explained by assuming that there is no step of heat regulation in both cases and that the time between turning on and off the stove is longer in Case 7 than that in Case 13. The variation trend of CO concentration in Case 19 is similar to that in Case 1, which can be attributed to the step to lowering the heat in both cases. This series of observations suggests that the emission of CO is mainly related to the burning of the gas, and the CO concentration presents a smooth fluctuation when the heat is turned to low. In Figure 7, it is also recognized that the

CO concentration in Case 7 has the maximum peak value and is about double the initial background level at 3600 s. Part of the reason might be due to the largest consumed gas volume used in Case 7. At 3600 s, the CO concentration in Case 1 is the only one that approaches the corresponding initial background value. One reason for this could be that the average consumed gas of boiled eggs is the lowest, and the accumulative CO in the kitchen can be exhausted in a timely manner after the stove is turned off.

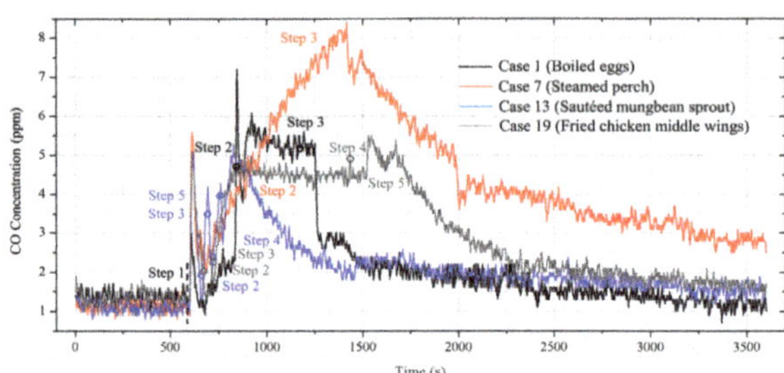

Figure 7. Variations in the concentrations of CO in Cases 1, 7, 13, and 19.

The concentrations of CO at the sampling point against time in cases with the dish of steamed perch are illustrated in Figure 8. It can be seen that CO concentrations are all lower than the acceptable value of 10 ppm. During the cooking procedures, the CO concentrations in cases with an idle exhaust hood are apparently higher than those in cases with a working exhaust hood. It appears that the working exhaust hood plays an important role in reducing indoor CO concentration. By comparing the data in cases with an idle exhaust hood, the CO concentrations in Cases 7 and 9 both increase first and then decrease, while those in Case 10 have a greater fluctuation and no obvious regularity, which bears similarity to the study of the fluctuation indices in chemistry [33,34]. This may be attributed to the stronger convection when the door and window are opened simultaneously. At 3600 s, the concentrations of CO in Cases 7 and 9 are higher and close to the initial background values, whereas those in Case 10 are lower than the initial background CO concentrations. This could mean that natural ventilation can promote a decrease in CO concentration and that cross-ventilation is more effective. The dramatic impact of increasing ventilation by opening the window and door on lowering the level of CO can also be seen in the study by Grabow et al. [20]. Moreover, the above results suggest that the effect of controlling CO concentration is more susceptible to the weather outside when the exterior window is opened. For the cases with a working exhaust hood, CO concentrations increase slightly during cooking and are even lower than the initial background values when the exhaust hood is turned off. After the exhaust hood stops working, the CO concentrations increase rarely and then reach stable values. The CO concentrations at 3600 s are nearly equal to the initial background level. The peak values of the CO concentrations from highest to lowest order the cases with a working exhaust hood in the sequence of Case 11, Case 12, and Case 8, which has a similar condition with the temperature pattern. It may be that natural ventilation has little contribution to reducing CO concentration when the exhaust hood keeps working.

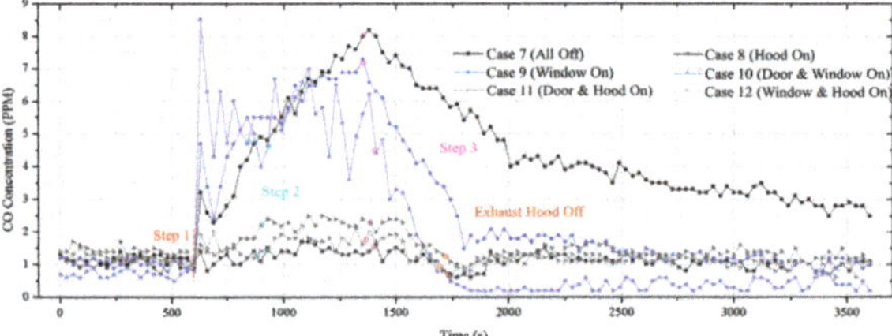

Figure 8. Variations in concentrations of CO while cooking the steamed perch.

3.4. Concentration of Carbon Dioxide

Figure 9 shows the concentrations of CO_2 versus time in Cases 1, 7, 13, and 19 with a closed interior door, exterior window, and exhaust hood. From Figure 9, it can be observed that the CO_2 concentration rises sharply once the fire is lit, which is similar to the findings of Zhao et al. [8]. Subsequently, the CO_2 concentrations gradually decline after the stove is turned off. Different from the results of the CO concentration, the CO_2 concentrations in these four cases rapidly exceed the acceptable value of the Indoor Air Quality Standard in China (1000 ppm) [32] when the stove is turned on and always higher than 1000ppm during cooking. These data imply that CO_2 concentrations have an adverse impact on the air quality in the kitchen during cooking without ventilation. The CO_2 concentration in Case 1 presents a decrease as soon as the stove is turned to low and shows a stationary fluctuation until the stove is turned off. This suggests that the magnitude of the heat is closely associated with the change in the CO_2 concentration. In Case 7, the CO_2 concentration strikes the peak value of 5000 ppm many times and drops slowly after the stove is turned off. The CO_2 concentration in Case 13 grows rapidly before the stove is turned off and decreases sharply later. The CO_2 concentration in Case 19 exhibits an obvious decrease after the heat is turned to be low and then rises gradually. When the stove is turned off, the CO_2 concentration declines. A comparison of the data for different cases shows that the CO_2 concentration with the maximum value of 5000 ppm in Case 7, which uses the most gas, stays at a high level for the longest time among these cases and that the CO_2 concentration in Case 1 with the least consumed gas has the lowest peak value. After 3600 s, the CO_2 concentration in Case 1 is the only one that is lower than the acceptable level. These results suggest that the emission of CO_2 has a positive association with the consumed gas.

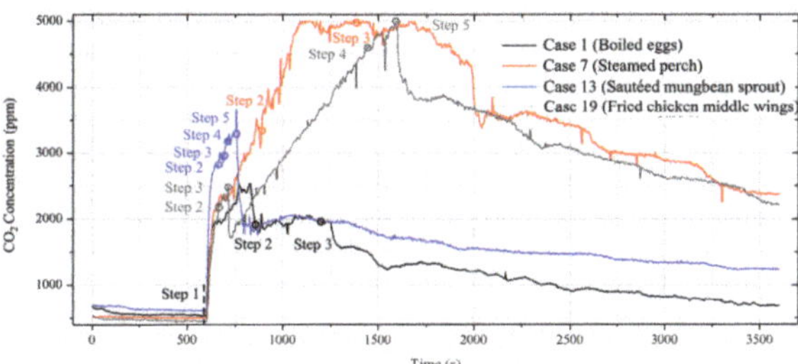

Figure 9. Variations in concentrations of CO_2 in Cases 1, 7, 13, and 19.

The concentrations of CO_2 at the sampling point against time in cases with the dish of fried chicken middle wings are presented graphically in Figure 10. The CO_2 concentrations in cases with a working exhaust hood are significantly lower than those in cases with an idle exhaust hood during cooking. This finding is consistent with the results reported by Zhao et al. [8]. Also, the CO_2 concentrations in cases with an idle exhaust hood all exceed the acceptable value of 1000 ppm when the cooking is carried out in the kitchen, whereas most of the CO_2 concentrations in cases with a working exhaust hood are below the acceptable level. The CO_2 concentrations in cases with a working exhaust hood fluctuate around a constant level during the cooking procedures and drop to the initial background values before the exhaust hood is turned off. These indicate that the exhaust hood is a valid tool for reducing the CO_2 generated from cooking and that the CO_2 concentration is basically lower than the acceptable level when the exhaust hood is working. For cases with an idle exhaust hood, the maximum CO_2 concentrations from large to small order the cases by ventilation patterns: Case 19, Case 21, and Case 22. The CO_2 concentration in Case 19 is still higher than 2000 ppm at 3600 s. Until about 3000 s from the beginning, the CO_2 concentration in Case 21 dips below 1000 ppm. The CO_2 concentration in Case 22 is relatively stable between Step 3 and Step 5, lower than 1000ppm at about 1650 s, and close to the initial background concentration at about 1700 s from the beginning. These data suggest that natural ventilation measures contribute to the elimination of CO_2 during cooking and that the effect is limited. If the exhaust hood keeps working, the case dependence of the gap between the maximum CO_2 concentration and the initial background value appears from large to small in the following order: Case 23, Case 24, and Case 20. It seems that the measure of opening the door or window merely cannot make the effectiveness of the CO_2 elimination better when the exhaust hood is turned on.

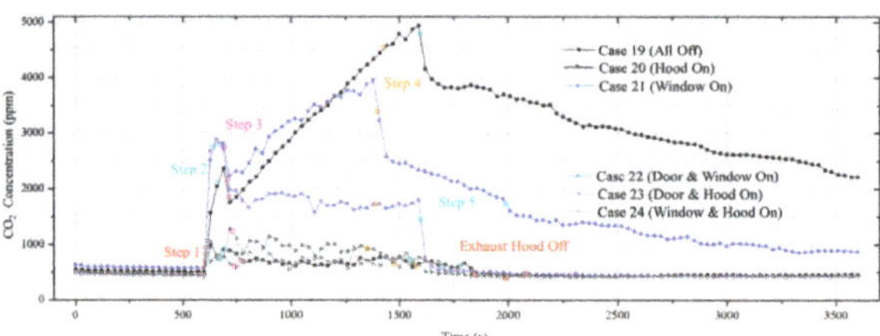

Figure 10. Variations in concentrations of CO_2 while cooking the fried chicken middle wings.

Based on the above discussions, we suggest that chefs should be aware of the impact of the kitchen environment on their own health during the cooking process and keep ventilation devices (such as an exhaust hood) working to maintain good indoor air quality. In addition, wearing a mask during cooking activities is also a way to protect people's health.

4. Conclusions

In the present work, the air temperature, relative humidity, and concentrations of CO and CO_2 during cooking procedures are measured in a Chinese residential kitchen. The influences of four cooking procedures and different ventilation patterns in the kitchen on these parameters are discussed. According to the data generated from the experiment and the above analysis, the main conclusions are as follows:

(1) When the stove is on, the temperature and CO and CO_2 emissions increase, and when the stove is turned off, the temperature and CO and CO_2 emissions decrease. The heat used in the cooking is closely related to the temperature and concentrations of CO and CO_2.

(2) With the reduction in gas consumption, the temperature and CO and CO_2 emissions during the cooking are reduced, and the relative humidity varies irregularly.

(3) Some cooking procedures, such as boiling water, are related to changes in the air temperature and relative humidity in the kitchen.

(4) The temperature and concentrations of CO and CO_2 are maintained at a high level in the kitchen without ventilation measures, and the working exhaust hood can effectively reduce the temperature and concentrations of CO and CO_2 during cooking.

(5) Natural ventilation, especially cross-ventilation, has a positive effect on controlling the air temperature, relative humidity, and concentrations of CO and CO_2 in kitchens with idle exhaust hoods.

(6) Opening the interior door or the exterior window only has no significant impact on the control of air temperature, relative humidity, and concentrations of CO and CO_2 during cooking in kitchens with working exhaust hoods.

The current research has some limitations. This paper does not discuss all cooking methods and dishes and only selects four dishes. Furthermore, other pollutants such as COF produced during different cooking procedures have not been studied. In the future, we will upgrade related instruments to measure other pollutants in the kitchen. In addition, the indoor air quality during some other cooking procedures is also well worth investigating.

Author Contributions: Conceptualization, X.C.; methodology, X.C. and Y.G.; formal analysis, Y.G., L.Y. and Y.L.; writing—original draft preparation, X.C., Y.G. and M.Q.; writing—review and editing, X.C., J.X. and P.W.; funding acquisition, X.C. and L.Y. All authors have read and agreed to the published version of the manuscript.

Funding: This research was funded by the National Natural Science Foundation of China (no. 51708180); the Key Scientific Research Project for Universities in Henan, China (no. 23A560010); the Youth Backbone Teacher Training Program of Henan University of Technology (no. 21420099); and the funding support from The Department of Science and Technology of Henan Province, China (no. 222103810076 and no. 232102111126).

Data Availability Statement: The data presented in this study are available on request from the corresponding author.

Conflicts of Interest: The authors declare no conflict of interest.

References

1. Lai, C. Assessment of side exhaust system for residential kitchens in Taiwan. *Build. Serv. Eng. Res. Technol.* **2005**, *26*, 157–166. [CrossRef]
2. Vu, T.V.; Ondracek, J.; Zdímal, V.; Schwarz, J.; Delgado-Saborit, J.M.; Harrison, R.M. Physical properties and lung deposition of particles emitted from five major indoor sources. *Air Qual. Atmos. Health* **2016**, *10*, 1–14. [CrossRef]
3. Chiang, C.; Lai, C.; Chou, P.; Li, Y. The influence of an architectural design alternative (transoms) on indoor air environment in conventional kitchens in Taiwan. *Build. Environ.* **2000**, *35*, 579–585. [CrossRef]
4. Ryhl-Svendsen, M.; Clausen, G.; Chowdhury, Z.; Smith, K.R. Fine particles and carbon monoxide from wood burning in 17th–19th century Danish kitchens: Measurements at two reconstructed farm houses at the Lejre Historical–Archaeological Experimental Center. *Atmos. Environ.* **2010**, *44*, 735–744. [CrossRef]
5. Singh, A.; Chandrasekharan Nair, K.; Kamal, R.; Bihari, V.; Gupta, M.K.; Mudiam, M.K.R.; Satyanarayana, G.N.V.; Raj, A.; Haq, I.; Shukla, N.K.; et al. Assessing hazardous risks of indoor airborne polycyclic aromatic hydrocarbons in the kitchen and its association with lung functions and urinary PAH metabolites in kitchen workers. *Clin. Chim. Acta* **2016**, *452*, 204–213. [CrossRef]
6. Sidhu, M.K.; Ravindra, K.; Mor, S.; John, S. Household air pollution from various types of rural kitchens and its exposure assessment. *Sci. Total Environ.* **2017**, *586*, 419–429. [CrossRef] [PubMed]
7. Amouei Torkmahalleh, M.; Naseri, M.; Nurzhan, S.; Gabdrashova, R.; Bekezhankyzy, Z.; Gimnkhan, A.; Malekipirbazari, M.; Jouzizadeh, M.; Tabesh, M.; Farrokhi, H.; et al. Human exposure to aerosol from indoor gas stove cooking and the resulting nervous system responses. *Indoor Air* **2022**, *32*, e12983. [CrossRef] [PubMed]
8. Zhao, Y.; Li, A.; Gao, R.; Tao, P.; Shen, J. Measurement of temperature, relative humidity and concentrations of CO, CO_2 and TVOC during cooking typical Chinese dishes. *Energy Build.* **2014**, *69*, 544–561. [CrossRef]
9. Andrey, L. The Effect of Supply Air Systems on Kitchen Thermal Environment. Available online: https://www.researchgate.net/publication/242516657_The_Effect_of_Supply_Air_Systems_on_Kitchen_Thermal_Environment (accessed on 1 July 2023).
10. Zhou, J.; Kim, C.N. Numerical Investigation of Indoor CO_2 Concentration Distribution in an Apartment. *Indoor Built Environ.* **2010**, *20*, 91–100. [CrossRef]
11. Liu, S.; Cao, Q.; Zhao, X.; Lu, Z.; Deng, Z.; Dong, J.; Lin, X.; Qing, K.; Zhang, W.; Chen, Q. Improving indoor air quality and thermal comfort in residential kitchens with a new ventilation system. *Build. Environ.* **2020**, *180*, 107016. [CrossRef]
12. Liu, S.; Dong, J.; Cao, Q.; Zhou, X.; Li, J.; Lin, X.; Qing, K.; Zhang, W.; Chen, Q. Indoor thermal environment and air quality in Chinese-style residential kitchens. *Indoor Air* **2020**, *30*, 198–212. [CrossRef]
13. Deng, N.; Fan, M.; Hao, R.; Zhang, A.; Li, Y. Field Measurements and Analysis on Temperature, Relative Humidity, Airflow Rate and Oil Fume Emission Concentration in a Typical Campus Canteen Kitchen in Tianjin, China. *Appl. Sci.* **2022**, *12*, 1755. [CrossRef]
14. Liu, Y.; Li, C.; Ma, H.; Dong, J. Investigation on the indoor environment during a whole cooking process under constant make-up air organization in a Chinese-style residential kitchen. *Indoor Built Environ.* **2023**, *32*, 1170–1186. [CrossRef]
15. World Health Organization. WHO Guidelines for Indoor Air Quality: Selected Pollutants, Geneva, Switzerland. Available online: https://www.who.int/publications/i/item/9789289002134 (accessed on 5 July 2023).
16. Reboul, C.; Thireau, J.; Meyer, G.; André, L.; Obert, P.; Cazorla, O.; Richard, S. Carbon monoxide exposure in the urban environment: An insidious foe for the heart? *Respir. Physiol. Neurobiol.* **2012**, *184*, 204–212. [CrossRef] [PubMed]
17. Jacobson, T.A.; Kler, J.S.; Hernke, M.T.; Braun, R.K.; Meyer, K.C.; Funk, W.E. Direct human health risks of increased atmospheric carbon dioxide. *Nat. Sustain.* **2019**, *2*, 691–701. [CrossRef]
18. Lim, K.; Lee, C. A numerical study on the characteristics of flow field, temperature and concentration distribution according to changing the shape of separation plate of kitchen hood system. *Energy Build.* **2008**, *40*, 175–184. [CrossRef]
19. Li, A.; Zhao, Y.; Jiang, D.; Hou, X. Measurement of temperature, relative humidity, concentration distribution and flow field in four typical Chinese commercial kitchens. *Build. Environ.* **2012**, *56*, 139–150. [CrossRef]
20. Grabow, K.; Still, D.; Bentson, S. Test Kitchen studies of indoor air pollution from biomass cookstoves. *Energy Sustain. Dev.* **2013**, *17*, 458–462. [CrossRef]
21. Zhou, B.; Chen, F.; Dong, Z.; Nielsen, P.V. Study on pollution control in residential kitchen based on the push-pull ventilation system. *Build. Environ.* **2016**, *107*, 99–112. [CrossRef]
22. Nejat, P.; Hussen, H.M.; Fadli, F.; Chaudhry, H.N.; Calautit, J.; Jomehzadeh, F. Indoor Environmental Quality (IEQ) Analysis of a Two-Sided Windcatcher Integrated with Anti-Short-Circuit Device for Low Wind Conditions. *Processes* **2020**, *8*, 840. [CrossRef]

23. Lu, S.; Zhou, B.; Zhang, J.; Hou, M.; Jiang, J.; Li, F.; Wang, Y. Performance evaluation of make-up air systems for pollutant removal from gas stove in residential kitchen by using a push-pull ventilation system. *Energy Build.* **2021**, *240*, 110907. [CrossRef]
24. Wang, L.; Xiang, Z.; Stevanovic, S.; Ristovski, Z.; Salimi, F.; Gao, J.; Wang, H.; Li, L. Role of Chinese cooking emissions on ambient air quality and human health. *Sci. Total Environ.* **2017**, *589*, 173–181. [CrossRef] [PubMed]
25. Chithra, V.S.; Shiva Nagendra, S.M. Impact of outdoor meteorology on indoor PM10, PM2.5 and PM1 concentrations in a naturally ventilated classroom. *Urban Clim.* **2014**, *10*, 77–91. [CrossRef]
26. Martins, V.; Moreno, T.; Mendes, L.; Eleftheriadis, K.; Diapouli, E.; Alves, C.A.; Duarte, M.; de Miguel, E.; Capdevila, M.; Querol, X.; et al. Factors controlling air quality in different European subway systems. *Environ. Res.* **2016**, *146*, 35–46. [CrossRef]
27. TSI. IAQ-Calc Indoor Air Quality Meter Model 7545 Operation and Service Manual. Available online: https://tsi.com/getmedia/812721b4-f84d-43aa-8b04-19a4ccab2bd7/7545-IAQ-Calc-1980576D?ext=.pdf (accessed on 9 July 2023).
28. Fengfeng, Y.; Zhang, W.; Dai, Z.; Xu, N. Study on ensuring indoor fresh air volume in winter. *Measurement* **2010**, *43*, 406–409. [CrossRef]
29. Wyon, D. Individual Microclimate Control: Required Range, Probable Benefits and Current Feasibility. Available online: https://www.researchgate.net/publication/310403150_Individual_microclimate_control_Required_range_probable_benefits_and_current_feasibility (accessed on 15 July 2023).
30. Giwa, S.O.; Nwaokocha, C.N.; Sharifpur, M. An appraisal of air quality, thermal comfort, acoustic, and health risk of household kitchens in a developing country. *Environ. Sci. Pollut. Res.* **2021**, *29*, 26202–26213. [CrossRef] [PubMed]
31. Giwa, S.O.; Nwaokocha, C.N.; Odufuwa, B.O. Air pollutants characterization of kitchen microenvironments in southwest Nigeria. *Build. Environ.* **2019**, *153*, 138–147. [CrossRef]
32. General Administration of Quality Supervision, Inspection and Quarantine Health of People's Republic of China; Ministry of Health of People's Republic of China; Ministry of Environmental Protection of People's Republic of China. Technical Specifications for Monitoring of Indoor Air Quality (HJ/T 167-2004). Available online: https://www.mee.gov.cn/ywgz/fgbz/bz/bzwb/jcffbz/200412/W020110127390160056847.pdf (accessed on 2 September 2023).
33. Raza, Z.; Akhter, S.; Shang, Y. Expected value of first Zagreb connection index in random cyclooctatetraene chain, random polyphenyls chain, and random chain network. *Front. Chem.* **2023**, *10*, 1067874. [CrossRef]
34. Tokmakoff, A. Chemical Dynamics in Condensed Phases-Fluctuations and Randomness—Some Definitions. Available online: https://chem.libretexts.org/@go/page/107300 (accessed on 25 July 2023).

Article

Optimization of Cost–Carbon Reduction–Technology Solution for Existing Office Parks Based on Genetic Algorithm

Zhenlan Dou [1], Lu Jin [2], Yinhui Chen [3],* and Zishuo Huang [3],*

[1] State Grid Shanghai Municipal Electric Power Company, Shanghai 200122, China; douzhenlan@epri.sgcc.com.cn
[2] China Electric Power Research Institute, Beijing 100192, China; jinlu@epri.sgcc.com.cn
[3] College of Architecture and Urban Planning, Tongji University, Shanghai 200092, China
* Correspondence: 17196695359@163.com (Y.C.); huangzish@tongji.edu.cn (Z.H.)

Abstract: With limited investment costs, how to fully utilize the carbon-reduction capacity of a campus in terms of buildings, equipment, and energy is an important issue when realizing the low-carbon retrofit of office parks. To this end, this paper establishes a mathematical optimization model for the decarbonization-based retrofit of existing office parks, based on the genetic algorithm, taking into account the relationship between cost, energy-consumption, and carbon-emissions, and taking the maximum carbon reduction of the park over its whole life as the optimization goal. The validity of the model was verified in conjunction with a case study of an office park in Nanchang, China. The case study shows that, compared with current typical parks, the carbon reduction through an office park's decarbonization retrofit has a non-linear correlation with the investment cost, and when the total investment cost of the park is above CNY 60 million, the increase in carbon reduction with the increase in the investment cost is gradually weakened, and the park achieves the maximum carbon reduction of 236,087 t when the investment cost reaches CNY 103 million. Under the current technical and economic conditions, the investment-cost–carbon-reduction benefits of different carbon-reduction technologies are different, the carbon-reduction benefit of increasing renewable energy utilization is the best, and the carbon-reduction benefit of upgrading the energy efficiency of the park's supply-and-use system is lower than that of renewable energy utilization, but better than that of upgrading the performance of the building envelope system. In addition, the configuration of the parameters of the same low-carbon technology in different forms of buildings varies significantly, due to differences in the building form and daily use. The model established in this paper is able to give a comprehensive optimized building–equipment–energy configuration plan for existing office parks, when maximizing carbon reduction under different investment costs, which guides the park's decarbonization retrofit.

Keywords: genetic algorithm; existing office parks; whole life cycle; cost benefits; carbon reduction; retrofit

Citation: Dou, Z.; Jin, L.; Chen, Y.; Huang, Z. Optimization of Cost–Carbon Reduction–Technology Solution for Existing Office Parks Based on Genetic Algorithm. *Processes* **2023**, *11*, 2452. https://doi.org/10.3390/pr11082452

Academic Editors: Yabin Guo, Zhanwei Wang, Yunpeng Hu and João M. M. Gomes

Received: 20 July 2023
Revised: 5 August 2023
Accepted: 10 August 2023
Published: 15 August 2023

1. Introduction

The rapid increase in carbon emissions has made environmental issues too prominent to be ignored [1]; global carbon emissions are expected to increase to 30% above the 2010 level by 2030 [2]. In China, CO_2 emissions generated by parks account for 31% of the country's total carbon emissions [3], so carbon reduction in parks has become an inevitable requirement for China's low-carbon transition. According to the IPCC, the main sources of carbon emissions on the park scale are energy and buildings [4].

In terms of energy, most of the existing research focuses on the optimization of the configuration method and operation strategy of the integrated energy system in the park. For example, Song, Z. et al. developed a multi-objective optimization model to synergistically optimize the configuration and operation strategy of a combined cooling, heating, and power (CCHP) system, with the objectives of minimizing the cost, primary energy consumption, and carbon emissions [5]. Wu, D. et al. proposed a multi-parameter synergistic optimization method aimed at cost reduction, carbon reduction, and independence in the low-carbon park energy system, with photovoltaics, wind power, lithium batteries, and heat storage tanks, and explored the equipment configuration and operating parameters [6]. Wang, Y. et al. developed a multi-objective optimization model for an integrated energy system to optimize the capacity allocation of the energy system, using the Strength Pareto Evolutionary Algorithm 2 (SPEA2), and through a sequential preference technique similar to the ideal solution (TOPSIS), to minimize the total annual cost and carbon dioxide emissions under different investment cost constraints [7]. Guo, W. et al. constructed an integrated energy system, including combined heat and power (CHP), a heat pump (HP), and energy storage (ES), and considered the optimization objectives of the operating cost, energy efficiency, and renewable energy consumption rate, taking into account the carbon emission and demand response, and used the multi-objective particle swarm optimization algorithm (MOPSO) to optimize the operation strategy of the integrated energy system [8].

In terms of buildings, research has focused on passive energy-saving design, such as envelope thermal insulation design, and the optimization of the building form and layout. For example, Luo, Z. et al. took the optimization objective of maximizing the carbon-reduction benefit and cost-effectiveness of the whole life cycle of the building, to obtain the optimal configurations of seven design parameters for the thermal performance of the envelope, and two parameters related to the users' willingness to save energy (cooling and heating temperatures) [9]. Fesanghary, M. et al. proposed a multi-objective optimization model based on the harmony search algorithm (HS). The minimization the life cycle cost (LCC) and CO_2-equivalent (CO_2-eq) of the building was used as the objective function, and the building envelope parameters were used as the design variables [10]. Ferrara, M. et al. used a combination of TRNSYS and GenOpt, with a global cost function as the objective function for optimization, and a particle swarm optimization algorithm was used to minimize the objective function and identify the cost-optimal building configuration [11]. Gerber, D.J. et al. chose ModelCenter as the process integration and design optimization software, and used a genetic algorithm to develop a multi-objective optimization model [12]. Yigit, S. et al. developed a software package that combines customized thermal simulation software with Matlab's Optimtool [13]. Tuhus-Dubrow, D. et al. combined a genetic algorithm with a building energy simulation engine, to build an optimization model [14]. Carli, R. et al. developed a multi-objective optimization algorithm, to improve, in an integrated and holistic way, the building energy efficiency and comfort, by efficiently allocating the budget to the buildings [15].

In addition, many scholars consider the coupled effect of the building and energy system to integrate and optimize the building envelope and air-conditioning system. Chantrelle, F. et al. developed a multi-objective optimization tool for building renewal, MultiOpt, that focuses on optimizing the building envelope, air conditioning loads, and control strategies, using a genetic algorithm (NSGA-II) coupled with TRNSYS, and economic and environmental databases [16]. Petkov, I. et al. proposed a multi-stage multi-objective scalable optimization framework (called MANGOret) to provide optimal configurations for multi-energy systems and the envelope retrofits of existing buildings, using the multi-objective building optimization tools Mobo and TRNSYS [17]. Abdou, N. carried out a multi-objective optimization, using the multi-objective building optimization tool Mobo, in conjunction with TRNSYS, to find the optimal building envelope design, and the optimal sizing of the renewable energy system for a net-zero energy building in Tetouan (Morocco) [18]. Lin, Y.H. et al. established a multi-objective optimization decision model (MOBELM) for the energy performance of the building envelope and air-conditioning system with the help of MATLAB R2021a (9.10.1602886), with the optimization objectives of minimizing the building cost and carbon emissions [19]. Bichiou, Y. et al. demonstrated a comprehensive energy simulation environment to optimize the building envelope characteristics and HVAC system design and operation strategies, with the optimization objective of minimizing the whole-life-cycle cost, and compared the robustness and effectiveness of three algorithms, namely, the genetic algorithm, particle swarm algorithm, and sequential search algorithm, in the simulation environment [20]. Hashempour, N. et al. reviewed the literature related to energy efficiency optimization for existing buildings, and pointed out that GA ranked first (41%) in terms of contribution among the studies analyzed, with the NSGA-II algorithm receiving the most attention [21]. Mela, K. compared the functionality and the results provided by six multiple criteria decision; unfortunately, the best MCDM method was not discovered [22].

It can be seen that previous studies mainly focus on a single dimension, such as energy or buildings, and the selection of carbon-reduction technologies is mostly limited to the selection of one or two carbon-reduction technologies for optimization. In addition, the research on the optimization of carbon-reduction technologies in parks mainly focuses on the planning and design of new parks, offering few optimizations for retrofitting existing parks, and most of them are qualitative technical guidelines [23–26], lacking quantitative analysis. The office park is a systematic integration of various types of single buildings, equipment, and energy sources, and the form and function of different buildings are different, so when the investment cost is given, how can we configure multiple low-carbon technologies among buildings, to maximize the benefits of carbon reduction in the park? This is one of the questions that need to be answered.

Therefore, in this paper, we will consider the relationship between cost, energy consumption, and carbon emissions, and establish an integrated optimal configuration model of building–equipment–energy, based on the genetic algorithm, with the goal of maximum carbon reduction in the whole life cycle of an existing office park retrofit. The innovation of this paper is to maximize the whole carbon emission reduction of the park under limited cost constraints, and to make full use of the carbon-reduction potential of buildings, equipment, and renewable energy in all aspects, through a reasonable configuration solution, which is of great guiding significance for the low-carbon transformation of the established park.

2. Optimization Model

Currently, there are several mature low-carbon technologies for building bodies, equipment, and energy, but the costs and carbon-reduction potential of different low-carbon technologies vary greatly, and they are applicable at different stages. Under limited cost constraints, first of all, according to the region, climate, form, function, and other characteristics of the specific project, the low-carbon benefits of various types of resource inputs should be comprehensively optimized, and compared and weighed against the low-carbon benefits of buildings, equipment, and renewable energy use, etc., so as to filter out the optimal configuration of the various low-carbon technologies that can be utilized in the target park. The establishing process of the optimization model is shown in Figure 1, and is mainly divided into three steps. The first step is to select low-carbon technologies. Based on the carbon-reduction benefits and investment costs of relevant low-carbon technologies, we identify low-carbon technologies applicable to the retrofit of the target parks. The second step is to establish the optimization model. Based on the low-carbon technologies selected to form the optimization variables for the park retrofit, furthermore, we obtain the quantitative relationship between each optimization variable and carbon emissions and investment costs, then we define the objective function, and set the constraints and initial values of the optimization model. The third step is model solving. Using the genetic algorithm, we obtain the calculation results of the model, i.e., the maximum life-cycle carbon reduction achieved by the park's decarbonization retrofit under different investment cost constraints, and the corresponding optimal configuration scheme.

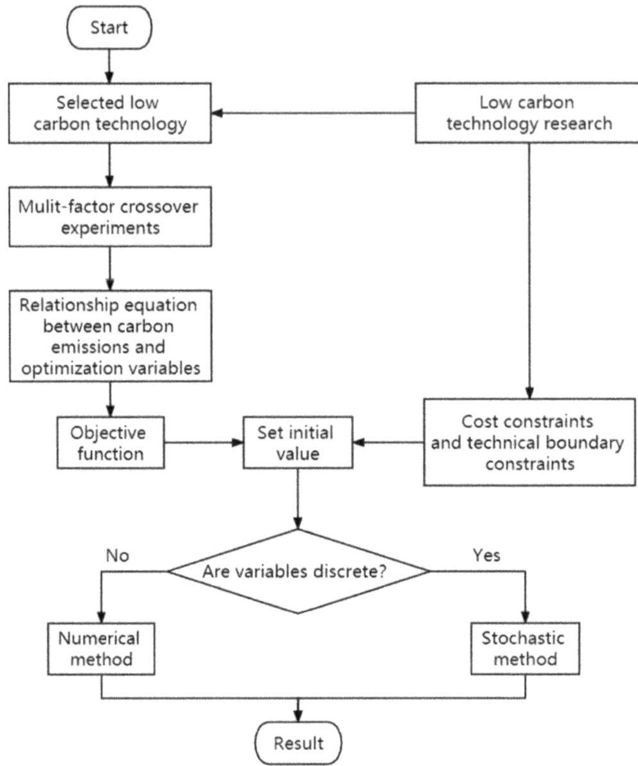

Figure 1. Optimization flow chart.

2.1. Objective Function

In this paper, the maximum life cycle carbon reduction of the park's decarbonization retrofit is taken as the objective, assuming that the park contains n buildings with a total of m low-carbon technologies available. For a given parameter of a low-carbon technology applied to a building in the park, the corresponding carbon reduction and investment cost of the technology can be obtained, as shown in Equations (1) and (2), respectively:

$$E_{i,j}^{k} = f(k_{i,j}) \tag{1}$$

$$C_{i,j}^{k} = g(k_{i,j}) \tag{2}$$

where $E_{i,j}^{k}$ is the whole-life-cycle reduction in carbon emissions from the use of low-carbon technology j on building i; $k_{i,j}$ is the technical parameter for the use of low-carbon technology j on building i; and $C_{i,j}^{k}$ is the investment cost of applying low-carbon technologies j to the building i.

The objective function for maximizing the whole-life-cycle carbon reduction in the park is established as Equation (3):

$$E = \sum_{i=1}^{n} \sum_{j=1}^{m} E_{i,j}^{k} = \sum_{i=1}^{n} \sum_{j=1}^{m} E_{OP,i,j}^{k} - \sum_{i=1}^{n} \sum_{j=1}^{m} E_{CO,i,j}^{k} \tag{3}$$

where $E_{OP,i,j}^{k}$ is the reduction in carbon emissions during the operational phase of the use of low-carbon technology j on building i; and $E_{CO,i,j}^{k}$ is the increase in carbon emissions during the construction phase of the use of technology j on building i.

2.2. Constraints

2.2.1. Technical Boundary Constraints

Different low-carbon technologies have different parameter-change characteristics, and technical boundaries. For example, for low-carbon technologies such as the roof and external wall insulation thickness, external window heat transfer coefficient, and photovoltaic installation area, their technical parameter changes can be regarded as continuous, with upper and lower bounds. For carbon-reduction technologies such as heat-reflective coatings, air-conditioning intelligent operation and control systems, and energy-saving lighting retrofits, there are only two scenarios in which the change in the technical parameters takes place: when the technology is used, the parameter takes the value of 1; when it is not used, the parameter takes the value of 0. Then, the constraints of the technical boundary are shown in Equation (4):

$$k_{i,j}^{L} \leq k_{i,j} \leq k_{i,j}^{U} \tag{4}$$

2.2.2. Cost Constraints

The application of each low-carbon technology to each building will generate the corresponding investment cost; based on the accumulation of the investment cost of all the carbon-reduction technologies in the park, we get the total cost of the park's decarbonization retrofit, as shown in Equation (5):

$$C_0 = \sum_{i=1}^{n} \sum_{j=1}^{m} C_{i,j}^{k} \leq C \tag{5}$$

where n is the number of individual buildings in the park; m is the number of low-carbon technologies adopted; k is the value of the parameter taken for the use of technology j on building i; and $C_{i,j}^{k}$ is the cost of adopting the j-th technology in building i with technical parameters taking the value k. The total cost C_0 is obtained by superimposing all the sub-costs $C_{i,j}^{k}$.

2.2.3. Model Solving

Optimization models include linear optimization models, nonlinear optimization models, mixed integer linear or nonlinear optimization models, multi-objective optimization models, and many other types. Efficient and accurate solution algorithms can be selected according to the model characteristics, such as the gradient descent method, Newton method, genetic algorithm, and so on.

The optimization model established in this paper is a single-objective optimization model, with the objective of minimizing the total carbon emissions in the whole life cycle of the park. According to whether the optimization variables are continuously reachable or not, it is divided into nonlinear optimization programming and mixed integer linear programming, which corresponds to the choice of calculus or stochastic methods for solving the model using MATLAB2021a (9.10.1602886) software.

3. Case Study

3.1. Park Overview

The research object of this paper is a selected office park in Nanchang, Jiangxi Province, which covers an area of 25,000 square meters, with a total of 11 single buildings: respectively, a restaurant (No. R01), four enclosed office buildings with the same structural form (No. W01), and another enclosed office building (No. W02). In addition, there are four "E"-type office buildings with the same structural form (No. E01), and another "E"-type office building (No. E02).

The effect diagram of the park is shown in Figure 2, and the main technical indexes of each individual building in the park are shown in Table 1.

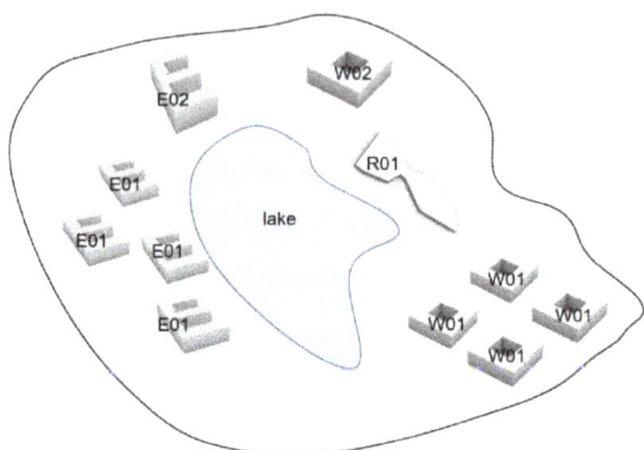

Figure 2. Effective diagram of the park.

Table 1. Technical indicators for each type of building in the park.

Building	Number	Individual Building Area/m^2	No. of Floors	Window–Wall Ratio	Roof Area/m^2	External Wall Area/m^2	External Window Area/m^2	The Area of East, West, and South External Wall/m^2
R01	1	23,150	2	0.8	11,575	1200	4800	1200
W01	4	2088	6	0.3	3348	5223	2712	2717
W02	1	33,480	10	0.3	3348	8705	4687	4528
E01	4	22,902	6	0.3	3817	7160	3856	4086
E02	1	38,170	10	0.3	3817	11,933	6427	6809

3.2. Selection of Optimization Variables

Based on the considerations of carbon-reduction potential and investment cost, Sun, J. categorized various types of low-carbon technologies. Firstly, there are high-potential and high-cost, low-carbon technologies represented by a distributed energy supply and ultra-low energy buildings. Secondly, there are high-potential, low-cost, low-carbon technologies represented by building photovoltaics and various types of heat pump technologies. Thirdly, the carbon-reduction technologies represented by a building's photovoltaic technology and by arbor and shrub configuration, solid waste recycling, and wastewater treatment are low-potential and low-cost low-carbon technologies and, fourthly, there are low-potential and high-cost, low-carbon technologies represented by a garbage pneumatic recycling system and three-dimensional greening [27]. Xiao, H. et al. pointed out that air-conditioning systems, household appliances, and lighting systems are the three technologies with the highest potential for carbon reduction in the long term, accounting for 72.9% of the total carbon-reduction potential in the building sector in China [28]. The literature [29,30] points out that heat-reflective insulation coatings on external walls can effectively reduce the surface temperature of external walls in hot-summer and cold-winter regions, with good carbon-reduction benefits. In addition, photovoltaics have excellent carbon-reduction benefits [31,32]. For the optimization of the layout of the building form, various types of heat pump technology, and other types of low-carbon technology, despite having significant carbon-reduction benefits, are not suitable for the retrofit of existing parks.

In this paper, for the low-carbon retrofit of an existing office park in Nanchang City, considering economy and feasibility, the seven low-carbon technologies of exterior wall insulation (abbreviated as "WAIT"), roof insulation (abbreviated as "RIT"), thermal performance of exterior windows (abbreviated as "WDIT"), heat-reflective coatings (abbreviated as "HRIT"), lighting energy-saving retrofit (abbreviated as "LEST"), intelligent control system of HVAC (abbreviated as "ACIT"), and photovoltaics (abbreviated as "PVs") are finally selected as the optimization variables to establish a comprehensive optimization model for the low-carbon optimization of the park's retrofit. The optimization variables of this park are set as shown in Table 2.

Table 2. Optimized variable settings.

| Building | Building Envelope | | | Use or Not of Heat Reflective Coating Technology $k_{i,4}$ | Use or Not of HVAC Intelligent Control Technology $k_{i,5}$ | Use or Not of Lighting Energy-Saving Technology $k_{i,6}$ | PV Area k_7 |
	Thickness of Roof Insulation $k_{i,1}$	Thickness of External Wall Insulation $k_{i,2}$	Heat Transfer Coefficient of Windows $k_{i,3}$				
R01	$k_{1,1}$	$k_{1,2}$	$k_{1,3}$	$k_{1,4}$	$k_{1,5}$	$k_{1,6}$	
W01	$k_{2,1}$	$k_{2,2}$	$k_{2,3}$	$k_{2,4}$	$k_{2,5}$	$k_{2,6}$	
W02	$k_{3,1}$	$k_{3,2}$	$k_{3,3}$	$k_{3,4}$	$k_{3,5}$	$k_{3,6}$	k_7
E01	$k_{4,1}$	$k_{4,2}$	$k_{4,3}$	$k_{4,4}$	$k_{4,5}$	$k_{4,6}$	
E02	$k_{5,1}$	$k_{5,2}$	$k_{5,3}$	$k_{5,4}$	$k_{5,5}$	$k_{5,6}$	

3.3. Optimizing Variable Constraints

(1) Low-carbon technologies for building envelopes

We examine the performance limits of low-carbon technologies for envelope structures in China. For the exterior envelope insulation performance retrofit in this park, the parameter-setting constraints for roof insulation, external wall insulation, and external window heat transfer coefficients are shown in Equations (6)–(8), respectively:

$$0 \leq k_{i,1} \leq 120 \tag{6}$$

$$0 \leq k_{i,2} \leq 120 \tag{7}$$

$$0.5 \leq k_{i,3} \leq 2.6 \tag{8}$$

where $k_{i,1}$ is the thickness of the exterior wall insulation, mm; $k_{i,2}$ is the thickness of the roof insulation, mm; $k_{i,3}$ is the heat transfer coefficient of the window, W m^{-2} K^{-1}.

(2) Heat-reflective coatings

The application effect of heat-reflective coatings is related to the climate zone in which the building is located, and the part of the application, generally in the tropics or hot-summer and cold-winter areas, where the application effect is better, and the application of reflective heat-insulating coatings on the north façade of the building is not obvious. The region in which this case is located is a hot-summer and cold-winter region, taking into account the economy and feasibility of the project implementation plan to optimize the application of this technology to the east, south, and west elevations of the building. The application area of the exterior heat-reflective coatings is not a continuous variable; rather, for a particular monolithic building, there are only two cases: when not in use, $k_{i,4}$ is 0; when in use, $k_{i,4}$ is 1; and the constraint relationship is shown in Equation (9):

$$k_{i,4} = \begin{cases} 0 \\ 1 \end{cases} \tag{9}$$

When $k_{i,4}$ is 0, it means that the building does not use this technology, and when $k_{i,4}$ is 1, it means that the building uses heat-reflective thermal insulation coatings.

(3)   HAVC Intelligent Control System

For the HAVC intelligent control system, there are two cases; i.e., when a building is set up with this control system, it takes the value of 1, and when the building is not set up with this control system, it takes the value of 0. The constraint relationship is shown in (10):

$$k_{i,5} = \begin{cases} 0 \\ 1 \end{cases} \tag{10}$$

(4)   Lighting-energy-saving retrofit

Lighting-energy-saving retrofit technology is only considered for spaces with high occupancy rates, such as offices and conference rooms, to which the economic benefits of lighting-energy-saving retrofits are relatively obvious. For some auxiliary equipment rooms, stairwells, and other areas with a low level of space occupancy, the technology is not considered. The constraint relationship is shown in Equation (11):

$$k_{i,6} = \begin{cases} 0 \\ 1 \end{cases} \tag{11}$$

When a building adopts a lighting-energy-saving retrofit, then $k_{i,6}$ is 1, and when the technology is not adopted, $k_{i,6}$ takes the value of 0.

(5)   PV-paved

There is a limited area in the park and, at the same time, there is the requirement for a green space rate, so the available photovoltaic area in the park is concentrated on the roof of the building, and the photovoltaic pavement area in the park is constrained, as shown in Equation (12):

$$0 \leq k_{i,7} \leq A_P \tag{12}$$

where A_P is the area of the park where PV can be installed, which, in this paper, is taken as 70% of the total building roof area.

3.4. Investment Cost Constraints

The total investment cost of the decarbonization of the park is the sum of the investment costs of implementing each low-carbon technology in individual buildings, as shown in Equation (13). Among them, each individual investment cost is shown in Table 3.

$$C_0 = \sum_{j=1}^{6} C_{1,j}^k + 4 \times \sum_{j=1}^{6} C_{2,j}^k + \sum_{j=1}^{6} C_{3,j}^k + 4 \times \sum_{j=1}^{6} C_{4,j}^k + \sum_{j=1}^{6} C_{5,j}^k + C_7^k \tag{13}$$

The formula for calculating the investment cost of each type of low-carbon technology is shown in Table 3.

Table 3. The quantitative relationship between various low-carbon technology parameters and investment costs.

Low-Carbon Technology		Relational Equation for Cost and Parameter Configuration	Equation No.
Building thermal performance	Roof	$C_{i,1}^k = k_{i,1} \times AD_i \times P_1$	(14)
	External wall	$C_{i,2}^k = k_{i,2} \times AQ_i \times P_2$	(15)
	External window	$C_{i,3}^k = k_{i,3} \times AC_i \times P_3$	(16)
Heat-reflective coating technology		$C_{i,4}^k = (49.75 \times k_{i,4}^2 - 602.45 \times k_{i,4} + 1875) \times AW_i$	(17)
HVAC intelligent control technology		$C_{i,5}^k = k_{i,5} \times P_5$	(18)
Lighting-energy-saving technology		$C_{i,6}^k = k_{i,6} \times P_6$	(19)
Photovoltaic power technology		$C_{i,7}^k = k_{i,7} \times P_7$	(20)

Note: $k_{i,j}$ represents a technology configuration parameter; $P_i, i = 1, 2, \dots 7$ represents the investment cost unit price of each technology, respectively.

3.5. Quantitative Relationship between Carbon Reduction and Optimization Variables

In order to obtain the quantitative relationship between the parameter settings of low-carbon technologies, and the whole life cycle carbon reduction in each type of building, this paper establishes a model of each type of building in Rhino and Grasshopper, sets up different thicknesses for the roof and exterior wall insulation, and different heat transfer coefficients for the exterior windows (the variable settings are shown in Table 4), and simulates the calculation of the operational stage of different types of building with different envelope parameter settings for carbon reduction. Meanwhile, the quantitative relationship between the PV pavement area and PV power generation is calculated, according to ref. [26].

Table 4. Low-carbon technology variable settings.

Building	Optimization Variables	Unit	Range	Step
	Thickness of external wall insulation	mm	[0, 120]	10
	Thickness of roof insulation	mm	[0, 120]	10
R01, W01,	Heat transfer coefficient of windows	W m^{-2} K^{-1}	[0.5, 2.6]	0.5
W02, E01, E02	Use or not of heat-reflective coating technology	/	0/1	\
	Use or not of lighting-energy-saving technology	/	0/1	\
	Use or not of HVAC intelligent control technology	/	0/1	\
Park	PV area	m^2	[0, 33, 180]	1

Based on the modeling results, manner in which the carbon emissions increased in the construction stage, and reduced in the operation stage of each type of building envelope renovation in the park under different parameter settings was derived, and data fitting was performed to obtain the quantitative relationship equations between the thickness of the roof insulation layer, the thickness of the exterior wall insulation layer, the heat transfer coefficients of the exterior windows of each type of building, and the carbon emissions increased in the construction stage (CP-IN) and the carbon emissions reduced in the operation stage (OP-DE). In addition, for low-carbon technologies, such as heat-reflective coatings and lighting-energy-saving retrofits, the relationship between the parameter values taken and the carbon reduction can be given directly via a calculation. The quantitative relationship between each optimization variable and carbon reduction in the park is as follows.

(1)   Enhancement in the insulation performance of the envelope

The carbon reduction in the operational phase of the exterior wall, roof, and window insulation enhancement was calculated using the Grasshopper(Build 1.0.0007) software, the saved electricity was discounted according to the grid carbon emission factor, and the relationship equation between each optimization variable and its carbon reduction was obtained via fitting, which is summarized as shown in Table 5. The increased carbon emissions in the construction phase of each building's exterior wall, roof, and window insulation performance enhancement were calculated, according to Equations (21)–(23):

$$E_{co,i,1}^{k_{i,1}} = AQ_i \times \frac{k_{i,1}}{1000} \times 28.2 \tag{21}$$

$$E_{co,i,2}^{k_{i,2}} = AD_i \times \frac{k_{i,2}}{1000} \times 28.2 \tag{22}$$

$$E_{co,i,3}^{k_{i,3}} = AC_i \times (-4.84 \times k_{i,3} + 37) \tag{23}$$

Table 5. The quantitative relationship between each optimization variable and its carbon reduction in the operation stage.

Building	Low-Carbon Technology	Relationship Equation between Optimization Variables and Carbon Reduction in the Operational Phase	Equation No.
R01	External wall insulation	$E^{k_{1,1}}_{op,1,1} = \left(5 \times 10^{-5} \times k_{1,1}{}^2 - 0.0089 \times k_{1,1} - 17.91\right) \times A_1$	(24)
	Roof insulation	$E^{k_{1,2}}_{op,1,2} = \left(4.3 \times 10^{-3} \times k_{1,2}{}^2 - 0.62 \times k_{1,2} + 13.49\right) \times A_1$	(25)
	External window insulation	$E^{k_{1,3}}_{op,1,3} = \left(-30.32 \times k_{1,3} + 57\right) \times A_1$	(26)
W01	External wall insulation	$E^{k_{2,1}}_{op,2,1} = \left(5 \times 10^{-5} \times k_{2,1}{}^2 - 0.0089 \times k_{2,1} + 138.09\right) \times A_2$	(27)
	Roof insulation	$E^{k_{2,2}}_{op,2,2} = \left(0.01 \times k_{2,2}{}^2 - 0.28 \times k_{2,2} - 3.15\right) \times A_2$	(28)
	External window insulation	$E^{k_{2,3}}_{op,2,3} = \left(51.38 \times k_{2,3} - 140.02\right) \times A_2$	(29)
W02	External wall insulation	$E^{k_{3,1}}_{op,3,1} = \left(3.1 \times 10^{-3} \times k_{3,1}{}^2 - 1.09 \times k_{3,1} + 16.23\right) \times A_3$	(30)
	Roof insulation	$E^{k_{3,2}}_{op,3,2} = \left(0.01 \times k_{3,2}{}^2 - 0.29 \times k_{3,2} + 1.61\right) \times A_3$	(31)
	External window insulation	$E^{k_{3,3}}_{op,3,3} = \left(76.96 \times k_{3,3} - 148.04\right) \times A_3$	(32)
E01	External wall insulation	$E^{k_{4,1}}_{op,4,1} = \left(1.1 \times 10^{-3} \times k_{4,1}{}^2 - 0.39 \times k_{4,1} + 0.64\right) \times A_4$	(33)
	Roof insulation	$E^{k_{4,2}}_{op,4,2} = \left(4 \times 10^{-4} \times k_{4,2}{}^2 - 0.17 \times k_{4,2} - 0.17\right) \times A_4$	(34)
	External window insulation	$E^{k_{4,3}}_{op,4,3} = \left(13.39 \times k_{4,3} - 36.8\right) \times A_4$	(35)
E02	External wall insulation	$E^{k_{5,1}}_{op,5,1} = \left(1.4 \times 10^{-3} \times k_{5,1}{}^2 - 0.48 \times k_{5,1} + 12.59\right) \times A_5$	(36)
	Roof insulation	$E^{k_{5,2}}_{op,5,2} = \left(4 \times 10^{-4} \times k_{4,2}{}^2 - 0.16 \times k_{4,2} + 23.86\right) \times A_5$	(37)
	External window insulation	$E^{k_{5,3}}_{op,5,3} = \left(41.26 \times k_{5,3} - 48.76\right) \times A_5$	(38)

Note: $i = 1, 2, \ldots 5$ are denoted as restaurant R01, low-rise enclosed building W01, high-rise enclosed building W02, low-rise "E" building E01, and high-rise "E" building E02, respectively; $j = 1, 2, 3$ denotes the external wall insulation, roof insulation, and external window insulation, respectively.

(2) HVAC intelligent control, heat-reflective coatings, lighting-energy-saving retrofit

The carbon reduction by HVAC smart controls and heat-reflective coatings during the operational phase is calculated in Equation (39):

$$E^{k_{i,j}}_{op,i,j} = \left(E_{op,i,0} - \sum_{j=1}^{3} E^{k_{i,j}}_{op,i,j}\right) \times \eta_j \quad j = 5, 6 \tag{39}$$

where η_j is the energy-saving rate of the j-th technology; in this case, the energy-saving rate of intelligent control η_5 is taken as 8%; and the energy-saving rate of heat-reflective coating η_6 is taken as 2%.

The carbon reduction in the operational phase of the lighting-energy-saving retrofit is calculated in Equation (40):

$$E^{k_{i,4}}_{op,i,4} = 171.09 \times k_{i,4} \times A_i \tag{40}$$

The increased carbon emissions in the construction phase from lighting energy efficiency retrofits, HVAC smart controls, and heat-reflective coatings are calculated in Equation (41):

$$E^{k_{i,j}}_{co,i,j} = E^{k_{i,j}}_{op,i,j} \times 10\% \quad \begin{array}{l} i = 1, 2, \cdots, 5 \\ j = 4, 5, 6 \end{array} \tag{41}$$

(3) Photovoltaic

The carbon reduction by the park's PV during the operation phase is discounted, based on its power generation, with reference to the grid carbon emission factor. The power generation is calculated with reference to [33], to obtain the carbon reduction of PV in the operation phase as Equation (42), and the carbon emission increased in the construction phase is calculated as 10% of the carbon reduction in the operation phase, as Equation (43):

$$E_{op,pv,7}^{k_{pv,7}} = 4117.07 \times S_{pv} \tag{42}$$

$$E_{co,pv,7}^{k_{pv,7}} = E_{op,pv,7}^{k_{pv,7}} \times 10\% \tag{43}$$

4. Results and Discussion

Based on the basic information of the park, the quantitative relationship between each low-carbon technology parameter and its whole-life-cycle carbon emission and investment cost is substituted into the optimization model. to obtain the optimal configuration of the building-envelope–equipment–renewable-energy-system of the park under different investment-cost constraints.

4.1. Optimization of the Park under Infinite Cost

Firstly, the no-investment-cost constraint is set, and the optimal configuration of different carbon-reduction technologies for each type of building in the park is obtained through model calculations, as shown in Table 6.

Table 6. Parameter configurations for LCTs at infinite cost.

	Building	R01	W01	W02	E01	E02
Building envelope	The thickness of the roof insulation/mm	71	19	19	120	120
	The thickness of the external wall insulation/mm	0	120	120	120	120
	Heat transfer coefficient of the windows/W m^{-2} K^{-1}	2.6	0.5	0.5	0.5	0.5
Use or not of heat-reflective coating technology		0	1	1	1	1
Use or not of HVAC intelligent control technology		1	1	1	1	1
Use or not of lighting-energy-saving technology		1	1	1	1	1
Whole-life carbon reductions for different building types/t		8462	9580	16,064	7301	12,405
PV area/m^2				33,180		
Whole-life-cycle carbon reduction by PV/t				131,627		
Total life-cycle carbon reduction of the park/t				236,087		
Total investment/CNY 10,000				10,300		

From the above table, it can be seen that the maximum life-cycle carbon reduction that the park can achieve through the optimal configuration of the selected low-carbon technologies is 236,087 t, corresponding to a total investment cost of CNY 103 million. The average carbon-reduction benefit for the entire park is 2.29 kg CNY^{-1}; at this point, additional investments continue to be made, and carbon reduction is not increasing. It can also be seen that the equipment energy efficiency improvement technology has been applied to all the buildings in the park without investment cost constraints. It is worth noting that, when the maximum carbon reduction is achieved, the configuration parameters of the envelope thermal insulation technology of each type of building in the park differ significantly and, for the "E"-type of building, which has the highest requirements for the envelope thermal insulation technology, the thickness of the insulation layer of the roof and the external wall, and the heat-transfer coefficient of the external window, have reached the constraint boundaries, which are 120 mm, 120 mm, and 0.5 W m^{-2} K^{-1}, respectively. While the restaurant building's (R01) enclosure thermal insulation technology parameter configuration requirements are low, its roof and exterior wall insulation thickness, and exterior window heat-transfer coefficient were 71 mm, 0 mm, and 2.6 W m^{-2} K^{-1}.

4.2. Analysis of Carbon-Reduction Benefits under Infinite Cost

Based on the results of the model calculations, the carbon-reduction benefits of LCTs in various types of building are analyzed, as follows.

(1)   Carbon-reduction benefits of different LCTs in different buildings

Based on the model-optimization results, the carbon-reduction benefits (whole-life-cycle carbon reduction at unit cost) of each LCT in each building are calculated, as shown in Figure 3.

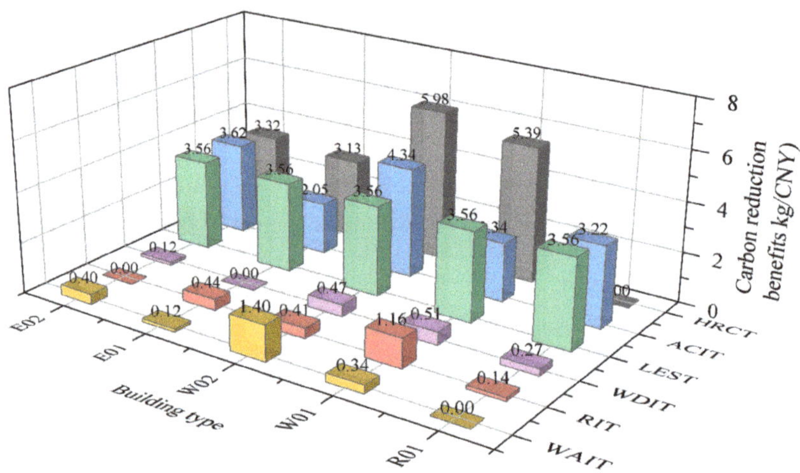

Figure 3. Carbon-reduction benefits of different LCTs in different types of building.

As can be seen from Figure 3, the carbon-reduction benefits of different LCTs in the same building vary greatly and, in general, compared with envelope thermal insulation technologies, equipment energy efficiency improvement technologies, such as ACIT and LEST, have higher carbon-reduction benefits. In addition, with the exception of LEST, the carbon-reduction benefits of the same LCT vary across buildings. For example, for the high-rise enclosed office building W02, the WAIT, ACIT, and HRCT have achieved the highest carbon-reduction benefits, compared with the other types of building. The applicability of the LEST is good in all types of building, with a carbon-reduction benefit of 3.56 kg CNY^{-1}, which is due to the low correlation between the lighting retrofit technology and the building thermal performance and building form, and the relatively low cost and linear relationship with the retrofit area.

It is worth noting that, for the restaurant building R01, the parameters of WAIT and HRCT take the value of 0. The carbon-reduction benefits of RIT and WDIT are very low, at 0.14 and 0.27 kg CNY^{-1}, respectively, which indicates that these LCTs are poorly adapted to the restaurant building. Unlike with the other types of building, the window-to-wall ratio of the restaurant is 0.8, and the building does not comply with the concept of an ultra-low-energy building, and it is more difficult and costly to improve the thermal performance of the envelope, so the measures related to the retrofit of the envelope have a very low carbon-reduction benefit.

(2)   Carbon-Reduction Benefits of Various Types of Building and Various Types of LCT

Based on the results of the model calculations, the carbon-reduction benefits of each type of building, and the carbon-reduction benefits of each type of LCT are shown in Figures 4 and 5, respectively.

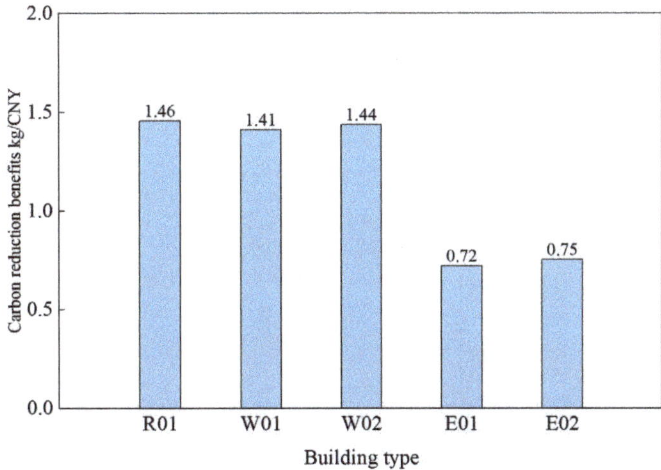

Figure 4. Carbon-reduction benefits of various building types.

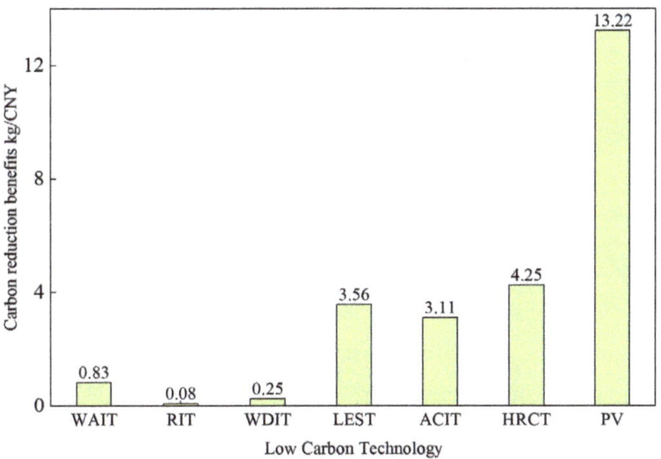

Figure 5. Carbon-reduction benefits of various low-carbon technologies.

As can be seen from Figure 4, the carbon-reduction benefits of restaurant building R01, enclosed building W01, and enclosed building W02 are higher, while the carbon-reduction benefits of "E"-type building E01 and "E"-type building E02 are lower.

As can be seen from Figure 5, the carbon-reduction benefits of LEST, ACIT, and HRCT are higher, while the carbon-reduction benefits of LCT related to the improvement of the thermal performance of the building envelope are lower.

4.3. Optimization of the Park at Limited Cost

According to the calculation results of the optimization model under an infinite investment cost, it can be seen that the cost corresponding to reaching the maximum carbon reduction in the park is CNY 103 million; based on this result, different limits of investment cost from CNY 0 to 100 million are set, and the optimal configuration scheme for the comprehensive decarbonization retrofit of the park under different cost constraints is calculated. The relevant calculation results are as follows.

(1) Variation in total carbon reduction in the park with the total investment cost

Based on the results of the model calculations, the variation in the total carbon reduction of the park with the increase in the investment cost is obtained, as shown in Figure 6.

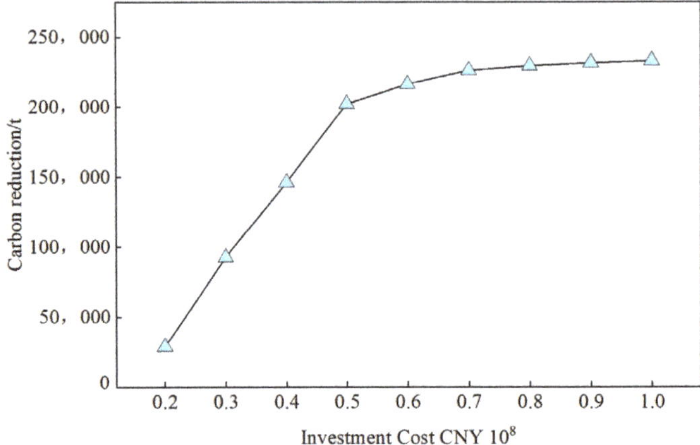

Figure 6. Variation in the total carbon reduction with the investment cost.

As can be seen from Figure 6, the total carbon reduction in the park is non-linear with the change in investment cost; when the investment cost is below CNY 60 million, the carbon reduction increases faster with the investment cost, which is approximately linear, and the carbon reduction per-unit cost is basically unchanged. However, as the investment cost continues to increase, the trend of carbon reduction increase is gradually weakened, and the carbon reduction per-unit cost gradually decreases. When the investment cost reaches CNY 103 million, the park achieves the maximum carbon reduction of 236,087 t; at this time, the marginal benefit is 0. Up to this point, the carbon reduction in the park will not be increased with the incremental increase in the investment.

(2) Variation in parameters with the investment costs for different LCTs in the same building

Based on the modeling results, the variation in the parameters of the carbon-reduction technologies with the investment cost in each building type was obtained separately, as shown in Figure 7.

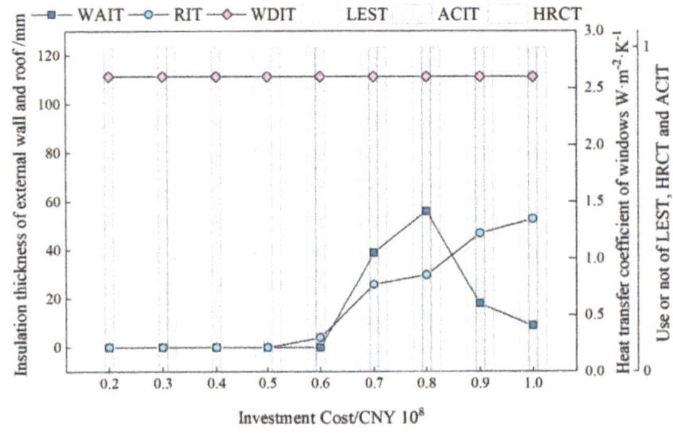

(**a**) Parameter changes in LCTs in R01

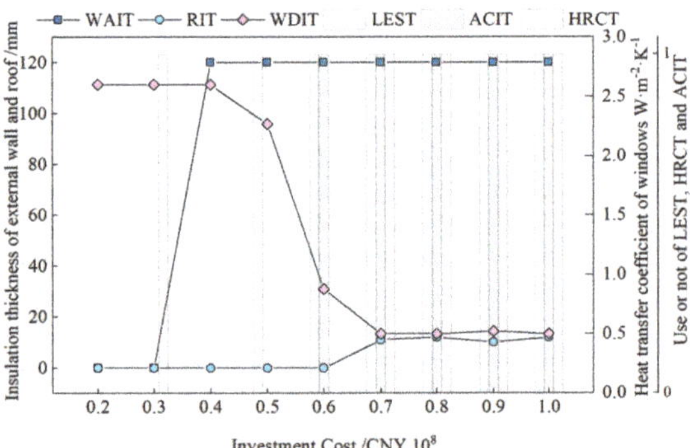

(**b**) Parameter changes in LCTs in W01

Figure 7. *Cont.*

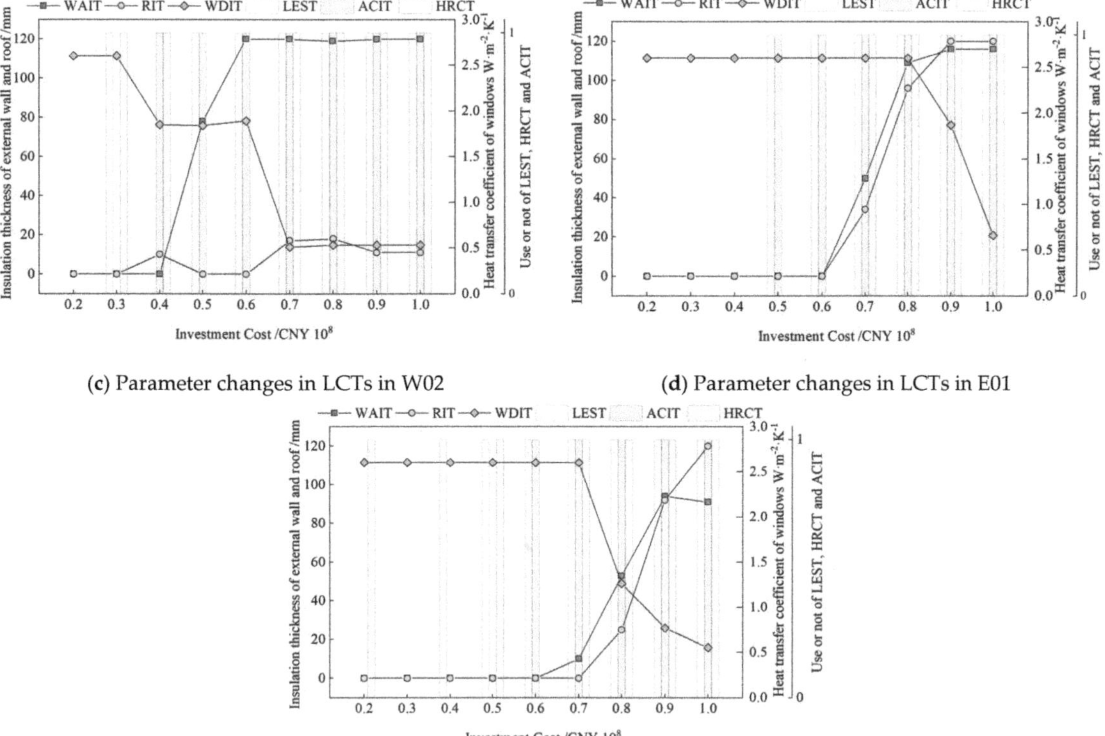

(**c**) Parameter changes in LCTs in W02

(**d**) Parameter changes in LCTs in E01

(**e**) Parameter changes in LCTs in E02

Figure 7. Variation in the parameters of different carbon-reduction technologies with the investment cost in each type of building.

As can be seen in Figure 7, there is a large difference in the variation in the parameters of the various types of LCT in the same type of building. With the increase in the investment cost, the priority of using various LCTs in different types of building is different. Changes in the technical parameters of the restaurant building R01 differ significantly from the other four types of office building. Changes in the technical parameters of the restaurant building R01 differ significantly from the other four types of office buildings. The other four types of office buildings have a relatively high priority for HRCT, but there are differences in the investment costs when the technology starts to be used.

For restaurant building R01, equipment energy efficiency upgrading technologies, such as ACIT and LEST, begin to be applied at an investment cost of CNY 30 million, indicating the high priority of this technology for restaurant buildings. In addition, the WDIT does not change with the investment cost, and remains at the initial value of 2.6. The RIT starts be applied when the investment reaches CNY 60 million.

For the low-rise enclosed office building W01, the first technologies configured are WAIT and HRCT and, when the investment cost reached CNY 40 million, the reduction in the window heat transfer coefficients begins. The parameters of the roof insulation technology do not change significantly with the investment cost.

For the high-rise enclosed office building W02, LEST, HRCT, and WDIT are the first technologies to be applied in the building, at a total investment cost of CNY 30 million. When the investment cost reaches CNY 70 million, with the exception of RIT, other LCTs are maximized.

For the low-rise "E" office building, the implementation of LCTs begins when the investment cost is CNY 50 million, with HRCT being the first technology to be implemented and, when the investment cost reached CNY 65 million, RIT and WAIT begin to be applied on a large scale, with all LCTs being maximized in the building when the total investment cost reaches CNY 10,000 million.

For the high-rise "E"-type office building, the first to be applied is the HRCT, and then, when the total investment reaches CNY 35 million, LEST and ACIT begin to be applied in the building, while the thermal performance improvement of the building envelope begins to be gradually configured when the total investment reaches CNY 70 million. By the time the total investment reaches CNY 105 million, all the LCTs are maximized in the building.

(3)   Variation in the configuration parameters of the same LCT in different buildings

According to the results of the model calculations, the variation in the configuration parameters of the same carbon-reduction technology in different buildings, with the investment cost is obtained, as shown in Figure 8.

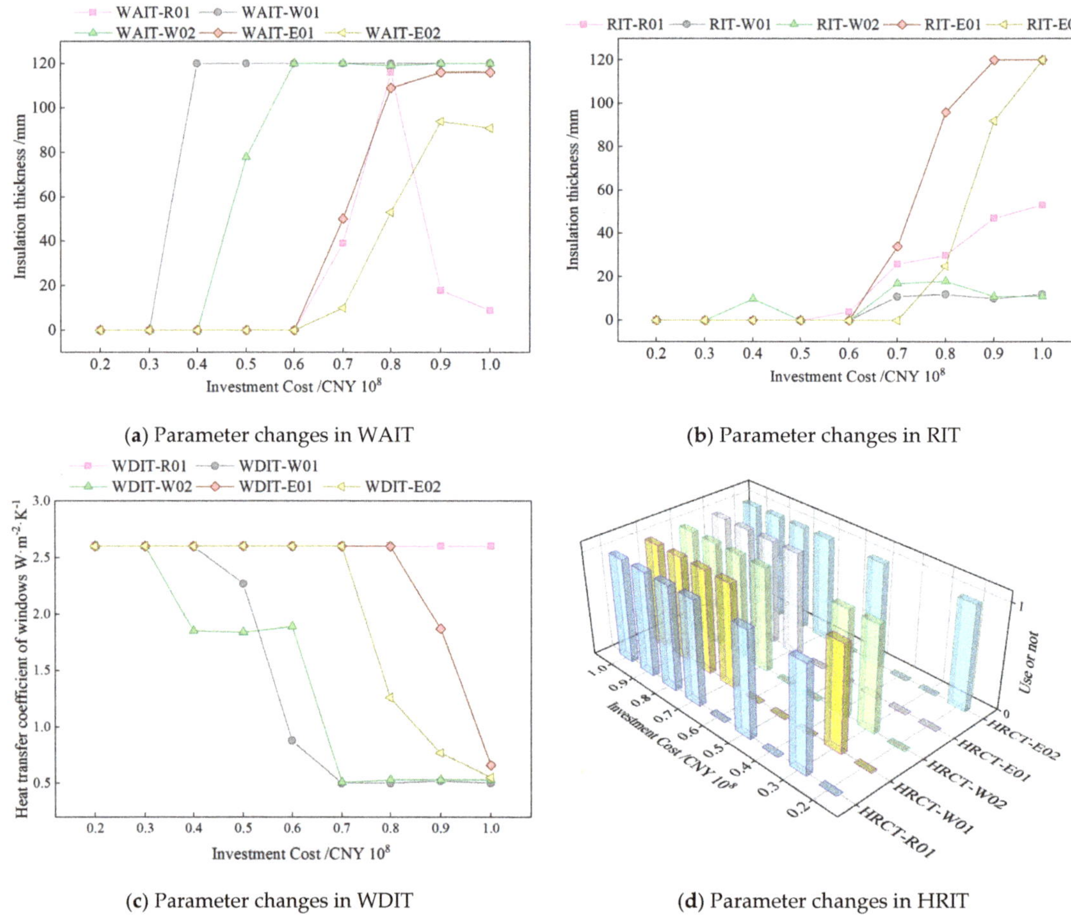

(**a**) Parameter changes in WAIT

(**b**) Parameter changes in RIT

(**c**) Parameter changes in WDIT

(**d**) Parameter changes in HRIT

Figure 8. *Cont.*

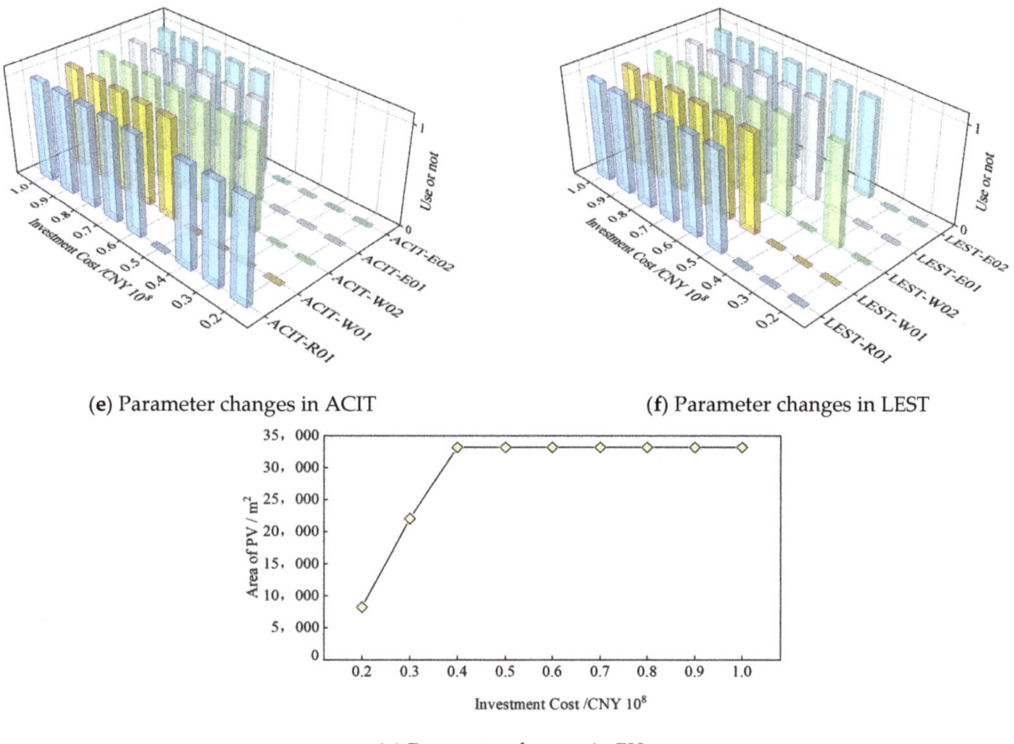

Figure 8. Parameter changes with investment costs for each LCT.

As can be seen in Figure 8, the same LCT is prioritized differently for use in different buildings, as investment costs increase.

For the WAIT, comparing its priority in different types of building, we can see that its application priority on enclosed office buildings is relatively higher, while its application priority on restaurant buildings is the lowest. Comparing W01 and W02, and E01 and E02, it can be found that, for the enclosed or "E" type of building, the priority of WAIT is higher for low-rise buildings than for high-rise buildings.

For the RIT, the technology only begins to be applied on a large scale when the investment cost is CNY 70 million, indicating that the technology has a low priority. From the comparison of its priority in different types of building, it can be seen that the technology in various types of building showed a priority of no significant difference. However, the technology in different types of building on the configuration parameters varies significantly; with the "E"-type office buildings E01 and E02 at the investment cost of CNY 100 million, the roof is set up with a thicker insulation layer, both 120 mm, while, at this point, the thickness of the roof insulation layer of the enclosed office buildings W01 and W02 is only 10 mm.

For the WDIT, comparing the priority of this technology on different types of building, it can be seen that the priority of the enclosed office buildings W01 and W02 is higher, the priority of the "E" buildings E01 and E02 is relatively lower, and there is no configuration of this technology for the restaurant building R01. Comparing W01 and W02, and E01 and E02, it can be found that, for enclosed or "E"-type buildings, the technology configuration priority of high-rise buildings is higher than that of low-rise.

For the HRCT, the technology begins to be applied when the total investment cost is CNY 20 million, indicating that the technology is prioritized higher. Through comparing its priority in different types of building, it can be seen that the difference in the priority of this technology in various types of building is small, and the priority in E02 buildings is relatively low.

For the ACIT, comparing its priority on different types of building, it can be seen that its priority is higher on the restaurant building R01, which starts to apply it when the total investment cost is CNY 20 million. Meanwhile, high-rise office buildings W02 and E02 start applying the technology at a total investment cost of CNY 50 million, and low-rise office buildings W01 and E01 start applying the technology at a total investment cost of CNY 60 million.

For the LEST, through comparing its priority in different types of building, it can be seen that there is no obvious difference in the priority of the technology in various types of building. At a total investment cost of about CNY 40 million, the technology starts to be applied in E02 and W02 and, when the total investment cost is CNY 50 million, the technology starts to be applied comprehensively in the park.

4.4. Analysis of Carbon-Reduction Benefits with Limited Cost

Based on the results of the optimization model calculations, the variation in the carbon-reduction benefits of each LCT with the investment cost in each type of building is obtained, as shown in Figure 9.

From Figure 9f, it can be seen that PVs have an outstanding carbon-reduction benefit of up to 13.22 kg CNY^{-1}. Comparing Figure 9a–e, it can be seen that, with the increase in investment cost, the changes in the carbon-reduction benefits of various types of LCT in different types of building vary greatly.

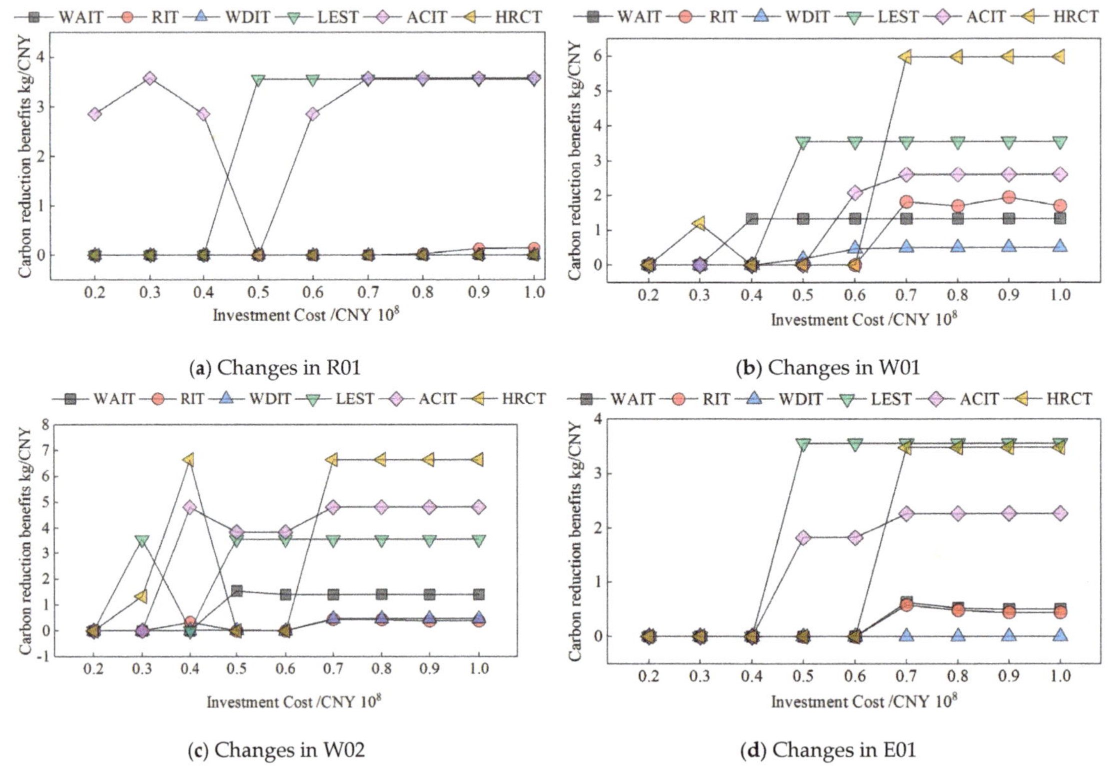

(**a**) Changes in R01

(**b**) Changes in W01

(**c**) Changes in W02

(**d**) Changes in E01

Figure 9. *Cont.*

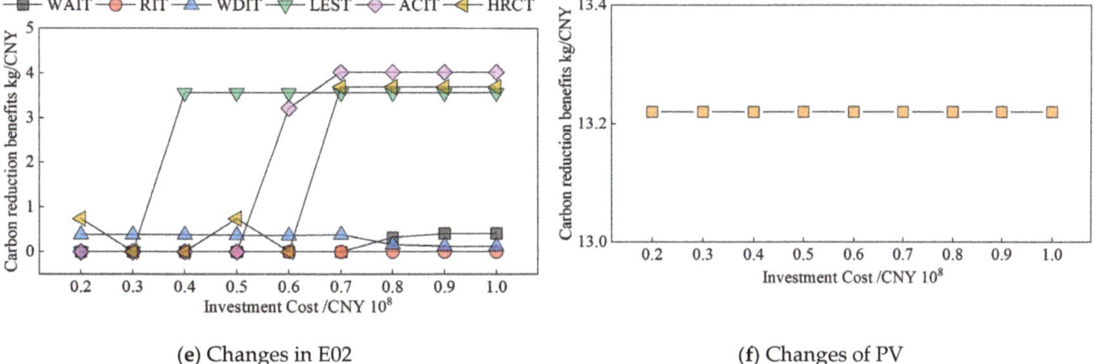

(**e**) Changes in E02

(**f**) Changes of PV

Figure 9. Carbon-reduction benefits of various LCTs in each building type changing with the investment cost.

Among them, there is a big difference in the change in the carbon-reduction benefit between restaurant building R01 and other types of building, as can be seen from Figure 9a: in the restaurant building R01, the ACIT and LEST have a high carbon-reduction benefit; at the investment cost of CNY 50–110 million, the building implements LEST, and the carbon-reduction benefit is 3.56 kg CNY^{-1}; at the investment cost of CNY 70–110 million, the carbon-reduction benefit of ACIT is 3.58 kg CNY^{-1}. The carbon-reduction benefit of the technology to enhance the thermal insulation performance of the envelope, such as WAIT and RIT, is 0 in this restaurant building.

For the four types of office buildings, HRCT, LEST, and ACIT have relatively higher carbon-reduction benefits, but the carbon-reduction benefits for different types of building have large differences. For example, the low-rise enclosed building W01, configured with HRCT, achieved a carbon-reduction benefit of 5.98 kg/CNY, and this result was 6.65, 3.48, and 3.69 kg CNY^{-1} for W02, E01, and E02, respectively. This shows that the carbon-reduction benefit of those technologies in enclosed buildings is higher than that in "E"-type buildings; the maximum carbon-reduction benefit achieved by the ACIT in low-rise buildings W01 and E01 is 2.6 and 2.27 kg CNY^{-1} respectively, while the maximum carbon-reduction benefit achieved in high-rise buildings W02 and E02 is 4.82, 4.02 kg CNY^{-1} respectively, indicating that the carbon-reduction benefit of high-rise buildings is higher than that of low-rise buildings.

5. Conclusions

This paper establishes a comprehensive optimization model for the retrofit of building-equipment-renewable-energy-systems in existing office parks, with the goal of maximizing the carbon reduction over the whole life cycle. Based on the modeling results, the maximum carbon reduction achievable in the park and the corresponding maximum investment cost are obtained. By setting different investment cost constraints, the maximum carbon reduction and the change in building-equipment-renewable-energy-system configuration parameters with the increase in investment cost are obtained. Based on the analysis of the carbon-reduction benefits and the prioritization of various LCTs in the different types of building in the park, the main conclusions are as follows:

① With the increase in investment cost, the carbon reduction of the park increases gradually, but non-linearly. When the total investment cost of the park is below CNY 60 million, the carbon reduction increases faster with the investment cost, but as the investment cost continues to increase, the trend of increasing carbon reduction gradually decreases, and when the investment cost reaches CNY 103 million, the park achieves the maximum carbon reduction of 236,087 t. Up to this point, the carbon reduction of the park will not increase with the investment.

② By analyzing the change in parameters of different LCTs with investment costs in various types of building, we can see that, overall, the changes in the technical parameters of restaurant buildings are significantly different from those of office buildings. The decarbonization retrofit of the restaurant is more suitable for the equipment energy efficiency improvement technology, not for the thermal performance improvement technology of the building envelope. In office buildings, both the equipment energy efficiency improvement and thermal performance improvement technologies for the building envelope are applicable, but equipment retrofitting is prioritized over the building envelope. It is worth noting that the same technology is prioritized differently in different types of building; e.g., WAIT is prioritized more highly in enclosed buildings than in "E" buildings.

③ An analysis of the carbon-reduction benefits of different LCTs in different building types found that, overall, the carbon-reduction benefits of all LCTs were higher in high-rise buildings than in low-rise buildings, higher in office buildings than in restaurants, and higher in enclosed buildings than in E-type buildings. In addition, the carbon-reduction benefits of different LCTs in the same type of building varied considerably and, overall, retrofitting equipment and renewable energy systems had better carbon-reduction benefits than retrofitting the thermal performance of the building envelope.

④ The optimal configuration of the building-equipment–renewable-energy-system retrofit in the park with the change in investment cost has a large difference. The paper gives the configuration parameters of each low-carbon technology in different types of building in an office park in Nanchang City with the increase in investment cost, but in other office parks, with different building compositions as well as climatic zones, the specific parameter settings of the optimal configuration scheme under different investment cost constraints should be determined based on the results of the modeling calculations.

This paper establishes an optimal configuration model for building-equipment–renewable-energy-system retrofits based on the relationship between investment cost, energy consumption, and carbon emissions in the park. It provides a reference for the design of low-carbon retrofits of existing parks. However, this paper does not take the guidance of building design on users' autonomous energy-saving behaviors into account in the establishment of the model, and the model needs to be further optimized in future research.

Author Contributions: Conceptualization, Z.D. and L.J.; methodology, Y.C.; software, Y.C.; validation, Z.H., Z.D. and L.J.; formal analysis, Z.H.; investigation, Z.H.; data curation, Y.C.; writing—original draft preparation, Y.C.; writing—review and editing, L.J.; visualization, Y.C.; funding acquisition, Z.D. All authors have read and agreed to the published version of the manuscript.

Funding: This research was funded by the science and technology project of State Grid Corporation of China: entitled "Research and Demonstration of Key Technologies for the Implementation of Zero Carbon Park Based on Flexible Resource Carbon Control of Cold, Heat and Electricity" (grant number 5400-202325227A-1-1-ZN).

Data Availability Statement: The data used to support the findings of this study are included in the article.

Acknowledgments: Many thanks to State Grid Corporation of China for providing the research funding.

Conflicts of Interest: The authors declare no conflict of interest.

References

1. Dong, F.; Yu, B.; Hadachin, T.; Dai, Y.; Wang, Y.; Zhang, S.; Long, R. Drivers of carbon emission intensity change in China. *Resour. Conserv. Recycl.* **2018**, *129*, 187–201. [CrossRef]
2. Liu, Q.; Zhang, W.; Yao, M.; Yuan, J. Carbon emissions performance regulation for China's top generation groups by 2020: Too challenging to realize? *Resour. Conserv. Recycl.* **2017**, *122*, 326–334. [CrossRef]
3. Yin, H.; Zhou, X. Research on Low-Carbon Development Model of Domestic and Overseas Typical Cities and Parks. *Shanghai Energy Sav.* **2022**, *04*, 363–369.

4. IPCC. *2019 Refinement to the 2006 IPCC Guidelines for National Greenhouse Gas Inventory*; IPCC: Geneva, Switzerland, 2019.

5. Song, Z.; Liu, T.; Liu, Y.; Jiang, X.; Lin, Q. Study on the optimization and sensitivity analysis of CCHP systems for industrial park facilities. *Int. J. Electr. Power Energy Syst.* **2020**, *120*, 105984. [CrossRef]

6. Wu, D.; Han, S.; Wang, L.; Li, G.; Guo, J. Multi-parameter optimization design method for energy system in low-carbon park with integrated hybrid energy storage. *Energy Convers. Manag.* **2023**, *291*, 117265. [CrossRef]

7. Wang, Y.; Li, R.; Dong, H.; Ma, Y.; Yang, J.; Zhang, F.; Li, S. Capacity planning and optimization of business park-level integrated energy system based on investment constraints. *Energy* **2019**, *189*, 116345. [CrossRef]

8. Guo, W.; Wang, Q.; Liu, H.; Desire, W.A. Multi-energy collaborative optimization of park integrated energy system considering carbon emission and demand response. *Energy Rep.* **2023**, *9*, 3683–3694. [CrossRef]

9. Luo, Z.; Lu, Y.; Cang, Y.; Yang, L. Study on dual-objective optimization method of life cycle energy consumption and economy of office building based on HypE genetic algorithm. *Energy Build.* **2022**, *256*, 111749. [CrossRef]

10. Fesanghary, M.; Asadi, S.; Geem, Z.W. Design of low-emission and energy-efficient residential buildings using a multi-objective optimization algorithm. *Build. Environ.* **2012**, *49*, 245–250. [CrossRef]

11. Ferrara, M.; Fabrizio, E.; Virgone, J.; Filippi, M. A simulation-based optimization method for cost-optimal analysis of nearly Zero Energy Buildings. *Energy Build.* **2014**, *84*, 442–457. [CrossRef]

12. Gerber, D.J.; Pantazis, E.; Wang, A. A multi-agent approach for performance based architecture: Design exploring geometry, user, and environmental agencies in façades. *Autom. Constr.* **2017**, *76*, 45–58. [CrossRef]

13. Yigit, S.; Ozorhon, B. A simulation-based optimization method for designing energy efficient buildings. *Energy Build.* **2018**, *178*, 216–227. [CrossRef]

14. Tuhus-Dubrow, D.; Krarti, M. Genetic-algorithm based approach to optimize building envelope design for residential buildings. *Build. Environ.* **2010**, *45*, 1574–1581. [CrossRef]

15. Carli, R.; Dotoli, M.; Pellegrino, R.; Ranieri, L. Using multi-objective optimization for the integrated energy efficiency improvement of a smart city public buildings' portfolio. In Proceedings of the 2015 IEEE International Conference on Automation Science and Engineering (CASE), Gothenburg, Sweden, 24–28 August 2015; IEEE: New York, NY, USA; pp. 21–26.

16. Chantrelle, F.P.; Lahmidi, H.; Keilholz, W.; El Mankibi, M.; Michel, P. Development of a multicriteria tool for optimizing the renovation of buildings. *Appl. Energy* **2011**, *88*, 1386–1394. [CrossRef]

17. Petkov, I.; Mavromatidis, G.; Knoeri, C.; Allan, J.; Hoffmann, V.H. MANGOret: An optimization framework for the long-term investment planning of building multi-energy system and envelope retrofits. *Appl. Energy* **2022**, *314*, 118901. [CrossRef]

18. Abdou, N.; El Mghouchi, Y.; Hamdaoui, S.; Mhamed, M. Optimal Building Envelope Design and Renewable Energy Systems Size for Net-zero Energy Building in Tetouan (Morocco). In Proceedings of the 2021 9th International Renewable and Sustainable Energy Conference (IRSEC), Tetouan, Morocco, 28 March 2022.

19. Lin, Y.H.; Lin, M.D.; Tsai, K.T.; Deng, M.J.; Ishii, H. Multi-objective optimization design of green building envelopes and air conditioning systems for energy conservation and CO_2 emission reduction. *Sust. Cities Soc.* **2021**, *64*, 102555. [CrossRef]

20. Bichiou, Y.; Krarti, M. Optimization of envelope and HVAC systems selection for residential buildings. *Energy Build.* **2011**, *43*, 3373–3382. [CrossRef]

21. Hashempour, N.; Taherkhani, R.; Mahdikhani, M. Energy performance optimization of existing buildings: A literature review. *Sustain. Cities Soc.* **2020**, *54*, 101967. [CrossRef]

22. Mela, K.; Tiainen, T.; Heinisuo, M. Comparative study of multiple criteria decision making methods for building design. *Adv. Eng. Inform.* **2012**, *26*, 716–726. [CrossRef]

23. Shan, W. Research on Low-Carbon Design of High-Tech Park Office Building under Ecological Concept. Master's Thesis, Nanchang Hangkong University, Nanchang, China, 2020.

24. Zhu, J. Study on the Low Carbon Design Strategy of the Office Building of High-Tech Park. Master's Thesis, Huazhong University of Science and Technology, Wuhan, Chian, 2012.

25. Huang, B.; Jiang, P.; Wang, S.; Zhao, J.; Wu, L. Low carbon innovation and practice in Caohejing high-tech industrial park of Shanghai. *Int. J. Prod. Econ.* **2016**, *181*, 367–373. [CrossRef]

26. Wang, Z.; Guo, H.; Wang, S.; Meng, F. Analysis of building energy consumption and discussion on energy-saving reform in an office park. *Archit. Technol.* **2020**, *51*, 670–672.

27. Sun, J. Methodology of Integrated Planning Toward Urban Block-Level Carbon Reduction. *Urban Plan. Forum* **2022**, *6*, 102–109.

28. Xiao, H.; Wei, Q.; Wang, H. Marginal abatement cost and carbon reduction potential outlook of key energy efficiency technologies in China's building sector to 2030. *Energy Policy* **2014**, *69*, 92–105. [CrossRef]

29. Guo, W.; Qiao, X.; Huang, Y.; Fang, M.; Han, X. Study on energy saving effect of heat-reflective insulation coating on envelopes in the hot summer and cold winter zone. *Energy Build.* **2012**, *50*, 196–203. [CrossRef]

30. Ye, X.; Chen, D. Thermal insulation coatings in energy saving. In *Energy-Efficient Approaches in Industrial Applications*; IntechOpen: London, UK, 2018.

31. Liu, Z.; Liu, Y.; He, B.J.; Xu, W.; Jin, G.; Zhang, X. Application and suitability analysis of the key technologies in nearly zero energy buildings in China. *Sustain. Energy Rev.* **2019**, *101*, 329–345. [CrossRef]

32. Liu, Y.; Xue, S.; Guo, X.; Zhang, B.; Sun, X.; Zhang, Q.; Dong, Y. Towards the goal of zero-carbon building retrofitting with variant application degrees of low-carbon technologies: Mitigation potential and cost-benefit analysis for a kindergarten in Beijing. *J. Clean. Prod.* **2023**, *393*, 136316. [CrossRef]
33. Tang, J.; Cai, X.; Li, H. Study on development of low-carbon building based on LCA. *Energy Procedia* **2011**, *5*, 708–712. [CrossRef]

 *processes*

Article

Experimental Study on Convection and Heat Conduction Heating of an Air-Conditioned Bed System under Winter Lunch Break Mode

Junjie Jin [1,*], Peiyao Duan [2], Yu Liu [1], Honglin Chen [1] and Tingting Yu [1]

[1] School of Energy and Building Environmental Engineering, Henan University of Urban Construction, Pingdingshan 467036, China; liuyu@huuc.edu.cn (Y.L.); 20172065@huuc.edu.cn (H.C.); 30040504@huuc.edu.cn (T.Y.)

[2] Aviation Engineering Institute, Anyang University, Anyang 455000, China; ayxyzsb@163.com

* Correspondence: junjiefx@huuc.edu.cn; Tel.: +86-15036853737

Abstract: In this paper, an experimental study of a system for heating an air-conditioned bed during a 2 h lunch was carried out. The results show that the power consumption of heat conduction heating was only 0.34 kW·h and that the average heat dissipation was 81.3 W, while the power consumption of convection heating was 1.43 kW·h, accompanied by an average heat dissipation of 748.7 W. Regardless of the power consumption or the heat dissipation, the convection heating was significantly higher than the heat conduction heating. As a result, the room air temperature increased from 12.3 °C to 17.3 °C under convection heating, but only increased from 14.4 °C to 15.2 °C under heat conduction heating. The study results indicate that when using heat conduction heating, water temperatures in the range of 38~40 °C could meet the thermal comfort needs of the human body; however, a higher temperature range was required when using convection heating. In contrast, the grade of the hot water required for heat conduction heating was lower. It was also found that the temperature under convection heating rises faster, but it tends to lead to a dry feeling after a long time, while the conductive heating showed a slower temperature rise. There was a cool feeling for 20 min when the heating started, and then the thermal comfort improved. The air-conditioning system in this paper was investigated in a heating experiment in the winter lunch break mode and compared with convection heating. The heat conduction heating resulted in better thermal comfort and higher energy efficiency. It is suggested to adopt the heat conduction heating mode in the winter heating operation of this system.

Keywords: air-conditioned bed system; heat conduction heating; convection heating; thermal comfort; energy conservation

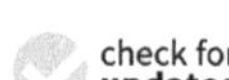
check for **updates**

Citation: Jin, J.; Duan, P.; Liu, Y.; Chen, H.; Yu, T. Experimental Study on Convection and Heat Conduction Heating of an Air-Conditioned Bed System under Winter Lunch Break Mode. *Processes* **2023**, *11*, 2391. https://doi.org/10.3390/pr11082391

Academic Editors: Yabin Guo, João M. M. Gomes, Zhanwei Wang and Yunpeng Hu

Received: 28 June 2023
Revised: 23 July 2023
Accepted: 25 July 2023
Published: 9 August 2023

1. Introduction

With increasing energy demand in the world, energy conservation and reductions in carbon emissions have become important issues. In order to reduce the energy consumption of air conditioning and improve personal comfort, some scholars focus on the study of personal comfort system (PCS) [1]. A PCS relaxes the restriction on the temperature of inactive areas, and only adjusts the thermal comfort of the area where the individual is located, which is conducive to reducing the temperature difference between indoor and outdoor spaces and, thereby, reducing energy consumption. PCS, namely, task ambient conditioning or personal conditioning system, is a personal heating system that can ensure human thermal comfort under cold conditions by adopting low energy consumption. It is a good supplement to district heating systems in Northern China [2]. In the published review papers [3–6], researchers mainly discuss the thermal comfort and energy efficiency of various PCS, and focus more on cooling and ventilation than heating. There is currently limited research on PCS for heating during winter sleep. In this paper, a capillary convection and heat conduction air-conditioned bed system is proposed, which creates a local thermal comfort environment when people sleep and rest. It not only provides heating in winter, but also cooling in summer. There are three heat transfer modes: heat conduction, radiation, and convection. The energy-saving effect is obvious, and it can be used in homes, hotels, hospitals, and other places, which has certain research significance. This paper mainly focuses on the experimental study of heating in the noon break mode.

2. Literature Review

At present, the research related to personal comfort system is mainly based on personalized air supply in summer [7]. There are few studies on heating in winter, and they mainly apply to office spaces. In some cases, the airflow is directed toward the head and other parts of the human body by setting an outlet or placing a fan near the desk [8–10]. There are also heating or cooling seats designed to meet the requirements of human thermal comfort [11,12]. Other designs focus more directly on a certain part of the human body, such as a heating device placed on the feet, and operate through the form of heat conduction or radiation heating for the human body [13,14]. Foda et al. [15] proposed a strategy to maximize the energy efficiency using local floor heating and found that local floor heating can meet the comfort requirements at 18 °C. Zeiler et al. [16] studied the thermal comfort in school classrooms, and their results show that radiant heating is more comfortable than traditional heating methods. Deng et al. [17] found that the overall thermal sensation and comfort resulting from personalized heating at 16 °C were rated as neutral and comfortable, respectively, which were significantly higher than the ratings obtained without the personalized heating at 18 °C. In addition, there are some differences in the thermal sensation of different parts of the human body, and the influence on the overall thermal sensation is also different. Arens et al. found that in colder environments, the thermal sensation of certain parts of the human body, such as the back, chest, etc., is greater than that of the hands and feet [18]. Lv et al. conducted the relevant experimental research on human thermal sensation and found that in a specific hot and humid environment, local heating or cooling of indoor personnel can produce a sense of thermal comfort [19]. Schellen et al. studied the differences in the perception of thermal comfort between genders under non-uniform environmental conditions, and found that local thermal sensation and skin temperature have a more pronounced effect on the overall thermal sensation in women than in men [20].

In recent years, some experts and scholars have conducted much research work on personal comfort system in the sleep environment. Wang et al. [21] developed a water-heated bed, and Yu et al. [22] developed a solar phase-change heat-storage heating bed system. Both of these designs provide a thermal comfort environment during sleep through heat conduction. To provide a thermal comfort environment under sleeping conditions through convection, Pan et al. [23] designed a personal sleep comfort system with a flexible air duct and a static pressure box, but the structure is more complex. Mao et al. [24–27] designed a simplified system on this basis, removing the flexible air duct and the static pressure box, which is conducive to its development. Du et al. [28] added a radiation plate on the basis of the design by Mao et al., which assists in solving the discomfort caused by the tuyere blowing on to the human body. In addition, radiant air conditioning also has several advantages. Radiation heating or cooling has a small vertical temperature gradient and can maintain a high level of thermal comfort [29,30]. Xin et al. [31] designed a local space capillary radiation air-conditioning system for the sleeping environment. The capillary radiation plate was set above the horizontal human body and surrounded by a baffle and a bed curtain. The thermal comfort during heating was good, and the system was energy efficient. However, condensation tends to occur in radiant air-conditioning refrigeration, which is a key factor restricting its development. The air-conditioning bed system proposed in this article can provide both heating and cooling, and can avoid condensation during refrigeration. The system provides a high level of thermal comfort, has a simple structure, and results in obvious energy savings, which is conducive to its future development.

3. Experiment Introduction

3.1. Experimental Room and Equipment

The laboratory room is selected from the student dormitory building of Henan University of Urban Construction, where the walls are all made of 240 mm brick walls, the inside is painted with 20 mm white plaster, the outside is painted with 20 mm white plaster or cement mortar, the external windows are installed with 6 mm thick single-layer ordinary glass, and the room height is 3.1 m. During the lunch break heating experiment, none of the adjacent rooms (up and down, left and right) were heated. The size of the wooden bed in the room is 1860 × 1000 mm, a layer of 20 mm thick thermal insulation cotton is laid on the bed, a layer of capillary tube network is laid on top of the thermal insulation cotton, and a sponge-padded mattress is placed on the capillary tube. The capillary tube adopts the third-type of Polypropylene Pipe (PP-R), the main pipes for water supply and return are both 20 × 2.0 mm, the size of capillary diameter is 4.3 × 0.8 mm size, and the spacing is 40 mm. The relevant layout plan is shown in Figure 1.

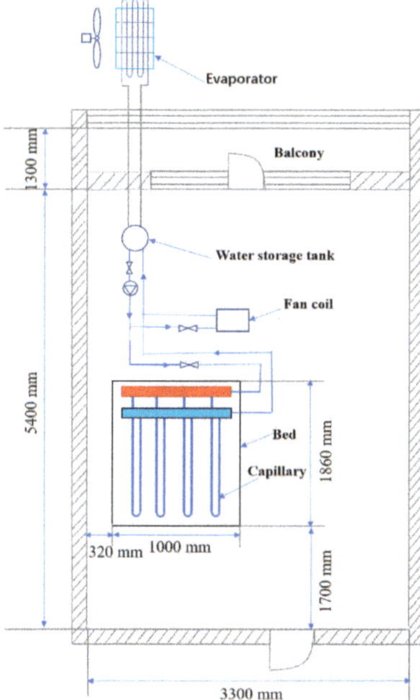

Figure 1. Layout of experimental room and air-conditioned bed system.

In the experiment, 1.8 kg of sponge-padded mattress and 1.9 kg of cotton silk quilt are selected. It adopts the FLIR T540 infrared thermal imager, the T-type thermocouple, the COS-04 temperature and humidity recorder, the GTLWGY10ALC2SSN turbine flowmeter, and the UNI-T anemometer, with the YP5000 multi-channel temperature recorder, to record data. The air duct adopts a telescopic PVC aluminum foil pipe with a diameter of 183 mm, and the heat source unit adopts a fixed frequency air source heat pump to provide the hot water for heating. A 280 W desktop computer is placed in the laboratory to collect data. As shown in Figures 2–5.

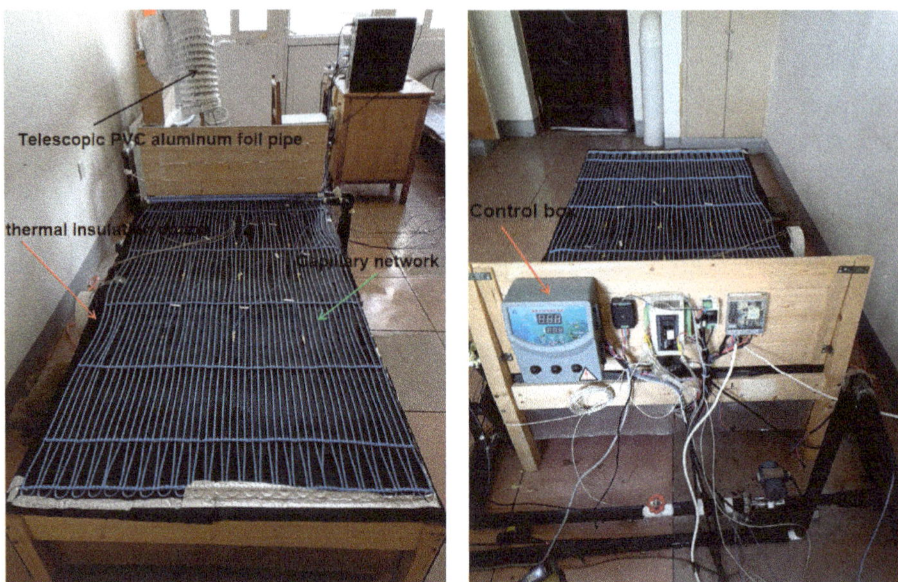

Figure 2. Physical diagram of capillary convection and heat conduction air-conditioned bed system.

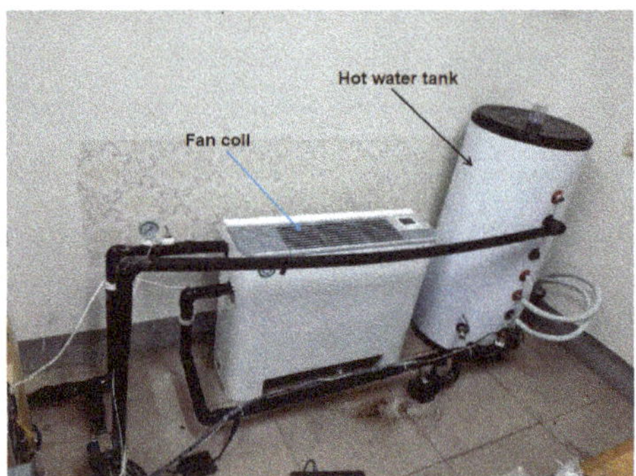

Figure 3. Fan coil and hot water tank.

Figure 4. Air source heat pump.

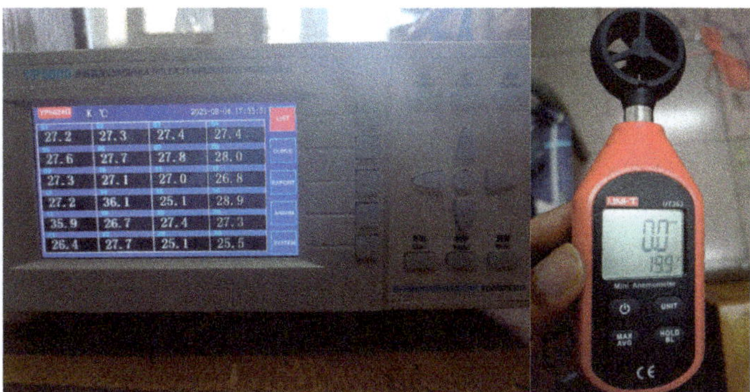

Figure 5. YP5000 multi-channel temperature recorder and anemometer.

3.2. Subjects and Measurement Point Arrangement

The subjects are four male college students and four female college students, aged 20–23 years, with a height of (1.70 ± 0.12) m, a body mass of (60 ± 14) kg, and a body mass index (BMI) of (20.4 ± 2.0) kg/m^2. In the experiment, everyone is required to wear thin long sleeves on top and thin pants on bottom, and the thermal resistance of the clothes is about 0.4 clo [32]. All subjects are in good physical condition, and do not exercise vigorously before the start of the experiment, and once the experiment begins, the subjects are required to lie flat and remain still throughout the entire process. During the experiment, we should know the subject's evaluation of thermal comfort at all times. Figure 6 shows the field photo of the subject during the experiment.

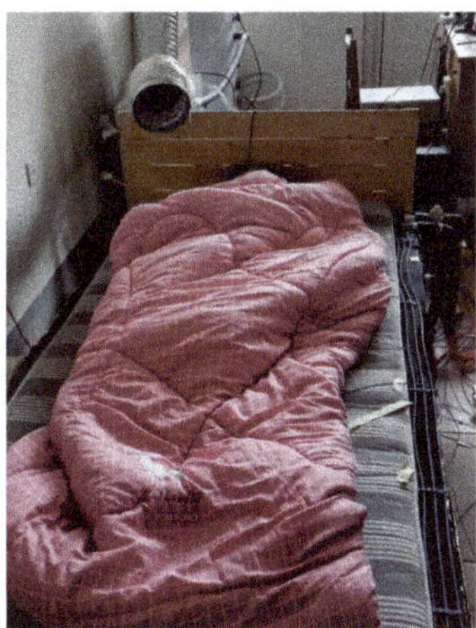

Figure 6. Field view of the subject during the experiment.

During the experiment, the inlet main pipe and outlet main pipe of the capillary network are arranged with T-type thermocouples, respectively, and the measured data is used as the supply water temperature (SWT) and return water temperature (RWT) for the experiment. The temperature of the thermocouple arranged on the surface of the mattress directly below the back of the human body is considered as the temperature of the contact surface between human body and bed (TSHB). Three thermocouples are arranged on the surface of the capillary water supply pipe, and six thermocouples are arranged on the surface of the return pipe, and the average value of their measurements is considered as the surface temperature of supply water capillary (STSWC) and the surface temperature of return water capillary (STRWC). The indoor and outdoor temperature and humidity changes are recorded with the temperature and humidity recorder.

3.3. Experimental Research Method

Choosing the time between 12:00–14:00 on 21 February 2023, and 26 February 2023, as the experimental time, the heating experiment mainly based on heat conduction and the heating experiment mainly based on convection were carried out. The external weather conditions during these times are relatively similar, which is conducive to comparing and analyzing the experimental results. During the experiment focused on heat conduction for heating purposes, the water pump will be running throughout the entire process, and while the heat pump unit will be started at 12:00 and stopped at 12:30, the supply water temperature will be controlled within a certain range using the control system, and the hot water will be sent into the capillary tube for the heat conduction experiments. During the experiment focused on convection, hot water will be also provided by the air source heat pump, the control system will maintain the supply water temperature within a certain range, the hot water will be sent into the fan coil unit, and the convective heat exchange will occur through the fan, and then the hot air will be directly delivered near the human body.

The Predicted Mean Vote (PMV) and Predicted Percent Dissatisfied (PPD) proposed by Fanger et al. [33] are commonly used to evaluate human thermal sensation, but this evaluation index is applicable for the evaluation of human thermal comfort in a steady-state thermal environment, not suitable for a dynamic environment. In addition, the other methods include ASHRAE 7-level scale method for thermal sensation voting in the form of questionnaire survey and the classification method of thermal comfort proposed by He et al. [34], but most of them are evaluated for a certain temperature rather than a certain temperature range. This paper adopts the subjective evaluation method to evaluate the thermal comfort of the human body, which is mainly based on the subjective feelings of the testers about the thermal environment, and it is also a voting method to find different temperature ranges of thermal comfort under the conditions of non-uniform environment and dynamic environment. The adopted method is based on the thermal comfort classification method proposed by He et al. [34] and makes appropriate adjustments on this basis, dividing thermal comfort into five grades: very uncomfortable, uncomfortable, unfeeling, comfortable, and very comfortable. In addition, during the experiment, the thermal comfort of eight subjects was tracked and investigated at all times, and the subjective voting method was used to find different thermal comfort temperature ranges according to the number of votes cast by subjects at different TSHB, and the thermal comfort temperature range with a turnout rate exceeding 75% was recognized.

4. Results and Analysis

4.1. Experimental Results of Thermally Conductive Heating

4.1.1. Analysis of Data Changes at Different Measurement Points

As shown in Figures 7 and 8, the indoor air temperature (IAT) and the indoor humidity are relatively stable, with the temperature ranging from 14.4 to 15.2 °C, with an average of 14.9 °C, and the relative humidity in the range of 39.7 to 41.2%, with an average of 40.6%. The outdoor air temperature (OAT) and the outdoor humidity also varied little, with the temperature ranging from 6.9 to 7.6 °C, with an average of 7.2 °C, and the relative humidity in the range of 47.5 to 56.2%, with an average of 51.4%.

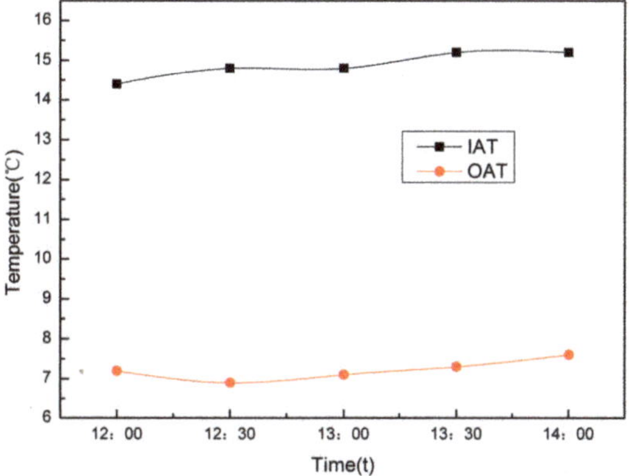

Figure 7. Change graph of indoor and outdoor air temperature.

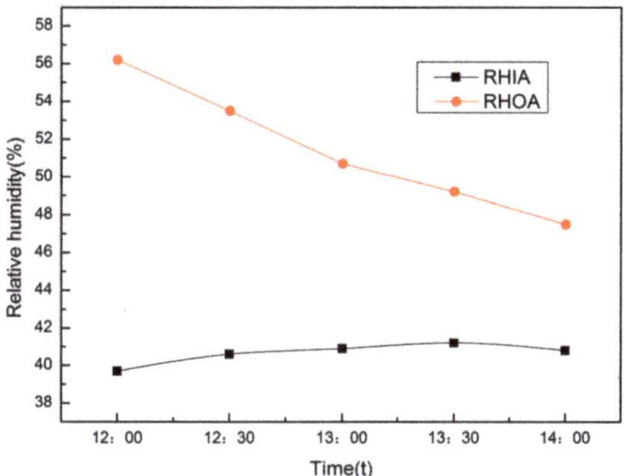

Figure 8. Variation curve of indoor and outdoor air relative humidity.

As shown in Figure 9, the unit is turned on at 12:00, the water supply temperature rises from 28.6 to 42.8 °C within 40 min, while the TSHB also rises rapidly from 24.5 to 37.5 °C. Referring to the ASHRAE seven-level scale method and making appropriate adjustments, the thermal sensation is divided into five levels. During the experiment, the thermal sensation of eight subjects is tracked and investigated at all times, and using subjective voting method, the temperature range of different thermal sensations is determined based on the number of votes cast by subjects at different TSHB. The specific results are shown in Table 1. When people lie on the bed, they feel obvious coldness at the beginning, then feel the surface temperature of the sponge-padded mattress is rising, and finally the cold feeling gradually disappears and a good warm feeling appears, which is consistent with the survey results in Table 1. Hot water enters into the capillary tube, first convection heat exchange with the inner wall of the capillary, then the heat is transferred to the outer surface of the capillary tube by heat conduction, and then to the outer surface of the sponge-padded mattress through thermal conduction. Part of the heat is transferred to the human body and quilt in contact with it through thermal conduction, while the remaining heat is mainly transferred to the quilt and room through heat radiation. During the period from 12:00 to 12:20, it is found that the temperature of the contact surface between human body and bed increased rapidly, and obviously exceeded the temperature of the supply and return water. It is obvious that it cannot be completely caused by the heat transfer from the heating hot water, and the main reason is that the human body is in contact with the surface of the mattress, and the heat of the human body is transferred to the contacted mattress, resulting in a significant increase in temperature at that area.

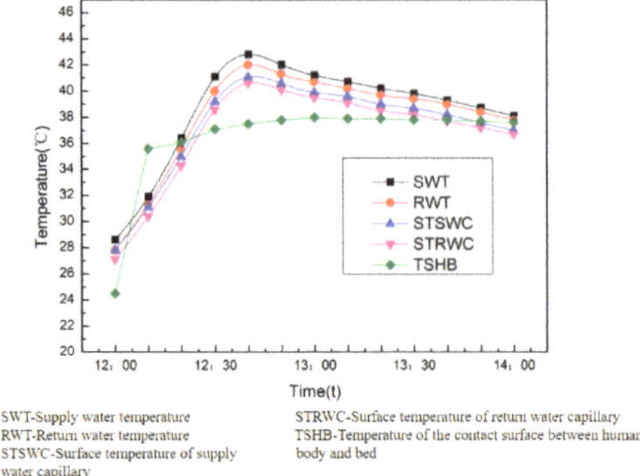

SWT-Supply water temperature
RWT-Return water temperature
STSWC-Surface temperature of supply water capillary
STRWC-Surface temperature of return water capillary
TSHB-Temperature of the contact surface between human body and bed

Figure 9. Temperature variation curve of different measuring points during thermal conduction heating.

Table 1. Thermal sensation subjective voting questionnaire.

Temperature Range (°C)	Thermal Sensation (Number of Voters)				
	Cold	Cool	Neutral	Warm	Hot
(24.0–33.0)	8	0	0	0	0
(33.0–36.0)	2	6	0	0	0
(36.0–36.6)	0	1	7	0	0
(36.6–37.5)	0	0	1	7	0
(37.5–38.0)	0	0	0	5	3

As shown in Figure 10, after the unit has been running for 10 min, the temperature of the four positions increases significantly. Subsequently, the temperature of the contact surface between human body and bed tends to be stable, as does the air temperature near the chest in the quilt (ATCQ). While the air temperature near the legs in the quilt (ATLQ) and the surface temperature of the sponge-padded mattress in the middle of the legs (STM) show a slow increasing trend. The air temperature near the chest in the quilt is always higher than the air temperature near the legs in the quilt, with an average difference of 1.8 °C. On the one hand, this is related to the heat dissipation of the human body, as the chest area has greater heat dissipation, while the leg area has less. On the other hand, it is also because the quilt is covering the human body, sinking with gravity, and tightly adhering to the surface of the human body and the mattress, which hinders the circulation of the air inside the quilt. This tends to make the temperature of the legs rise slower, making people feel cold in their feet at first.

From the temperature curve change rule in Figure 9, it can be seen that after the unit is shut down, the supply water temperature still rises for about 10 min. The reason is that the temperature of the condenser will not decrease immediately, and the water in the hot water tank will still be heated until condenser achieves the same temperature as the water in the hot water tank. In addition, the temperature of the supply and return water and the surface temperature of the supply and return capillary, will gradually decrease after reaching the maximum value, and the water supply temperature will still remain at a high temperature of 38.1 °C at 14:00. In this process, the temperature of the contact surface between human body and bed reaches the maximum temperature of 38.0 °C, and then remains basically unchanged. Even at close to 14:00, it exceeds the surface temperature of the capillary for the supply and return water, and this is because as the supply and return water temperature decreases, the surface temperature of the capillary tube also decreases.

However, the sponge-padded mattress has a heat storage function, which greatly delays the attenuation of the temperature of the contact surface between human body and bed.

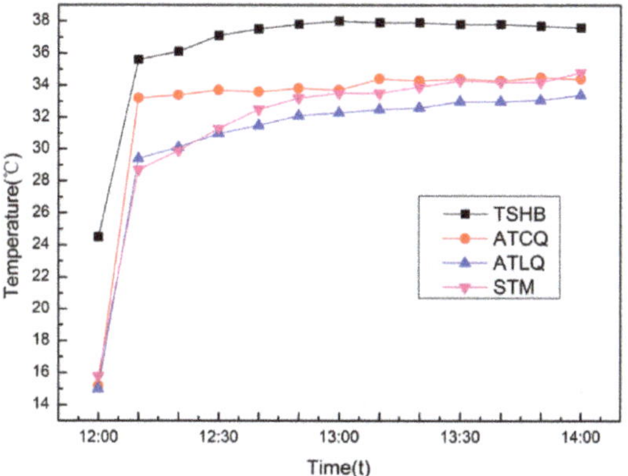

TSHB-Temperature of the contact surface between human body and bed
ATCQ-Air temperature near the chest in the quilt
ATLQ-Air temperature near the legs in the quilt
STM-Surface temperature of the sponge-padded mattress in the middle of the legs

Figure 10. Temperature change curve of measurement points in the quilt during thermal conduction heating.

4.1.2. Temperature Analysis of Infrared Thermal Imager

As shown in Figures 11–14, we have arranged measuring points at different locations on each graph, and from the temperature results shown at different measuring points, it can be seen that the surface temperature of the quilt is not high, and it changes very little over time, the average temperature in different time periods only changes within the range of 16.0 to 17.7 °C, which indicates that the quilt has a good heat insulation function. Additionally, it can be observed that the surface temperature of the quilt is lower at the edges compared to the central area. This is because the surface temperature of human body is higher, and the temperature of the quilt in contact with it is increased by heat conduction. It can also be seen that the outer surface temperature of the quilt at the foot position of the human body is slightly lower than that at other parts of the human body, resulting in a poor thermal comfort of the feet, which is also consistent with the above-mentioned law of air temperature distribution inside the quilt.

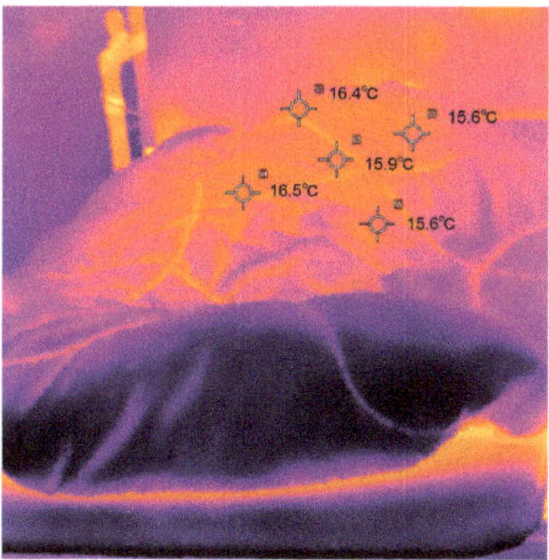

Figure 11. Temperature field distribution of infrared thermal imager at 12:30.

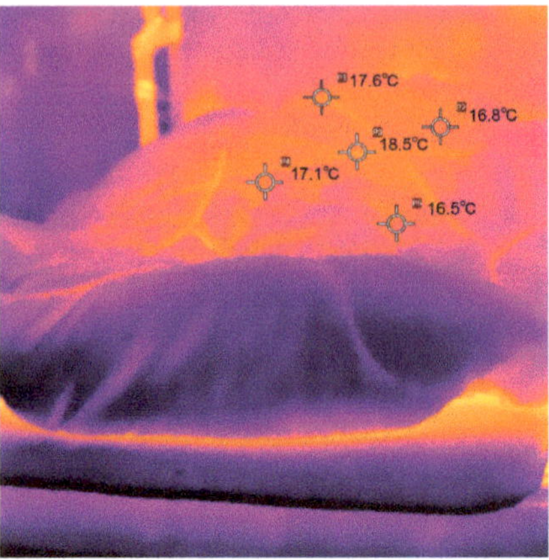

Figure 12. Temperature field distribution of infrared thermal imager at 13:00.

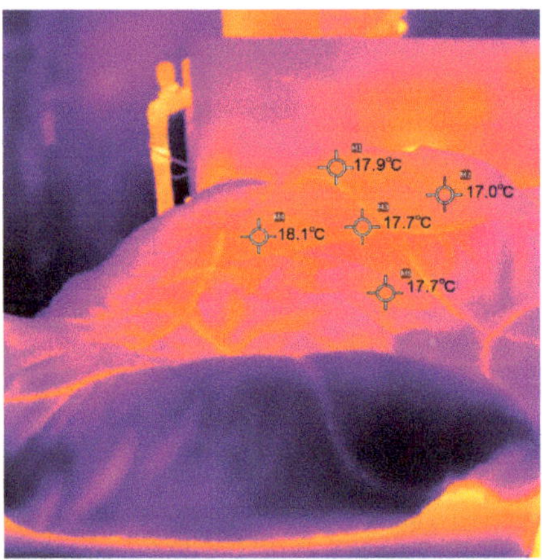

Figure 13. Temperature field distribution of infrared thermal imager at 13:30.

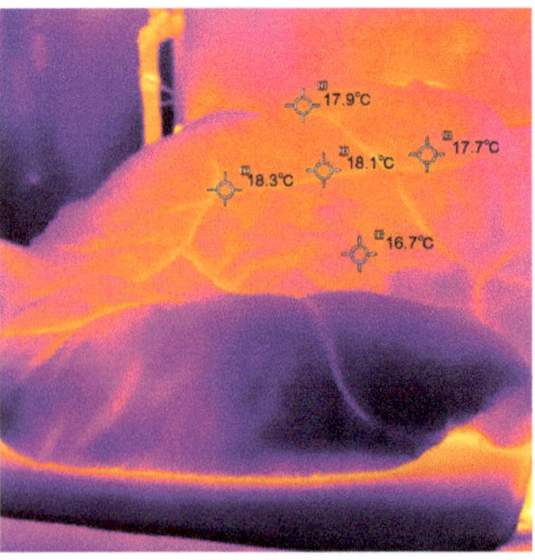

Figure 14. Temperature field distribution of infrared thermal imager at 14:00.

4.1.3. Analysis of Power and Heat Dissipation Data

As shown in Figure 15, the heat pump unit starts at 12:00 and stops at 12:30, operating for exactly half an hour. The average power of the unit during operation is 876.4 W, while the average power of the pump is 35.6 W. After the shutdown of heat pump unit, the pump continues to operate at an average power of 35.6 W, and the power consumption of the whole process is 0.52 kW·h. Through the experiment, it is found that if the lunch break mode experiment was conducted the day before and restarted the next day, the water supply temperature is often around 34 °C or even higher when the machine is turned on. The initial water supply temperature during the heat conduction heating experiment is only 28.6 °C, combining with the temperature change curve of supply and return water in Figure 9 and the power change curve in Figure 15, it can be seen that 0.18 kW·h energy consumption can be saved on this basis, that is to say, the heating in the lunch break mode only needs to consume 0.34 kW·h to meet the thermal comfort of the human body.

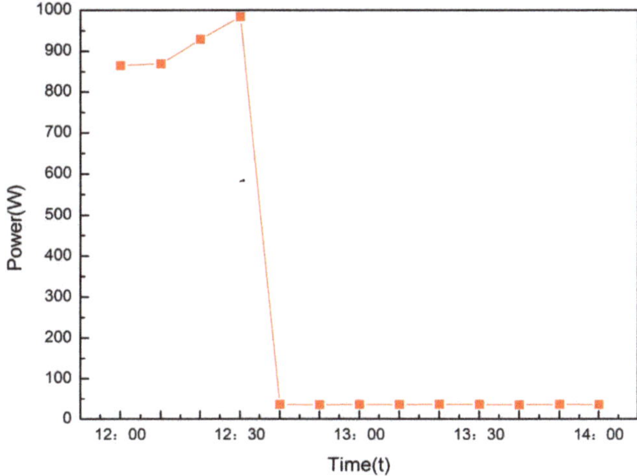

Figure 15. Power variation curve of heat conduction heating with time.

Through the water flow, the temperature difference between the supply and return water and the specific heat of the water, the heat dissipation of heat conduction heating can be calculated. During the whole experiment, the water flow rate changes slightly, and the average water flow rate is 2.0 L/min. As can be seen from Figure 16, the heat dissipation is large at the start, reaching 112.0 W, although the water supply temperature is only 28.6 °C. It is mainly because the ambient temperature of the room and the temperature of the mattress and quilt are relatively low when the unit is just started, which is easy to dissipate heat. With the increase in water supply temperature, the ambient temperature is also gradually increased, especially a sharp increase in the temperature of mattress and quilt, which reduces the temperature difference between them, but the heat dissipation is reduced, which reduces to 81.9 W at 12:10. With the further increase in the water supply temperature, it becomes more conducive to the surrounding heat dissipation, and its heat dissipation is also increased, especially reaching a maximum of 152.5 W at 12:30. And then as the temperature difference between the water supply and mattresses, quilt, etc., further decreases, the heat dissipation gradually decreases, and after 13:40, it tends to be flat. The average heat dissipation during the whole lunch break heating time is only 81.3 W, which is much lower than that of the conventional air conditioning system. In addition, it can also be seen that during the entire lunch break process, the time period of large heat dissipation is mainly concentrated in the period from the start of the unit to the 20 min after the unit stops.

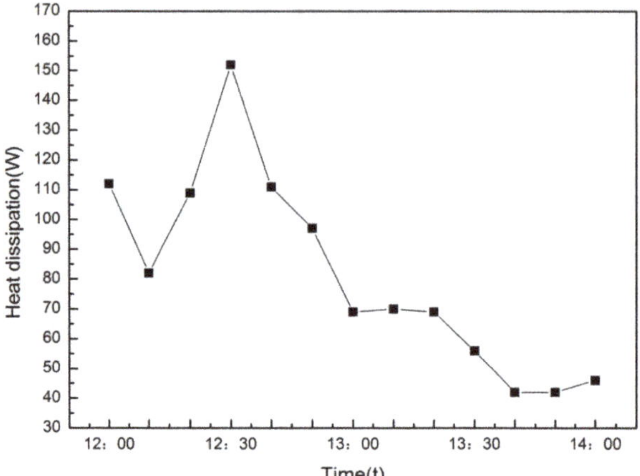

Figure 16. Curve of heat dissipation with time during heat conduction heating.

4.2. Experimental Results of Convective Heat Exchange Heating

4.2.1. Analysis of Data Changes at Different Measurement Points

From Figure 17, it can be seen that the indoor air temperature changes in the range of 12.3 to 17.3 °C, and the increase is more obvious. After half an hour, it tends to be stable, with an average of 16.2 °C. It can be seen that in the convective heating experiment, a large part of the heat is lost in the air, resulting in a significant increase in indoor air temperature. The outdoor air temperature ranges from 4.7 to 7.9 °C, and the average temperature is 6.4 °C, which is close to the weather conditions on the 21st.

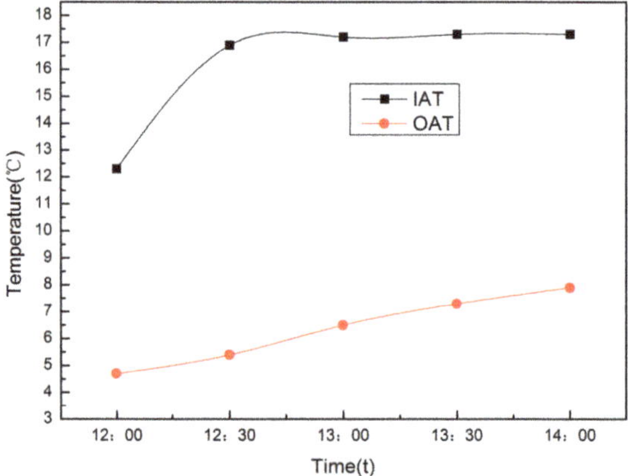

Figure 17. Change graph of indoor and outdoor air temperature.

From Figure 18, it can be seen that the indoor relative humidity ranges from 43.5 to 46.6%, with an average of 44.9%, which is also relatively stable. The outdoor relative humidity is relatively high, ranging from 60.3 to 80.6%, with an average of 71.8%, showing a downward trend.

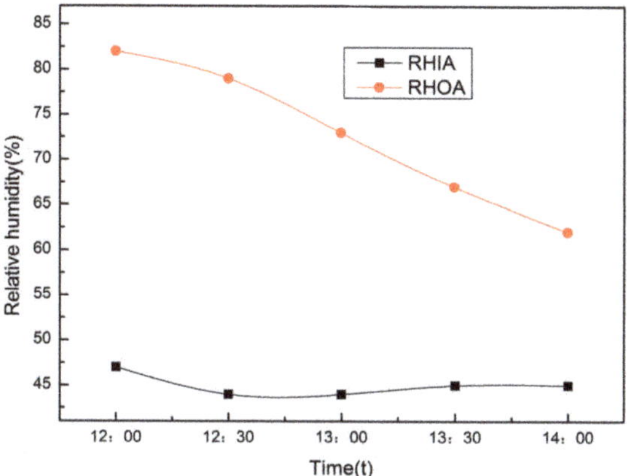

Figure 18. Variation curve of indoor and outdoor air relative humidity.

From Figure 19, it can be seen that when doing the convective heating experiment, the water supply temperature changes in the range of 39.9 to 48.8 °C, and the average temperature difference between the supply and return water is 6.7 °C, However, the average temperature difference between water supply and return in the heat conduction heating mode is only 0.6 °C. In addition, the air temperature near the legs in the quilt is still significantly lower than the air temperature near the chest in the quilt, and the average difference between the two is 1.6 °C. This is due to the greater heat dissipation of the chest than the legs, and the main reason is that the air outlet blows near the head of the person, and the temperature near the chest is obviously higher than the temperature in other parts. Throughout the experiment, it is found that the wind speed at the tuyere and the wind speed near the human face is uneven, and the maximum value and the minimum value are very different. In this paper, the maximum value is expressed. During the experiment, the wind speed in the air outlet and near the person's face has been stable over time, and the maximum wind speed of air outlet has been maintained at about 2.0 m/s, while the temperature of air outlet changes with the change in water supply temperature, and remains in the range of 29.6 to 33.3 °C. The maximum wind speed near human face has been maintained at about 0.8 m/s, and the temperature varies between 24.2 °C and 27.1 °C, which gives people a weak sense of blowing. The temperature is relatively comfortable, but there is a slight sense of dryness.

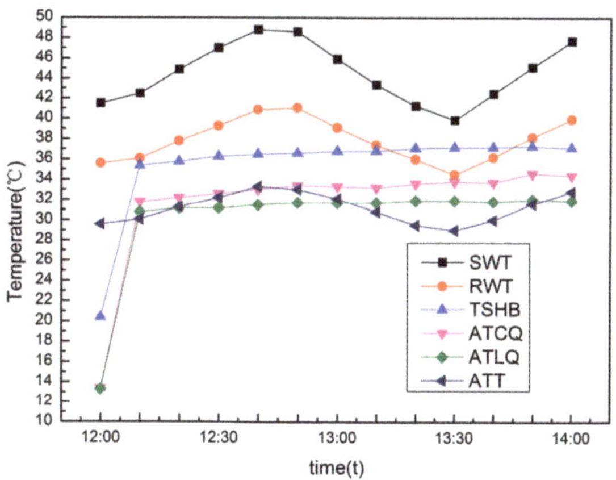

Figure 19. Temperature variation curves at different measuring points during convective heating.

4.2.2. Temperature Analysis of Infrared Thermal Imager

From Figures 20–24, we have also arranged measuring points at different locations on each graph, and from the temperature results shown at different measuring points, it can be seen that at the beginning of the experiment, the outer surface temperature of the quilt is significantly lower, with an average temperature of 15.3 °C. Subsequently, the temperature rises significantly and remains stable thereafter, with an average temperature ranging from 21.0 to 22.7 °C. The surface temperature of the quilt is significantly higher than that of the heat conduction heating experiment, resulting in a large amount of heat being lost into the air, which is also the fundamental reason why the room temperature is significantly higher than that of the heat conduction heating mode. The temperature is obviously higher in the area near the air outlet, while the temperature is lower in the farther place, especially near the human foot, which is also an important reason why the air temperature near the chest in the quilt is higher than the air temperature near the legs in the quilt. Therefore, at the beginning of lying on the bed, there will be a phenomenon of cold feet, until the phenomenon of cold feet gradually disappears and warm feet appears.

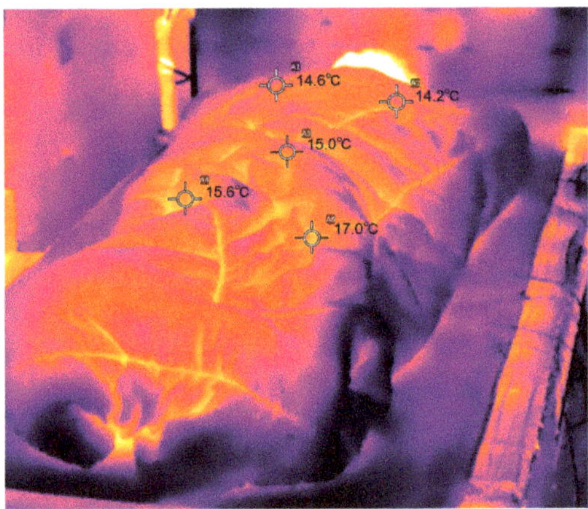

Figure 20. Temperature field distribution of infrared thermal imager at 12:00.

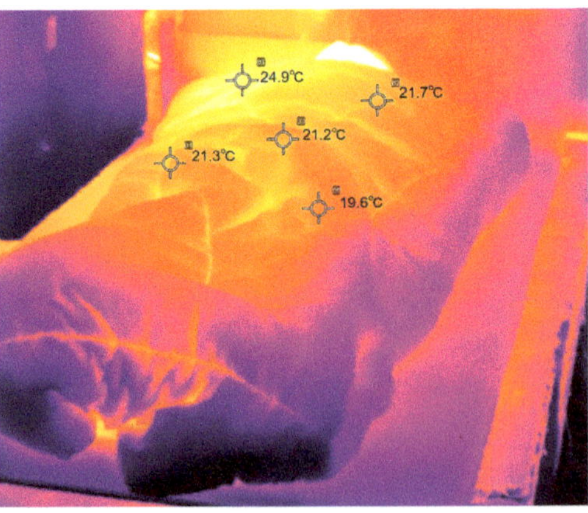

Figure 21. Temperature field distribution of infrared thermal imager at 12:30.

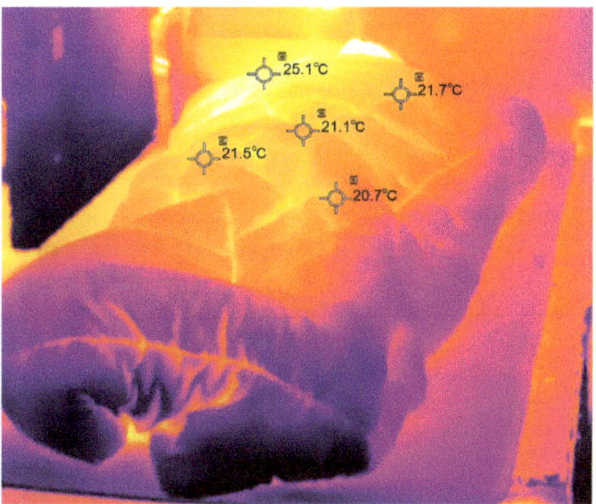

Figure 22. Temperature field distribution of infrared thermal imager at 13:00.

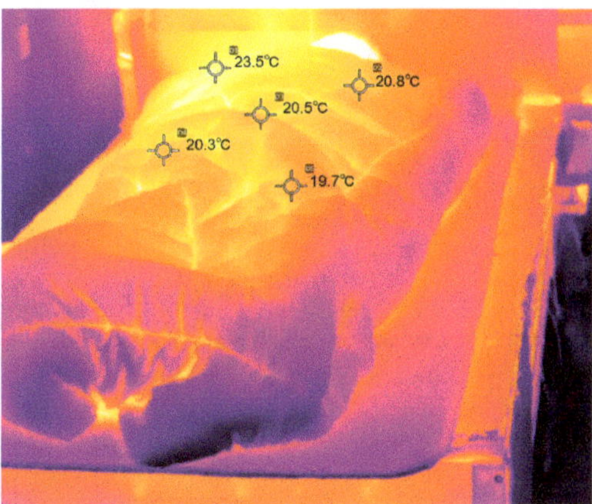

Figure 23. Temperature field distribution of infrared thermal imager at 13:30.

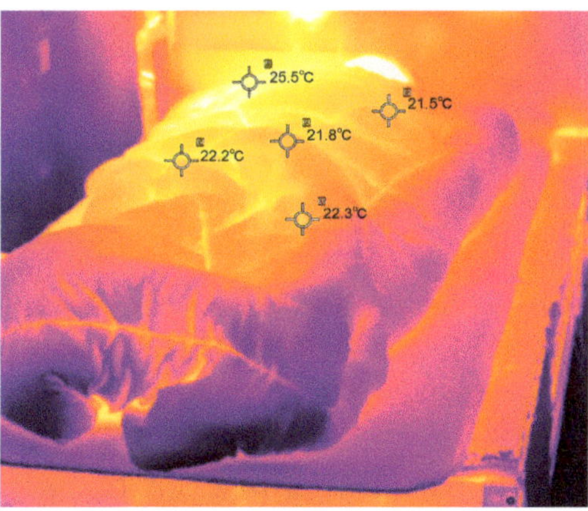

Figure 24. Temperature field distribution of infrared thermal imager at 14:00.

4.2.3. Analysis of Power and Heat Dissipation Data

As shown in Figure 25, the pump heat unit starts at 12:00 for convection heating. The initial water supply temperature is 41.5 °C. When the water supply temperature is close to 50 °C, the unit stops and reopens when it is reduced to close to 40 °C, so it changes periodically during operation. The heat pump unit has been running for two periods of time, with an average power of 928.5 W and a total of 76 min of operation, while the average power of the pump and fan is 105.9 W. The power consumption during the whole lunch break is 1.43 kW·h, which is significantly higher than that of heat conduction heating mode.

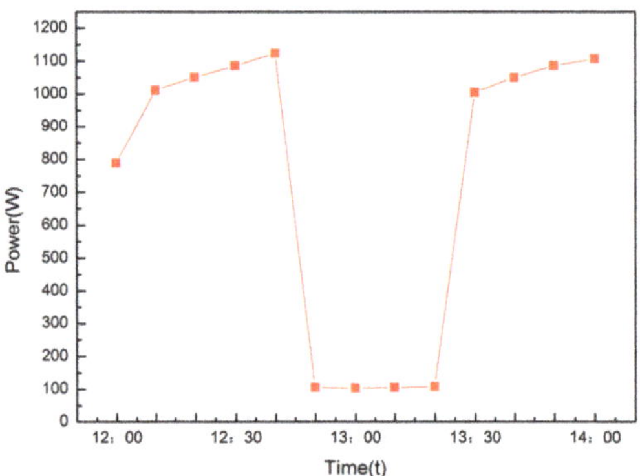

Figure 25. Power variation curve of convection heating with time.

As shown in Figure 26, the amount of heat dissipation varies with the operation law of the heat pump unit. The amount of heat dissipation increases when the heat pump unit is running, and begins to decrease when the unit stops. The average heat dissipation of the whole process is 748.7 W, which is much larger than the average heat dissipation of 81.3 W in the conductive heating mode, and the former is more than nine times the latter.

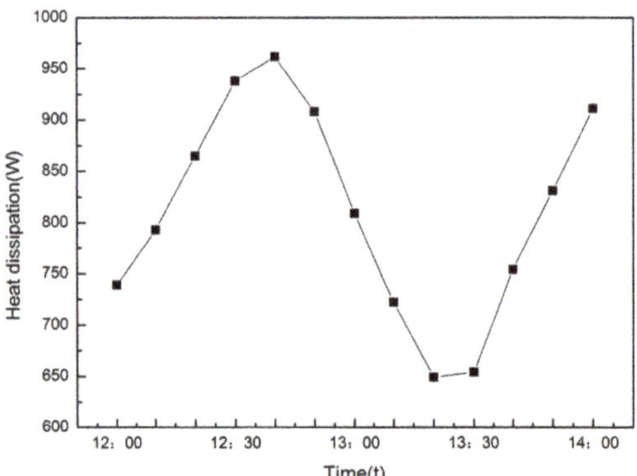

Figure 26. Curve of heat dissipation with time during convection heating.

5. Discussion

5.1. Thermal Comfort Analysis

In the study, it is found that in the heat conduction heating experiment, the human body basically reached a state of no cold feeling after 20 min of operation, and the discomfort basically disappeared. After that, the human body feels slightly warm and gradually warmer, and the sense of comfort gradually strengthens. Different individuals may have different levels of thermal comfort. This article adopts a subjective voting method to evaluate human thermal comfort. For the evaluation of thermal comfort in non-uniform and dynamic environmental conditions with direct contact, it is divided into five grades: very uncomfortable, uncomfortable, unfeeling, comfortable, and very comfortable. During the experiment, the subjects were continuously asked about their thermal comfort and were asked to vote, similar to the survey form in Table 1. By adopting the identification method with a voter turnout of more than 75%, finding different thermal comfort temperature ranges. Finally, it is concluded that the temperature of the contact surface between human body and bed is comfortable in the range of 36.7 to 37.2 °C and very comfortable in the range of 37.2 to 37.8 °C.

In the convection heating experiment, if the air velocity at the air outlet is too high, people will feel the obvious sense of blowing, thus feeling uncomfortable. When the air velocity is too low, the temperature of the blown air declines too quickly. When it reaches the head of human body, the air temperature is already low, which will also affect the thermal comfort. Therefore, during the experiment, the air velocity at the air outlet was adjusted to around 2.0 m/s, and the temperature near the human face could be maintained in the range of 24.2 to 27.1 °C. The blowing feeling was weak, and the overall thermal comfort was good, but for a long time, there would be a slight sense of dryness. The overall thermal comfort is not as good as the heat conduction heating.

5.2. Energy Saving Analysis

In the heat conduction heating experiment, the supply water temperature varied within the range of 28.6 to 42.8 °C, and the average heat dissipation during the whole lunch break heating time is only 81.3 W. If we exclude the heat loss caused by the exposed capillary tubes, this value would be even lower. If the initial water supply temperature is set at 34 °C according to the experimental experience, the power consumption during the lunch break is 0.34 kW·h, and a large part of the power consumption is used for heating the water in hot water tank and storing energy in the water. We choose a day with similar outdoor weather conditions, using a KFR-35GW air conditioner to perform heating experiments in the experimental room, also 2 h at noon, and found that when the indoor temperature is maintained around 20.3 °C, the power consumption is 2.21 kW·h. In comparison, the energy-saving effect of the air-conditioned bed system is very obvious, and its power consumption is only 15.4% of the ordinary air conditioner. In addition, it is found that the heat conduction heating water temperature is roughly in the range of 38 to 40 °C, which can makes people more comfortable. Compared with radiators, floor heating, etc., the required water temperature is lower, which is more conducive to energy saving. And the heat pump unit only needs to run for 30 min to meet the heating needs of the two hours of lunch break.

In the convective heating experiment, the water supply temperature varied within the range of 39.9 to 48.8 °C, and the heating effect is relatively rapid. However, the average heat dissipation during the lunch break is 748.7 W, which is more than nine times that of the heat conduction heating experiment. During the whole lunch break heating period, the power consumption is 1.43 kW·h, which is 64.7% of the ordinary air conditioner, and more than four times that of the heat conduction heating experiment. In addition, compared with the heat conduction heating experiment, the water supply temperature of the convection heating experiment is significantly higher, which is not conducive to energy saving. The air-conditioned bed system, especially the heat conduction heating mode, has an obvious energy-saving effect and strong thermal comfort, and its prospect for future application is good. However, the factors that constrain its development include the requirement for people to believe that it has obvious advantages along with its cost, installation, and other issues, especially the cooling effect. These will be solved in future research, so as to promote the possibility of widespread use of the air-conditioned system in the future.

6. Conclusions

In this paper, the heat conduction heating experiment and convection heating experiment of the capillary convection and heat conduction air-conditioned bed system are carried out, respectively. The experimental results show that during a 2 h lunch break, the heat conduction heating only consumes 0.34 kW·h of electrical energy, with an average heat dissipation of only 81.3 W, and the power consumption is 15.4% of the ordinary air conditioner. The convection heating needs to consume 1.43 kW·h of electric energy, with an average heat dissipation of 748.7 W, and the power consumption is 64.7% of the ordinary air conditioner. These two heating modes are both relatively energy-saving, especially the heat conduction heating mode, which has a more obvious effect.

From the images captured by the infrared thermal imager, it can be seen that compared to thermal conduction heating, the outer surface temperature of the quilt is significantly higher during convective heating, resulting in a higher heat dissipation of the quilt. In addition, there is a continuous flow of hot air from the air outlet, resulting in the room air temperature rising from 12.3 to 17.3 °C, while the air temperature of the heat conduction heating room only rises from 14.4 to 15.2 °C.

Compared to heat conduction heating, convective heating is effective in heating quickly, but it is prone to dryness for a long time, while heat conduction heating heats up slowly and may even feel cool in the first 20 min. But its thermal comfort has been better, making the overall thermal comfort of convective heating inferior to heat conduction heating. In addition, heat conduction heating requires a lower range of water temperature and has lower requirements for the quality of hot water, which is advantageous for energy saving.

Regardless of heat conduction heating or convective heating, the air temperature near the chest in the quilt is significantly higher than that near the legs in the quilt, causing the upper body to feel hotter than the lower body. This is also the reason why cold feet often occur at the beginning of the experiment. After comparison, it is found that the heat conduction heating provides better overall thermal comfort and is more energy efficient, but there is a phenomenon of cold feeling in the first 20 min of heating. It is suggested that when heating in winter, the system should use two heating modes simultaneously for the first 20 min, and then only heat conduction heating mode should be used, or the heat conduction heating mode should be started 20 min before the lunch break.

Author Contributions: Conceptualization, J.J. and T.Y.; methodology, J.J. and P.D.; software, H.C.; validation, J.J., Y.L., P.D., H.C. and T.Y.; formal analysis, P.D.; investigation, Y.L.; resources, H.C.; data curation, J.J. and T.Y.; writing—original draft preparation, J.J. and P.D.; writing—review and editing, J.J. and Y.L. All authors have read and agreed to the published version of the manuscript.

Funding: This research received no external funding.

Data Availability Statement: The data that support the findings of this study are available upon reasonable request from the corresponding author, Junjie Jin.

Conflicts of Interest: The authors declare no conflict of interest.

Nomenclature

IAT	Indoor air temperature (°C)
OAT	Outdoor air temperature (°C)
RHIA	Relative humidity of indoor air (%)
RHOA	Relative humidity of outdoor air (%)
SWT	Supply water temperature (°C)
RWT	Return water temperature (°C)
STSWC	Surface temperature of supply water capillary (°C)
STRWC	Surface temperature of return water capillary (°C)
TSHB	Temperature of the contact surface between human body and bed (°C)
ATCQ	Air temperature near the chest in the quilt (°C)
ATLQ	Air temperature near the legs in the quilt (°C)
STM	Surface temperature of the sponge-padded mattress in the middle of the legs (°C)
ATT	Air temperature of tuyere (°C)

References

1. Zhang, H.; Arens, E.; Zhai, Y. A review of the corrective power of personal comfort systems in non-neutral ambient environments. *Build. Environ.* **2015**, *91*, 15–41. [CrossRef]
2. Tian, X.Y.; Liu, W.W.; Liu, J.W.; Yu, B.; Zhang, J. Effects of personal heating on thermal comfort: A review. *J. Cent. South Univ.* **2022**, *29*, 2279–2300. [CrossRef]
3. Shahzad, S.; Calautit, J.K.; Calautit, K.; Hughes, B.R.; Aquino, A.I. Advanced Personal Comfort System (APCS) for the workplace: A review and case study. *Energy Build.* **2018**, *173*, 689–709. [CrossRef]
4. Godithi, S.B.; Sachdeva, E.; Garg, V.; Brown, R.; Kohler, C.; Rawal, R. A review of advances for thermal and visual comfort controls in personal environmental control (PEC) systems. *Intell. Build. Int.* **2019**, *11*, 75–104. [CrossRef]
5. Warthmann, A.; Wölki, D.; Metzmacher, H.; Van Treeck, C. Personal Climatization Systems—A Review on Existing and Upcoming Concepts. *Appl. Sci.* **2018**, *9*, 35. [CrossRef]
6. Rawal, R.; Schweiker, M.; Kazanci, O.B.; Vardhan, V.; Duanmu, L. Personal Comfort Systems: A review on comfort, energy, and economics. *Energy Build.* **2020**, *214*, 109858. [CrossRef]
7. Li, N.; He, D.; He, Y.; He, M.; Zhang, W. Experimental Study on Thermal Comfort with Radiant Cooling Workstation and Desktop Fan in Hot—Humid Environment. *J. Huanan Univ. Nat. Sci.* **2017**, *44*, 198–204. [CrossRef]
8. Zhai, Y.; Zhang, H.; Zhang, Y.; Pasut, W.; Arens, E.; Meng, Q. Comfort under personally controlled air movement in warm and humid environments. *Build. Environ.* **2013**, *65*, 109–117. [CrossRef]
9. Huang, L.; Ouyang, Q.; Zhu, Y.; Jiang, L. A study about the demand for air movement in warm environment. *Build. Environ.* **2013**, *61*, 27–33. [CrossRef]
10. Zhang, H.; Arens, E.; Kim, D.; Buchberger, E.; Bauman, F.; Huizenga, C. Comfort, perceived air quality, and work performance in a low-power task–ambient conditioning system. *Build. Environ.* **2010**, *45*, 29–39. [CrossRef]
11. He, Y.; Wang, X.; Li, N.; He, M.; He, D. Heating chair assisted by leg-warmer: A potential way to achieve better thermal comfort and greater energy conservation in winter. *Energy Build.* **2018**, *158*, 1106–1116. [CrossRef]
12. Pasut, W.; Zhang, H.; Arens, E.; Zhai, Y. Energy-efficient comfort with a heated/cooled chair: Results from human subject tests. *Build. Environ.* **2015**, *84*, 10–21. [CrossRef]
13. Oi, H.; Yanagi, K.; Tabata, K.; Tochihara, Y. Effects of heated seat and foot heater on thermal comfort and heater energy consumption in vehicle. *Ergonomics* **2011**, *54*, 690–699. [CrossRef] [PubMed]
14. Zhang, H.; Arens, E.; Taub, M.; Dickerhoff, D.; Bauman, F.; Fountain, M.; Pasut, W.; Fannon, D.; Zhai, Y.; Pigman, M. Using footwarmers in offices for thermal comfort and energy savings. *Energy Build.* **2015**, *104*, 233–243. [CrossRef]
15. Foda, E.; Sirén, K. Design strategy for maximizing the energy-efficiency of a localized floor-heating system using a thermal manikin with human thermoregulatory control. *Energy Build.* **2012**, *51*, 111–121. [CrossRef]
16. Zeiler, W.; Boxem, G. Effects of thermal activated building systems in schools on thermal comfort in winter. *Build. Environ.* **2009**, *44*, 2308–2317. [CrossRef]
17. Deng, Q.; Wang, R.; Li, Y.; Miao, Y.; Zhao, J. Human thermal sensation and comfort in a non-uniform environment with personalized heating. *Sci. Total Environ.* **2017**, *578*, 242–248. [CrossRef]
18. Arens, E.; Zhang, H.; Huizenga, C. Partial- and whole-body thermal sensation and comfort—Part II: Non-uniform environmental conditions. *J. Therm. Biol.* **2006**, *31*, 60–66. [CrossRef]
19. Lv, B.; Su, C.; Yang, L.; Wu, T. Effects of stimulus mode and ambient temperature on cerebral responses to local thermal stimulation: An EEG study. *Int. J. Psychophysiol.* **2017**, *113*, 17–22. [CrossRef]
20. Schellen, L.; Loomans, M.G.; de Wit, M.H.; Olesen, B.W.; van Marken Lichtenbelt, W.D. The influence of local effects on thermal sensation under non-uniform environmental conditions--gender differences in thermophysiology, thermal comfort and productivity during convective and radiant cooling. *Physiol. Behav.* **2012**, *107*, 252–261. [CrossRef] [PubMed]
21. Wang, D.; Chen, P.; Liu, Y.; Wu, C.; Liu, J. Heat transfer characteristics of a novel sleeping bed with an integrated hot water heating system. *Appl. Therm. Eng.* **2017**, *113*, 79–86. [CrossRef]
22. Yu, G.; Gao, S.; Qiang, j. Modeling of heartbeat for sleep in integrated with PCM thermal storage and solar heating. *J. Chongqing Univ.* **2011**, *34*, 89–92.
23. Pan, D.; Chan, M.; Xia, L.; Xu, X.; Deng, S. Performance evaluation of a novel bed-based task/ambient conditioning (TAC) system. *Energy Build.* **2012**, *44*, 54–62. [CrossRef]
24. Mao, N.; Pan, D.; Chan, M.; Deng, S. Experimental and numerical studies on the performance evaluation of a bed-based task/ambient air conditioning (TAC) system. *Appl. Energy* **2014**, *136*, 956–967. [CrossRef]
25. Mao, N.; Song, M.; Deng, S.; Pan, D.; Chen, S. Experimental and numerical study on air flow and moisture transport in sleeping environments with a task/ambient air conditioning (TAC) system. *Energy Build.* **2016**, *133*, 596–604. [CrossRef]
26. Mao, N.; Pan, D.; Li, Z.; Xu, Y.; Song, M.; Deng, S. A numerical study on influences of building envelope heat gain on operating performances of a bed-based task/ambient air conditioning (TAC) system in energy saving and thermal comfort. *Appl. Energy* **2017**, *192*, 213–221. [CrossRef]
27. Mao, N.; Hao, J.; Cui, B.; Li, Y.; Song, M.; Xu, Y.; Deng, S. Energy performance of a bedroom task/ambient air conditioning (TAC) system applied in different climate zones of China. *Energy* **2018**, *159*, 724–736. [CrossRef]
28. Du, J.; Chan, M.; Deng, S. An experimental study on the performances of a radiation-based task/ambient air conditioning system applied to sleeping environments. *Energy Build.* **2017**, *139*, 291–301. [CrossRef]

29. Catalina, T.; Virgone, J.; Kuznik, F. Evaluation of thermal comfort using combined CFD and experimentation study in a test room equipped with a cooling ceiling. *Build. Environ.* **2009**, *44*, 1740–1750. [CrossRef]
30. Mikeska, T.; Svendsen, S. Study of thermal performance of capillary micro tubes integrated into the building sandwich element made of high performance concrete. *Appl. Therm. Eng.* **2013**, *52*, 576–584. [CrossRef]
31. Xin, Y.; Cui, W.; Zeng, J. Experimental study on thermal comfort in a confined sleeping environment heating with capillary radiation panel. *Energy Build.* **2019**, *205*, 109540. [CrossRef]
32. Jiang, P.; Huang, Y.; Wei, X.; Mu, Y. Effect of Local Radiant Heating on Thermal Comfort in Cold Environment. *J. BEE* **2022**, *50*, 52–55. [CrossRef]
33. Fanger, P.O.; Melikov, A.K.; Hanzawa, H.; Ring, J. Air turbulence and sensation of draught. *Energy Build.* **1988**, *12*, 21–39. [CrossRef]
34. He, Y.; Li, N.; Huang, Q. A field study on thermal environment and occupant local thermal sensation in offices with cooling ceiling in Zhuhai, China. *Energy Build.* **2015**, *102*, 277–283. [CrossRef]

processes

MDPI

Article

Fault Diagnosis Based on Fusion of Residuals and Data for Chillers

Zhanwei Wang [1,2,*], Boyang Liang [1,2], Jingjing Guo [1,2], Lin Wang [1,2,*], Yingying Tan [1,2], Xiuzhen Li [1,2] and Sai Zhou [1,2]

[1] Institute of Building Energy and Thermal Science, Henan University of Science and Technology, Luoyang 471023, China; liangby2020@163.com (B.L.); gjjhaust@163.com (J.G.); tyyhaust@163.com (Y.T.); lixiuzhenvip@126.com (X.L.); zhousaiangel@163.com (S.Z.)

[2] Henan Provincial Engineering Research Center of Building Environmental Control and Safety, Luoyang 471023, China

* Correspondence: wzhanweisun@163.com (Z.W.); wlhaust@163.com (L.W.)

Abstract: Feature data refer to direct measurements of specific features, while feature residuals represent the deviations between these measurements and their corresponding benchmark values. Both types of information offer unique insights into the system's behavior. However, conventional diagnostic systems often struggle to effectively integrate and utilize both types of information concurrently. To address this limitation and improve diagnostic performance, a hybrid method based on the Bayesian network (BN) is proposed. This method enables the parallel fusion of feature residuals and feature data within a unified diagnostic model, and a comprehensive framework for developing this hybrid method is also given. In the hybrid BN, the symptom layer consists of residual nodes representing feature residuals and data nodes representing measured feature data. By applying the proposed method to two chillers and comparing it with state-of-the-art existing methods, we demonstrate its effectiveness and superiority. The results highlight that the proposed method not only accommodates the absence of either type of information but also leverages both of them to enhance diagnostic performance. Compared to using a single type of node, the hybrid method achieves a maximum improvement of 24.5% in diagnostic accuracy, with significant enhancements in F-measure observed for refrigerant leakage fault (34.5%) and excessive lubricant fault (32.8%), respectively.

Keywords: chillers; fault diagnosis; fusion; residual; data

Citation: Wang, Z.; Liang, B.; Guo, J.; Wang, L.; Tan, Y.; Li, X.; Zhou, S. Fault Diagnosis Based on Fusion of Residuals and Data for Chillers. *Processes* **2023**, *11*, 2323. https://doi.org/10.3390/pr1108

Academic Editor: Jie Zhang

Received: 25 June 2023
Revised: 21 July 2023
Accepted: 28 July 2023
Published: 2 August 2023

1. Introduction

Buildings play a critical role as the primary consumers of energy, accounting for around 40% of global energy consumption [1]. Among the energy-consuming systems in buildings, chillers are major contributors. However, when a chiller malfunctions, it can lead to reduced efficiency, energy waste, shortened lifespan, and compromised indoor comfort. In fact, a faulty chiller can result in an additional 30% increase in energy consumption [2]. Therefore, fault diagnosis plays a vital role in detecting and resolving issues with chillers, leading to significant energy savings.

After decades of development, fault diagnosis of chillers has been widely studied, and a large number of methods have been proposed and applied. Generally, these methods can be categorized into two types: model-driven and data-driven methods.

Model-driven methods usually establish a model that can identify and assess deviations (known as residuals) between actual operating levels and predefined normal (benchmark) operating levels. One of the main applications of model-driven methods is the acquisition of feature residuals. These residuals represent the differences between measured feature data and their corresponding benchmark values. The benchmark values are derived from a benchmark model and signify the system's normal operating state. In the realm of model-based approaches, Browne and Bansal [3] summarized and analyzed the relevant literature on steady-state models of the chiller. Additionally, Zhao et al. [4] and Kim and Braun [5,6] developed fault diagnosis techniques based on simplified physical models and decoupling models, respectively.

Data-driven methods analyze a substantial amount of measured data related to specific features for each fault to extract patterns. These patterns are then utilized to diagnose faults by identifying similarities among them. Data-driven methods are particularly effective in utilizing feature data, which refers to directly measured data of features. In recent years, with the rapid advancements in computer technology and artificial intelligence, data-driven methods have gained significant popularity in the field of fault diagnosis [7]. Examples of such methods include support vector machine (SVM) [8], convolutional neural network (CNN) [9], global density-weighted support vector data description [10], association rule mining [11], and unsupervised clustering models such as K-means, Gaussian mixture model clustering, and spectral clustering [12], among others.

In recent years, there has been a growing interest among researchers in integrating multiple methods to enhance diagnostic performance. Various approaches have been explored to improve existing methods by combining different techniques. For example, Han et al. [13] proposed a fault diagnosis method based on least squares support vector machines (LS-SVM). Three popular machine learning algorithms, namely k-nearest neighbor, SVM, and random forest, have been combined for fault diagnosis in chillers [14]. In the field of CNNs, scholars have introduced methods like sparsely local embedding CNN (SLENet) [15] and combined a data self-production (SP) algorithm with CNN to develop diagnostic methods like SP-CNN [16], aiming to enhance diagnostic performance. Similarly, in deep neural network (DNN) based methods, researchers have utilized optimization algorithms such as simulated annealing (SA) to optimize model parameters, resulting in diagnostic methods like SA-DNN [17]. Additionally, Wang et al. [18] attempted to integrate rule-based knowledge with data, creating a hybrid diagnostic strategy that combines the strengths of both approaches.

Current research results suggest that integrating multiple models into a hybrid method is an effective approach to improving the performance of fault diagnosis. These hybrid methods, which combine multiple models, are considered to be among the best methods currently available. However, it should be noted that existing hybrid methods mainly focus on combining different models without effectively integrating different types of information. In other words, they still rely on either feature residuals or feature data alone for diagnosis without leveraging both types of information simultaneously.

Both feature residuals and feature data contribute valuable information to fault diagnosis and should be leveraged to enhance diagnostic performance. On the one hand, feature residuals can be obtained by constructing a benchmark model for practical field applications. When a chiller experiences a fault, the fault-sensitive feature undergoes noticeable changes, leading to a significant deviation between the actual feature value and the benchmark value. For instance, experimental results demonstrate that the normal value of the condenser inlet–outlet water temperature difference for a chiller under standard conditions is 4.5 °C. However, when there is an abnormal decrease in the cooling water flow rate, the corresponding inlet–outlet water temperature difference abruptly increases to values ranging from 5.1 °C to 7.8 °C, depending on the degradation severity. Similar findings are also reported in the experimental results presented in reference [19]. On the other hand, directly measured feature data are becoming more readily available with the increasing amount of collected data. These feature measurements provide real-time information about the chiller's performance.

However, while traditional model-driven methods excel at utilizing feature residuals, they face challenges in effectively leveraging large amounts of feature data. On the other hand, data-driven methods are unable to concurrently incorporate both feature residuals and feature data within a single diagnostic system. This limitation hampers their ability to harness the full potential of both types of information.

Therefore, the main motivation for this paper is to maximize the utilization of information and achieve significant improvements in diagnostic performance by simultaneously fusing feature residuals and feature data within a unified diagnostic framework. The major contributions of this word are as follows:

(1) An open topology structure based on BN is proposed, which enables the parallel fusion of feature residuals and feature data within a unified diagnostic model. A comprehensive framework is provided for the development of this hybrid method.
(2) The proposed method not only accommodates the absence of either type of information but also leverages both of them to enhance diagnostic performance.
(3) The proposed method showcases enhanced diagnostic performance, data utilization flexibility, and reduced training and application times when applied to two real-world chillers and compared with state-of-the-art existing methods.

2. Methodology

BN is a widely used probabilistic graphical model that has found numerous applications in fault diagnosis. For example, Li et al. [20] integrated expert knowledge into BN to develop a diagnostic network guided by expert insights. Wang et al. [21] employed virtual in situ calibration based on Bayesian inference and Markov chain Monte Carlo for photovoltaic thermal heat pump systems. Li et al. [22] utilized multiple linear regression to enhance in-situ sensor calibration strategies using Bayesian inference. Chen et al. [23] proposed a discrete Bayesian network-based method for diagnosing cross-level faults in HVAC systems. Hu et al. [24] introduced the Bayesian belief network into the fault diagnosis process of variable refrigerant flow air conditioning systems for diagnosing refrigerant leakage and overcharge. Wang et al. [25,26] developed a series of fault detection and diagnosis methods based on BN specifically for chillers.

The key benefit of employing BN is its flexible network topology, enabling the fusion of diverse information sources by incorporating different types of nodes within the BN structure. The methodology for merging each type of node into the BN is elaborated upon in the following sections.

2.1. BN Driven by Residuals

The steps to obtain feature residuals using a model-driven method are as follows:

Firstly, the construction of a benchmark model is essential. Depending on the modeling approach chosen, the benchmark model can be a precise or simplified physical model or a black-box model based on regression prediction. This model represents the normal operating behavior of the chiller and provides benchmark values for different features.

Secondly, the comparison between measured values and benchmark values is conducted. Feature residuals are calculated by subtracting the measured values from their corresponding benchmark values. When a fault occurs in the chiller, specific features, especially those sensitive to faults, exhibit noticeable deviations from their benchmark values. It is important to note that under normal operating conditions, the measured values of different features should ideally align with their respective benchmark values. However, due to factors such as data collection errors, model inaccuracies, and computational processes, slight deviations between measured values and benchmark values may occur. These deviations are typically small and can be considered negligible within a certain level of statistical confidence.

Thirdly, the analysis of feature residuals takes place. Substantial changes in feature residuals indicate the occurrence of faults or abnormalities within the chiller. By examining the patterns in the feature residuals, it becomes possible to diagnose specific faults.

The fault diagnosis process described above can be integrated into a BN framework. Figure 1 illustrates the structure and parameters of the residual-driven BN, highlighting the utilization of feature residuals in the diagnostic process. By combining the model-driven method with the BN framework, the diagnostic model can effectively leverage the information derived from the feature residuals to enhance its performance.

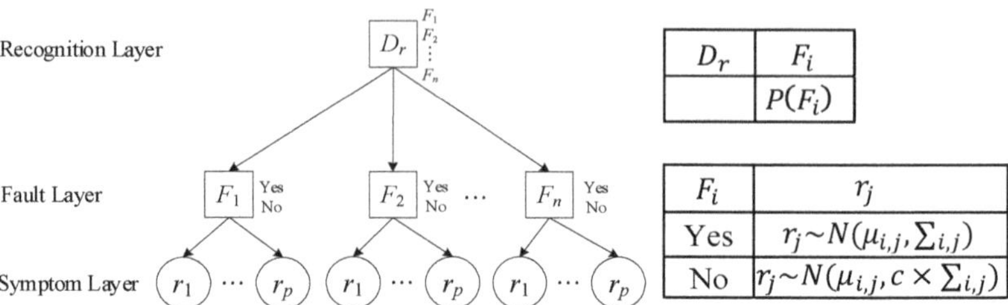

Figure 1. The structure and parameters of residual-driven BN.

The structure includes three layers, from top to bottom: the recognition layer, the fault layer, and the symptom layer. The recognition layer contains one top node D_r, the fault layer contains n fault nodes F_i ($i \in [1, n]$), and each fault node is connected to p residual nodes r_j ($j \in [1, p]$) in the symptom layer. The top node D_r has n states corresponding to n known faults. Each fault node F_i has two states: Yes and No, representing the occurrence and non-occurrence of faults, respectively. Each residual node r_j represents a continuous node consisting of the feature residuals. The residual node r_j enables the utilization of feature residuals.

Its parameters include the prior probabilities of the top node D_r and the conditional probabilities of each sub-node. For the BN shown in Figure 1, the prior probability of each state of the top node D_r ($P_i, i \in [1, n]$) can be determined by expert experience or statistical samples. The assignment principle of the conditional probabilities of the fault node F_i given the top node D_r state is shown in Equation (1):

$$\begin{cases} P(F_i = \text{Yes}|D_r = F_i) = 1; \\ P(F_i = \text{No}|D_r = F_i) = 0; \\ P(F_i = \text{Yes}|D_r = F_j) = 0; \\ P(F_i = \text{No}|D_r = F_j) = 1, i \neq j \end{cases} \tag{1}$$

The conditional probability of the residual node r_j is assumed to follow a Gaussian distribution. The two parameters that describe the distribution, mean ($\mu_{i,j}$) and covariance ($\Sigma_{i,j}$), given the parent node F_i state, need to be obtained through maximum likelihood estimation from historical data of feature residuals belonging to the fault F_i. The coefficient c in Figure 1 is used to determine the conditional probability distribution of the sub-node r_j when the state of the node F_i is No, and its calculation is shown in Equations (2) and (3). The detailed demonstration and validity of Equations (2) and (3) have been presented in the works of Wang et al. [25] and Verron et al. [27].

$$1 - c + \frac{pc}{CL} ln(c) = 0 \tag{2}$$

$$CL = \frac{p(N-1)(N+1)}{N(N-p)} F_\alpha(p, N-p) \tag{3}$$

In the equations, k represents the dimension of node r_j, $k = 1$. N represents the number of samples, and $F_\alpha(k, N - k)$ represents the α percentile of the Fisher distribution with degrees of freedom k and $N - k$. α is the significance level, which is determined through multiple attempts based on the principle of obtaining optimal diagnostic performance.

2.2. BN Driven by Data

The data-driven approach typically involves constructing a black-box model to establish the mapping relationship between input features and output faults. This approach requires a significant amount of data for training the models. The structure and parameters of the data-driven BN are illustrated in Figure 2.

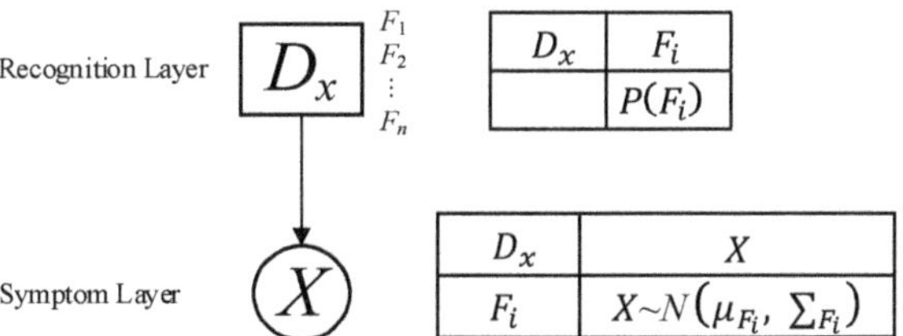

Figure 2. The structure and parameters of data-driven BN.

The structure includes two layers, from top to bottom: the recognition layer and the symptom layer, each of which includes one top node D_x and one data node X, respectively. The top node D_x has the same n states as the top node D_r in Figure 1; the data node X is a continuous node composed of m features. The data node X implements the use of feature data.

For the BN parameters, the prior probabilities of the top node D_x are exactly the same as those of the top node D_r in Figure 1. Assuming that the conditional probability of the data node X follows a multidimensional Gaussian distribution, the way to determine the distribution is exactly the same as that of the residual node r_j in Figure 1. Specifically, the two parameters that describe the distribution are the mean vector (μ_{F_i}) and the covariance matrix $(\sum_{F_i})$, which need to be obtained through maximum likelihood estimation from the feature measurement data belonging to the fault F_i.

2.3. BN Driven by the Fusion of Residuals and Data

The objective of this section is to combine the residual-driven model (Figure 1) with the data-driven model (Figure 2) into a unified BN. Typically, the BN structure is determined by establishing causal relationships between nodes. For complex systems with unclear internal mechanisms, it can be challenging to clarify these causal relationships, requiring the use of optimization algorithms for BN structure learning [28]. However, in the case of chillers, their thermodynamic principles are relatively well-defined, and the influence relationship between typical faults and features is generally understood. Therefore, there is no need to employ structure learning algorithms to determine the BN structure. The structure and parameters of the hybrid BN are depicted in Figure 3. It consists of four layers, namely the decision layer, recognition layer, fault layer, and symptom layer. The configuration of each layer is as follows:

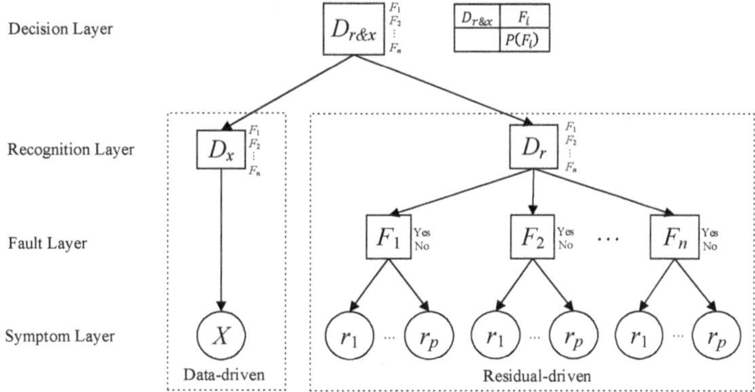

Figure 3. The structure and parameters of the hybrid BN driven by residual and data.

The function of the symptom layer is to acquire feature residuals and feature data, which serve as evidence for fault diagnosis. With the fusion of the residual-driven and data-driven components, the symptom layer now encompasses both residual nodes and data nodes.

The fault layer plays a critical role in evaluating the evidence obtained from the symptom layer and estimating the probabilities of each fault occurrence. It is specifically included in the residual-driven part of the diagnostic model. This distinction is made because the data-driven part involves numerous features, and almost all faults affect these features. Introducing individual fault nodes, as in the residual-driven part, for each fault in the data-driven part would substantially increase the complexity of model parameter configuration and reduce computational efficiency. Hence, fault nodes are only established separately in the residual-driven part, where the number of features is relatively smaller.

The recognition layer plays a crucial role in inferring the posterior probabilities of each fault by utilizing the posterior probabilities propagated from the fault layer or symptom layer. In the hybrid model combining the residual-driven and data-driven components, the recognition layer consists of two nodes, D_r and D_x, which represent the inference results from the respective parts.

In the diagnostic network presented in Figure 3, the two components involved in the fusion process perform parallel inference computations. Each component independently receives data from the corresponding nodes in its own section of the symptom layer and conducts fault diagnosis in parallel. This parallel inference leads to the generation of separate inference results in the identification layer. To complete the fusion of the identification results from both components, a new top node, $D_{r\&x}$, is introduced. The decision layer, as depicted in Figure 3, plays a critical role in combining and consolidating the diagnostic outcomes obtained from the parallel inference calculations performed by the two components. The proposed diagnostic process, which integrates residual-driven and data-driven models within a unified diagnostic framework, is referred to as a hybrid diagnostic method. The effectiveness of this framework has been confirmed in the study conducted by Atoui et al. [29].

The decision layer, represented by node $D_{r\&x}$, is a discrete node with the same states as nodes D_r and D_x. To ensure fairness and avoid bias towards any particular state, equal prior probabilities are assigned to each state of node $D_{r\&x}$. The principle for assigning conditional probabilities to nodes D_r and D_x in the identification layer follows the same approach. An example of this can be seen in Equation (4), which pertains to node D_r.

$$\begin{cases} P(D_r = F_i | D_{r\&x} = F_i) = 1 \\ P(D_r = F_i | D_{r\&x} = F_j) = 0, i \neq j \end{cases} \tag{4}$$

The assignment principles for the fault nodes F_i, residual nodes r_j, and data nodes X remain consistent with the principles used in the residual-driven and data-driven BN.

2.4. Working Mechanism of Residuals and Data in the Hybrid BN

In this paper, both feature residuals and feature data are utilized to diagnose faults by integrating them into a BN. BN is a probabilistic inference model that employs Bayesian inference to calculate posterior probabilities, specifically $P(Q|E = e)$, where Q represents an unobserved state, and E represents an observed state with the observed value of e. When an event occurs, it serves as evidence (an observed state). When evidence is fed into the BN, the information provided by the evidence is propagated throughout the network to update knowledge and obtain posterior probabilities of the unobserved states. This process is known as inference. BN uses predefined conditional probability distributions and observed evidence for inference.

In BN, the relationships between nodes are represented by the structure and conditional probability distributions. The structure of the BN is determined by the causal relationships between nodes, while the conditional probability distributions are derived from training data. Hence, when integrating feature residuals and feature data into the BN, there is no requirement to explicitly assign weights to these components. In practical applications, BN inherently considers the trade-off between feature residuals and feature data during the computation of posterior probabilities. This means that the diagnostic system automatically takes into account the relative importance and contribution of each type of information. Therefore, it avoids the uncertainty and potential bias that could arise from subjective weight assignments.

It is challenging to theoretically or mathematically show that the simultaneous use of both types of nodes is able to yield better diagnostic performance than using a single type of node, as it involves the complexity of the model and specific data distributions. However, it can be explained through the following reasonable inference.

For a newly observed sample, there are two possible outcomes when it is diagnosed separately by the two types of nodes. The first outcome is that both types of nodes diagnose the sample as the same fault. The second outcome is that the two types of nodes diagnose the sample as having different faults. Integrating the two types of nodes does not affect the first outcome but does impact the second outcome.

For example, let us consider a new observed sample, denoted as x, which is diagnosed by the residual node as fault F_1, while the data node diagnoses is as fault F_2. Moreover, the posterior probability of sample x being diagnosed as fault F_2 by the residual node is only slightly lower than the posterior probability of it being diagnosed as fault F_1. This indicates that the residual node has some ambiguity in diagnosing sample x. In this scenario, if the data node unequivocally diagnoses sample x as fault F_2, then when both types of nodes are used, sample x will be unequivocally diagnosed as fault F_2. If the sample x indeed belongs to fault F_2, then the diagnostic result is correct.

This inference process illustrates that when both types of nodes are simultaneously used, it captures more samples that are ambiguously diagnosed by the single-type nodes, thereby improving the diagnostic performance.

3. Framework Based on BN Driven by the Fusion of Residuals and Data

The diagnostic process, driven by the fusion of residuals and data, is illustrated in Figure 4, which consists of two main parts: construction of the hybrid BN model and online fault diagnosis.

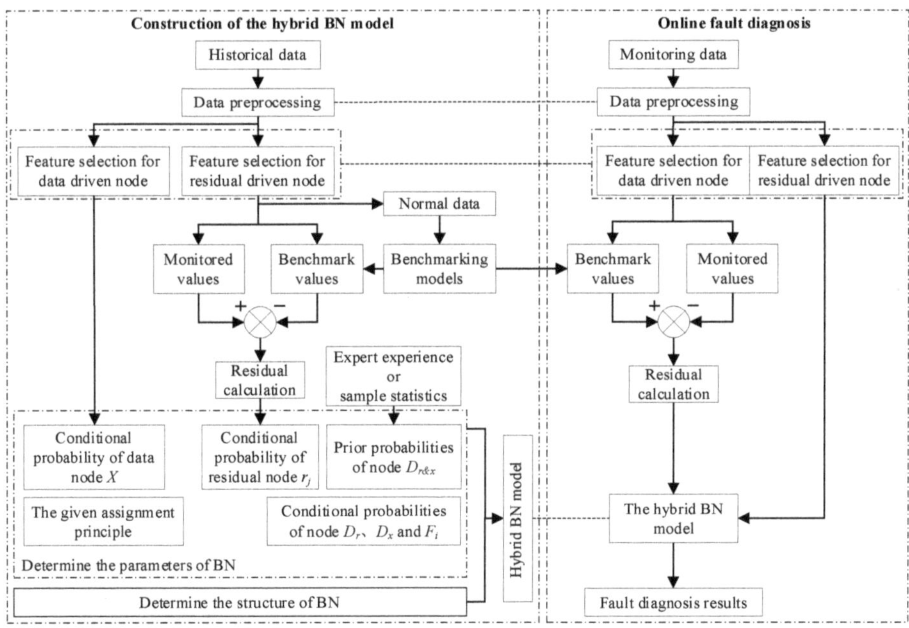

Figure 4. Framework of the fault diagnosis based on hybrid BN driven by residual and data.

3.1. Construction of the Hybrid BN Model

Construction of the hybrid BN model involves the following steps:

i   Data preprocessing and feature selection: The first step is to preprocess the historical data, which encompasses normal operating conditions and different fault types. This process entails removing any obvious transients and anomalies present in the data. Subsequently, appropriate features are selected for both the residual node r_j and the data node X.

ii   Development of the benchmark model and calculation of feature residuals: By using the normal data as a reference, a benchmark model is constructed for the selected features. Subsequently, the feature residuals are computed by quantifying the deviation between the measured values of each feature and their corresponding benchmark values for each fault scenario.

ii   Construction of the hybrid BN driven by residuals and data: Firstly, the structure of the hybrid BN, as depicted in Figure 3, is determined. Next, the prior probability of the top node $D_{r\&x}$ is established, taking into account expert knowledge or sample statistics. By using the predefined assignment principle, the conditional probabilities of nodes D_r, D_x, and the fault layer nodes F_i are sequentially determined. Finally, the conditional probability distributions of the residual node r_j and the data node X are estimated using maximum likelihood estimation, utilizing the feature residuals and the feature data.

By following these steps, the hybrid BN model is constructed, integrating the residual-driven and data-driven components. This model effectively merges the information from both feature residuals and feature data, allowing for accurate fault diagnosis and analysis.

3.2. Online Fault Diagnosis

In the practical application of the hybrid BN, the real-time monitored data undergo a two-step inference process.

In the first step, the data are input into the symptom layer, where they acquire the feature residuals from the residual-driven part and the feature data from the data-driven part. Through BN inference calculations, the posterior probabilities of the fault layer node F_i in the residual-driven part and the recognition layer node D_x in the data-driven part are obtained.

In the second step, based on the posterior probabilities of the fault layer node F_i, the BN inference calculation is performed again to obtain the posterior probabilities of the recognition layer node D_r in the residual-driven part. These posterior probabilities, along with the posterior probabilities of the recognition layer node D_x from the data-driven part, are propagated to the top layer. Finally, through further inference, the posterior probabilities of the decision layer node $D_{r\&x}$ are calculated.

According to the principle of maximum posterior probability, the state with the highest posterior probability of the node $D_{r\&x}$ is outputted as the fault diagnosis result. This ensures that the hybrid BN effectively combines the information from both the residual-driven and data-driven parts, enabling accurate fault diagnoses.

The BN inference algorithm includes two types: the exact inference algorithm and the approximate inference algorithm. Since the hybrid BN developed in this paper is not too complex, the exact inference algorithm, specifically the junction tree algorithm, is used.

4. Application and Performance Evaluation

In this section, the effectiveness and feasibility of the proposed hybrid method are evaluated by applying it to a real-world chiller system used in the ASHRAE RP-1043 project [19], as well as an actual maglev centrifugal chiller. The diagnostic performance of the proposed method is compared with that of existing advanced diagnostic methods.

4.1. Experimental Data

The ASHRAE RP-1043 project [19] used a centrifugal chiller with a cooling capacity of approximately 316 kW. Both the evaporator and condenser were shell-and-tube heat exchangers, with water flowing inside the tubes, and the refrigerant used was R134a, with a thermal expansion valve. The experiments were conducted under 27 operating conditions, and 64 parameters were measured and stored at 10 s intervals, including temperature, pressure, flow, power, etc. Through experiments, a large amount of data was obtained for the normal state of the unit and seven typical faults under four levels of degradation. These faults included reduced condenser water flow (RedCdW), reduced evaporator water flow (RedEvW), refrigerant leak (RefLeak), refrigerant overcharge (RefOver), condenser fouling (CdFoul), noncondensable gas in refrigerant (NcG), and excess oil (ExOil).

4.2. Data Preprocessing and Feature Selection

The data preprocessing method proposed in Ref. [30] was used to perform steady-state filtering on the original experimental data, filtering out any obvious dynamic and abnormal data. Three variables, namely the inlet and outlet temperatures of the chilled water and the inlet temperature of the cooling water, were chosen as the indicators for steady-state filtration.

After steady-state screening, for the normal samples and the samples including faults under four degradation levels, two-thirds of the steady-state data were randomly selected to form the training set, and the remaining one-third of the steady-state data were used for the testing set. For normal and each type of fault under each degradation level, there were approximately 800 and 400 samples in the training and testing sets, respectively. In total, the training set consisted of 23,200 samples, and the testing set consisted of 11,600 samples. This process of dividing the data into training and testing sets was repeated five times, resulting in five sets of training and testing data. Each of these five datasets was used separately to validate the diagnostic performance of the proposed method, providing a more comprehensive demonstration of its effectiveness. The training dataset was used to determine the model parameters, while the testing dataset was used to test and evaluate the diagnostic performance of the model.

Firstly, the features of the residual nodes r_j in the residual-driven part were selected. By referring to previous research findings [18,31,32], a set of fault-sensitive features and their corresponding calculations were determined. These features constitute the residual nodes r_j and are listed in Table 1. For detailed explanations of each feature, please refer to Table 2. Additionally, the association between features and faults can be found in Figure 5.

Table 1. Features selected by the residual node r_j in residual-driven part.

No.	Designation	Formulation	Corresponding Fault Types
1	Δt_c	$\Delta t_c = T_{co} - T_{ci}$	RedCdW
2	Δt_e	$\Delta t_e = T_{ei} - T_{eo}$	RedEvW
3	ε_{sc}	$\varepsilon_{sc} = \frac{T_{sub}}{T_c - T_{ci}}$	RefLeak/RefOver/CdFoul/NcG
4	$LMTD_c$	$LMTD_c = \frac{T_{co} - T_{ci}}{\ln((T_c - T_{ci})/(T_c - T_{co}))}$	RefLeak/RefOver/CdFoul/NcG
5	T_{oil}	Direct measurement	ExOil

Secondly, the features for the data nodes X in the data-driven part were selected. Considering the results of surveys conducted by Wang et al. [33] regarding the installation status of sensors in on-site chillers, features that are readily available are selected to form the data nodes X. The selected features are presented in Table 2.

Table 2. Features selected by the data node X in data-driven part.

No.	Designation	Description	Formulation
1	T_{ei}	Water temperature of evaporator inlet	Direct measurement
2	T_{eo}	Water temperature of evaporator outlet	Direct measurement
3	T_{ci}	Water temperature of condenser inlet	Direct measurement
4	T_{co}	Water temperature of condenser outlet	Direct measurement
5	T_e	Evaporating temperature	Direct measurement
6	T_c	Condensing temperature	Direct measurement
7	TEA	Evaporator approach temperature	$TEA = T_{eo} - T_e$
8	TCA	Condenser approach temperature	$TCA = T_c - T_{co}$
9	T_{sub}	Refrigerant sub-cooling temperature	Direct measurement
10	T_{suc}	Refrigerant suction temperature	Direct measurement
11	Tsh_{suc}	Refrigerant suction superheat temperature	$Tsh_{suc} = T_{suc} - T_e$
12	TR_{dis}	Refrigerant discharge temperature	Direct measurement
13	Tsh_{dis}	Refrigerant discharge superheat temperature	$Tsh_{dis} = T_c - TR_{dis}$
14	P_{in}	Compressor input power	Direct measurement
15	T_{oil}	Oil feed temperature	Direct measurement
16	P_{feed}	Oil feed pressure	Direct measurement

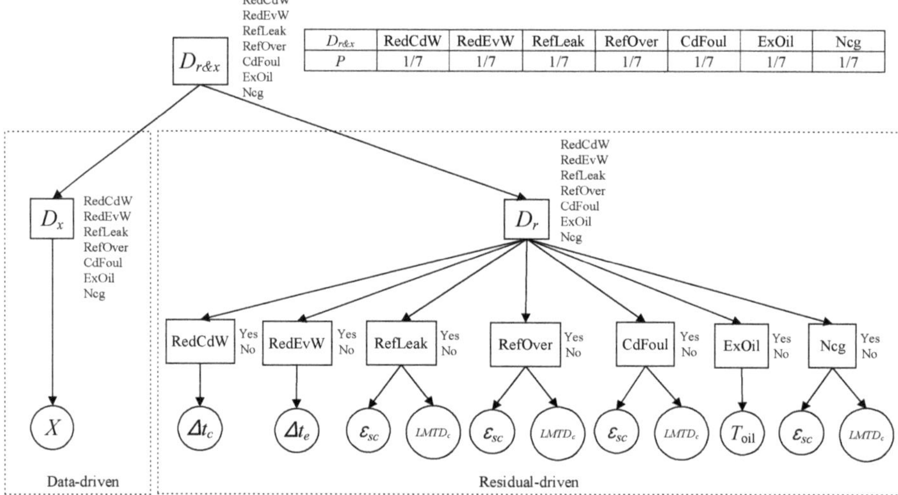

Figure 5. The structure and parameters of the hybrid method driven by residual and data.

4.3. Development of Benchmark Model

To determine the conditional probability distribution of the residual nodes r_j in the symptom layer, a benchmark model first needs to be constructed. Typically, for a fixed water flow rate system, the performance of a chiller can be represented as a relationship between Q_e (cooling capacity), T_{ci}, and T_{eo}. Therefore, the features listed in Table 1 are expressed as functions of these three parameters, as shown in Equation (5). This relationship has been proven effective in previous studies [31,32].

$$Y = f(Q_e, T_{ci}, T_{eo}) + \xi \tag{5}$$

where $Y = [\Delta t_c, \Delta t_e, \varepsilon_{sc}, LMTD_c, T_{oil}]$ represents the benchmark values of the features listed in Table 1, and $\xi \sim N(0, \sigma^2)$.

The task of determining benchmark values for these features can be transformed into a regression-based prediction problem. Previous studies have utilized various regression methods, including multiple linear regression, radial basis function, and support vector regression, for this purpose. In this study, the radial basis function was chosen as the regression method for building the benchmark models based on the comparative analysis conducted by Tran et al. [34]. The benchmark models were constructed with three layers: the input layer, the hidden layer, and the output layer. The input layer consists of three nodes corresponding to Q_e, T_{ci}, and T_{eo}. The number (h) of nodes in the hidden layer was determined by $h = 2u + 1$, where u is the number of nodes in the input layer. Therefore, in this case, $h = 7$. The output layer comprises five nodes representing the five features listed in Table 1. The weight range between the input layer and the hidden layer was set to [0, 1].

The benchmark models were trained using normal samples from the training set and tested using normal samples from the testing set. The goodness-of-fit of the models was evaluated based on the R-squared (R^2) value, where a value closer to 1 indicates better prediction performance. The test results are presented in Table 3, demonstrating the favorable overall prediction performance of the radial basis function-based benchmark models.

Table 3. Fitting accuracies of the benchmark models based on radial basis function.

Designation	Δt_c	Δt_e	ε_{sc}	$LMTD_c$	T_{oil}
R^2	0.979	0.986	0.883	0.958	0.962

4.4. Establishment of the Hybrid BN Driven by Residual and Data

The structure of the hybrid BN is depicted in Figure 5. In order to prevent any bias towards specific states, equal prior probabilities (1/7) were assigned to each state of the top node $D_{r\&x}$. The conditional probabilities of the nodes D_r and D_x were determined using Equation (4). The conditional probabilities of each fault node in the fault layer were established based on the assignment principle described in Equation (1). By referring to the relationship between faults and features presented in Table 1, the residual nodes connected to each fault node could be identified. For instance, the features ε_{sc} and $LMTD_c$ are sensitive to the RefLeak fault; hence, the residual nodes connected to the RefLeak fault node are the ε_{sc} and $LMTD_c$ nodes.

Firstly, the samples corresponding to faults from the training set were input into the trained benchmark models to obtain the benchmark values for each feature. Then, by comparing the benchmark values with the measured values of each feature, the feature residuals were obtained. Maximum likelihood estimation was applied to these feature residuals to derive the conditional probability distributions of the residual nodes. After several attempts, a significance level α of 0.025 was chosen, and the value of c was calculated using Equations (2) and (3) as 4.

The conditional probability distribution of the data node X in the data-driven part was obtained through maximum likelihood estimation using the feature data directly from the training set.

By completing the assignment of prior and conditional probabilities for all nodes in the hybrid BN, the construction of the hybrid BN model was finished.

4.5. Performance Evaluation Indexes

Multiple evaluation metrics were utilized to comprehensively assess the performance of the diagnostic model. These metrics include the confusion matrix, accuracy, precision, recall, and F-measure [35].

By taking the example of a confusion matrix (shown in Table 4) representing a binary classification problem, the calculation of these evaluation metrics is explained. In Table 4, TP represents the number of samples that are true positives (predicted as positive and are actually positive), TN represents the number of samples that are true negatives (predicted as negative and are actually negative), FP represents the number of samples that are false positives (predicted as positive but are actually negative), and FN represents the number of samples that are false negatives (predicted as negative but are actually positive).

Table 4. Confusion matrix explained by a binary classification problem.

True Class	Predicted (Diagnosed) Class	
	Positive	Negative
Positive	TP	FN
Negative	FP	TN

(1) Accuracy

$$Accuracy = \frac{TP + TN}{TP + TN + FP + FN} \tag{6}$$

(2) Precision

$$Precision = \frac{TP}{TP + FP} \tag{7}$$

(3) Recall

$$Recall = \frac{TP}{TP + FN} \tag{8}$$

(4) F-measure

$$F - measure = 2 \times \frac{Precision \times Recall}{Precision + Recall} \tag{9}$$

Precision measures the accuracy of classifying negative samples, indicating the proportion of predicted positive samples that are actually positive. Recall measures the accuracy of classifying positive samples, indicating the proportion of actually positive samples that are correctly predicted as positive. Both precision and recall provide insights into the diagnostic error. F-measure is a harmonic mean that takes into account both precision and recall.

4.6. Fault Diagnosis Results and Discussion

It is important to note that the fault diagnosis results for each scenario were calculated five times using the five datasets formed during the data preprocessing stage. The average of the five computations is presented and discussed as the final fault diagnosis result.

4.6.1. Diagnostic Results Using Only the Residual-Driven Part

In this section, only the feature residuals from the residual-driven part were used for fault diagnosis. The diagnostic performance was evaluated using the data from the testing set, and the test results represented by the confusion matrix are shown in Figure 6, along with the precision, recall, and F-measure. For instance, the precision, recall, and F-measure for RedCdW are determined to be 68.4%, 79.2%, and 73.4%, respectively.

The proposed method achieves a diagnostic accuracy of 73.1% in this scenario. Among the different fault types, the RedEvW and NcG faults exhibit the highest precisions exceeding 90%, while the RefLeak and ExOil faults have the lowest precision, below 60%. The high precision indicates a low false positive rate, meaning that once the fault is diagnosed, there is a high level of confidence that the fault has indeed occurred. The RedCdW, RedEvW, and RefOver faults demonstrate the highest recalls, close to 80%. The high recall indicates a low missed diagnosis rate, meaning that if a fault occurs, it can be accurately diagnosed. Only the RedEvW and NcG faults achieve F-measures above 80%, while the RefLeak fault has the lowest F-measure, at only 59.2%.

Only Residual	Diagnosed class						
True class	RedCdW	RedEvW	RefLeak	RefOver	CdFoul	ExOil	NcG
RedCdW	1268	10	23	68	22	209	0
RedEvW	12	1263	119	2	91	114	0
RefLeak	225	1	1009	91	111	162	0
RefOver	120	1	13	1275	52	121	19
CdFoul	138	1	203	102	1119	36	0
ExOil	8	0	443	27	27	1096	0
NcG	84	0	0	240	0	114	1161
Precision	68.4%	99%	55.8%	70.6%	78.7%	59.2%	98.4%
Recall	79.2%	79%	63.1%	79.7%	69.9%	68.5%	72.6%
F-measure	73.4%	87.8%	59.2%	74.9%	74.1%	63.5%	83.5%

Figure 6. Confusion matrix representing diagnosis results by using only residual nodes. (The row is the number of samples diagnosed by the method, while the column is the number of actually occurred samples. The correctly diagnosed samples are highlighted in green, and the falsely diagnosed samples are highlighted in red.)

These test results validate the effective fusion of the residual-driven model into the hybrid BN model and demonstrate its ability to independently fulfill the diagnostic task only using the feature residual node.

Indeed, the diagnostic performance of using only the residual-driven part for fault diagnosis is strongly affected by the fault-sensitive features listed in Table 1. In this study, the focus was on simplicity and ease of understanding rather than achieving optimal diagnostic performance for the residual-driven part. Therefore, only a small number of features were selected to form the residual nodes.

As shown in Table 1, the selection of the same feature (ε_{sc} and $LMTD_c$) for multiple faults (RefLeak, RefOver, CdFoul, and Ncg) has contributed to confusion between these faults. This overlap in feature selection has led to misdiagnosis among these fault types, as evident from the confusion matrix in Figure 6. Consequently, the diagnostic performance for these faults is not satisfactory. However, by incorporating additional fault-sensitive features, the diagnostic performance of the residual-driven part can be effectively enhanced. For example, introducing features such as the difference between the measured condensing temperature and the calculated condensing temperature based on the condensing pressure (which is sensitive to the Ncg fault [18]), including condensing pressure and subcooling (which are highly sensitive to the RefOver, RefLeak, and CdFoul faults [31,36]), or even considering the entropy efficiency of the compressor based on thermodynamic mechanisms and refrigerant flow rate [36,37], can significantly improve the diagnostic accuracy for these specific faults. Furthermore, by combining different types of nodes proposed in this paper, as discussed in Section 4.6.3, the overall fault diagnosis performance can also be further enhanced.

4.6.2. Diagnostic Results Using Only the Data-Driven Part

In this subsection, only feature data from the data-driven part were used for fault diagnosis. The test results, presented in the form of a confusion matrix in Figure 7, demonstrate that the proposed method achieved a high diagnostic accuracy of 94.5% in this case. Among the different fault types, six faults achieved precisions exceeding 90%, with the ExOil fault being the only exception at 86.4%. All seven faults achieved recalls of 80% or higher. Obviously, a low false positive rate and missed diagnosis rate were achieved. The F-measures for six faults are above 90%. The lowest F-measure is observed in the RefLeak fault at 86.1%, which is still considered effective for accurate fault diagnosis.

Diagnosed class

Only Data	RedCdW	RedEvW	RefLeak	RefOver	CdFoul	ExOil	NcG
RedCdW	1583	13	0	0	4	0	0
RedEvW	1	1592	2	0	5	0	0
RefLeak	2	1	1271	81	3	242	0
RefOver	1	0	46	1543	7	3	0
CdFoul	23	2	0	0	1433	0	143
ExOil	0	0	33	2	5	1560	0
NcG	0	0	0	0	1	0	1599
Precision	98.4%	99%	94%	94.8%	98.2%	86.4%	91.8%
Recall	98.9%	99.5%	79.4%	96.4%	89.5%	97.5%	99.9%
F-measure	98.7%	99.3%	86.1%	95.6%	93.7%	91.6%	95.7%

(True class shown along the vertical axis.)

Figure 7. Confusion matrix representing diagnosis results by using only data nodes (using the same representation as Figure 6).

These test results confirm that the data-driven model is successfully integrated into the hybrid BN model and demonstrate its ability to independently complete the diagnostic task only using the feature data node.

4.6.3. Diagnostic Results Using the Residual-Driven and Data-Driven Parts Together

In this section, feature residuals and feature data from both the residual-driven and data-driven parts were combined for fault diagnosis. The test results, depicted in the form of a confusion matrix in Figure 8, demonstrate that an exceptional diagnostic accuracy of 97.6% was achieved in this case. Among the different fault types, the precisions, recalls, and F-measures for all seven faults were above 90%, indicating an extremely low false positive rate and missed diagnosis rate.

Diagnosed class

Residual & Data	RedCdW	RedEvW	RefLeak	RefOver	CdFoul	ExOil	NcG
RedCdW	1587	10	0	0	2	0	0
RedEvW	2	1595	2	0	1	0	0
RefLeak	0	1	1459	54	3	82	0
RefOver	0	0	22	1570	3	4	0
CdFoul	22	0	2	3	1559	0	14
ExOil	0	0	31	1	3	1566	0
NcG	0	0	0	0	1	0	1599
Precision	98.5%	99.3%	96.3%	96.4%	99.1%	94.8%	99.1%
Recall	99.2%	99.7%	91.2%	98.1%	97.5%	97.9%	99.9%
F-measure	98.9%	99.5%	93.7%	97.3%	98.3%	96.3%	99.5%

(True class is shown along the row dimension.)

Figure 8. Confusion matrix representing diagnosis results by using residual and data nodes together (using the same representation as Figure 6).

When comparing the diagnostic results obtained from using only residual nodes or data nodes, it is evident that the diagnostic performance is significantly enhanced when both parts are used together. The diagnostic accuracy improves by up to 24.5%. For individual faults, there is an improvement in precision, recall, and F-measure. The largest improvement in precision is observed for RefLeak, with an increase of 40.5%, while the largest improvement in recall is seen for ExOil, with an increase of 29.4%. As a result, the F-measures for RefLeak and ExOil increased by up to 34.5% and 32.8%, respectively. This improvement can be attributed to the combination of evidence from both types of nodes, allowing for the utilization of more comprehensive information and ultimately enhancing diagnostic performance.

Fully leveraging all available information proves to be an effective approach to improving fault diagnosis performance. The results highlight the effectiveness of fusing residual and data nodes, enabling parallel fault diagnosis with each part independently completing the diagnostic task. When working together, this fusion leads to superior diagnostic performance.

4.7. Performance Comparison with the Latest Advanced Diagnostic Methods

The fault diagnosis performance of the proposed hybrid method was compared with that of the latest advanced methods proposed in similar studies. In order to ensure an impartial and effective comparison, the comparative methods were selected based on the following criteria: (i) they used the same ASHRAE RP-1043 experimental data, and (ii) they employed the latest improved algorithms for model development. As a result, four existing methods were chosen: SLENet-based [15], SP-CNN-based [16], SA-DNN-based [17], and LS-SVM-based [13] methods. These methods are considered to be the most advanced and have been reported to achieve superior diagnostic performance compared to conventional methods.

During the comparison, the proposed method used the same set of features as the comparative methods for fault indication and evaluated the results obtained by jointly utilizing residual and data nodes. The results of the comparison, represented by accuracies and F-measures, are shown in Figures 9 and 10. It is important to note that the performance of the comparative methods is directly sourced from the related literature. The reported performance in these literature sources should represent the best results achieved for the respective comparative methods. For instance, the diagnostic accuracies and F-measures of the SP-CNN-based method are based on the work by Guo et al. [16].

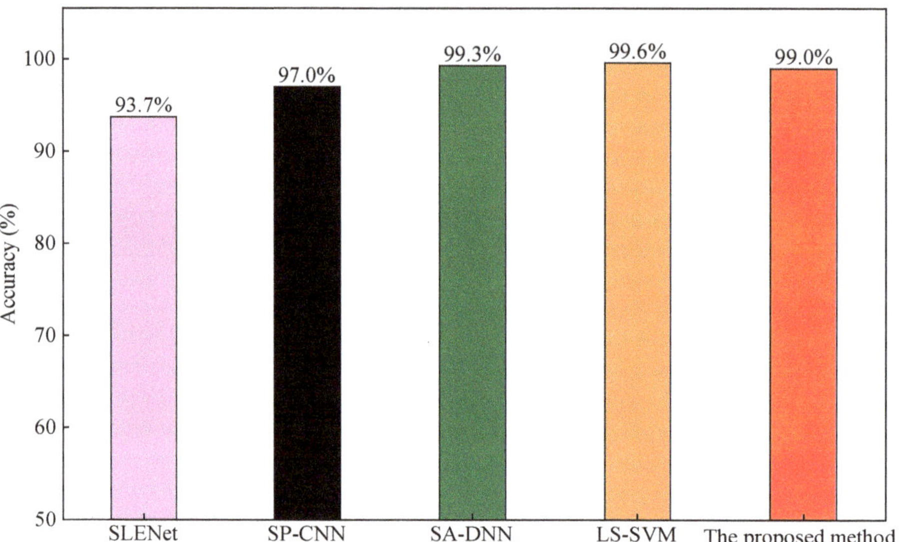

Figure 9. The comparisons of accuracies among the five diagnostic methods, SLENet from [15], SP-CNN from [16], SA-DNN from [17], LS-SVM from [13].

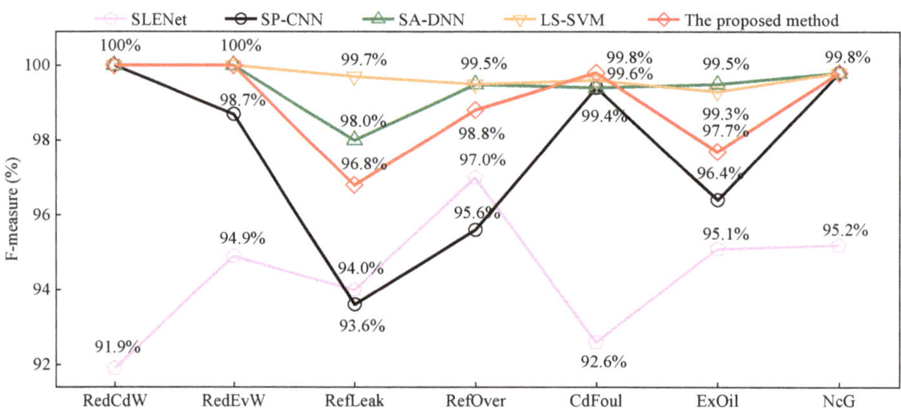

Figure 10. The comparisons of F-measures among the five diagnostic methods, SLENet from [15], SP-CNN from [16], SA-DNN from [17], LS-SVM from [13].

As depicted in Figures 9 and 10, the diagnostic accuracy of the proposed hybrid method surpasses that of the SLENet-based and SP-CNN-based methods and is on par with the LS-SVM-based and SA-DNN-based methods (with a difference of less than 6%).

Regarding F-measures, the proposed hybrid method achieves higher values for all faults compared to the SLENet-based and SP-CNN-based methods. When compared to the LS-SVM-based and SA-DNN-based methods, the proposed hybrid method demonstrates comparable or slightly higher F-measures for all faults, except for RefLeak and ExOil, where the difference is less than 3%.

The training and online diagnosis times of the SVM-based, CNN-based, DNN-based, and proposed methods were calculated and compared. For the model training, all samples from the training set were used, while 100 samples from the testing set were used to simulate an actual fault diagnosis process.

In the case of SVM, the radial basis function was used as the kernel function, and five-fold cross-validation and grid search algorithm were employed to optimize the penalty coefficient and kernel width. The grid searches were conducted within the region of $\left[2^{-4}, 2^4\right]$. The CNN architecture consisted of an input layer, two convolutional layers, two activation layers, two pooling layers, one fully connected layer, and an output layer. The rectified linear unit was used as the activation function. The DNN architecture included an input layer, five hidden layers, and an output layer, utilizing the hyperbolic tangent as the activation function. The proposed method incorporated feature residuals and feature data simultaneously for fault diagnosis.

The results are presented in Table 5, indicating that the proposed method requires shorter times for both model training and online diagnosis compared to the SVM-based, CNN-based, and DNN-based models. Particularly for model training, the proposed method demonstrates a significant reduction in time cost ranging from 56.2% to 76.9% compared to the other methods.

Table 5. The time cost of the model training and online diagnosis for the SVM-based, CNN-based, DNN-based, and proposed methods.

	SVM-Based	CNN-Based	DNN-Based	The Proposed Method
Time cost of model training	2164.7 s	1140.5 s	1382.5 s	499.5 s
Time cost of online diagnosis	5.8 s	6.4 s	6.8 s	5.6 s

Note: calculation time was evaluated in MATLAB 2014b environment installed on a computer with Intel Core i5-2430M (2.40 GHz) CPU and 8 GB of memory.

In summary, the proposed hybrid method has two distinct advantages compared to the comparison methods:

i. Method classification: The comparative methods are all data-driven approaches that rely solely on feature data for fault diagnosis. This is similar to the process of the proposed hybrid method when using only data nodes for fault diagnosis. However, the key advantage of the proposed hybrid method is its fusion of additional information, specifically feature residuals, during the fault diagnosis inference process. This fusion of multiple information sources is likely the main reason for the improved fault diagnosis performance.

ii. Model complexity and training cost: The comparative methods incorporate optimization algorithms to optimize model parameters or select the best features to enhance their performance. This, to some extent, increases the complexity of the models and the training cost. The shorter training and application times of the proposed hybrid method provide an additional advantage. The performance comparison results with existing advanced diagnostic methods further validate the effectiveness and superiority of the proposed hybrid method.

4.8. Application of the Proposed Method in Another Chiller

To further validate the effectiveness of the proposed hybrid method, it was applied to another magnetic centrifugal chiller with a cooling capacity of approximately 440 kW. This chiller also featured shell-and-tube heat exchangers for both the evaporator and condenser, with water flowing inside the tubes. The refrigerant used was R134a. The experimental setup involved various operating conditions, including different set points for the chilled water outlet temperature (5 °C, 7 °C, 8 °C, and 10 °C), cooling water inlet temperatures (25 °C, 27 °C, 30 °C, and 33 °C), and load ratios (40%, 50%, 60%, 70%, 80%, and 90%). Data were collected from a total of 51 operating conditions during the experiment, with a data acquisition interval of 10 s. A comprehensive dataset comprising measurements from 25 parameters was obtained. The five types of faults conducted in this experiment were RedCdW, RedEvW, RefLeak, RefOver, and CdFoul.

The experimental data from the magnetic centrifugal chiller underwent the same data preprocessing process as the ASHRAE RP-1043 data. However, since the magnetic centrifugal chiller was an oil-free system, parameters related to lubricating oil were excluded during the feature selection process. The test results demonstrate that the proposed hybrid method achieves an accuracy of 98.3% when both the residual and data nodes were used for fault diagnosis. The F-measures for fault diagnosis were calculated and presented in Table 6. The results reveal that all F-measures for the five faults exceed 95%, providing further evidence of the excellent diagnostic performance of the proposed hybrid method.

Table 6. The F-measures when both the residual and data nodes were used in the maglev chiller.

Fault	RedCdW	RedEvW	RefLeak	RefOver	CdFoul
F-measure	100%	99.4%	95.7%	95.7%	99.1%

4.9. Analysis of the Potential for Field Application of the Proposed Hybrid Method

By applying the proposed method to two actual chillers, it was demonstrated that the method achieves a high diagnostic accuracy of 97.6%. Moreover, when both feature residuals and feature data are used jointly, the method achieves precisions, recalls, and F-measures above 90% for all seven faults. These results indicate the effectiveness and reliability of the proposed method in fault diagnosis.

Furthermore, when compared to existing state-of-the-art methods, the proposed method shows comparable, and in some cases, even superior, diagnostic performance with higher accuracy and F-measure. This highlights the potential of the proposed method to outperform existing methods in practical applications.

The successful application of the proposed method on two actual chillers, the use of readily available features, and the shorter training and application times further demonstrate the feasibility and practicality of the method in real-world scenarios.

In addition to its strong diagnostic performance, the proposed hybrid method also exhibits great potential for practical application in the field. By utilizing feature residuals and feature data in parallel, the method is designed to be robust and tolerant towards any missing parts of the information. For instance, in situations where obtaining an accurate benchmark model is challenging or impractical, the evidence provided by the feature residuals may be unavailable. In such cases, the proposed hybrid method can still perform fault diagnosis by relying solely on the evidence from the feature data. Similarly, if feature data are unavailable or incomplete, the method can utilize only the feature residuals for diagnostic purposes. This flexibility in data utilization enhances the adaptability and robustness of the proposed hybrid method in practical scenarios.

Furthermore, when both feature residuals and feature data are available, the proposed hybrid method can leverage both sources of information simultaneously, resulting in optimal diagnostic performance and making it a promising solution for on-site fault diagnosis in chillers.

5. Conclusions

To effectively leverage the information from both feature residuals and feature data within a unified diagnostic system, a hybrid method based on BN is proposed. The effectiveness and superiority of the proposed hybrid method were validated through its application to two actual chillers. The main conclusions are as follows:

(1) The hybrid method not only enables independent diagnosis using either type of node but also allows for joint diagnosis using both feature residuals and feature data. This capability leads to improved diagnostic performance and enhances the field applicability of the method compared to approaches that solely rely on one type of node.

(2) The hybrid method demonstrates favorable diagnostic performance when utilizing either feature residuals or feature data alone. However, significant improvements in diagnostic performance are observed when both types of nodes are used together. For instance, the accuracy increases to 97.6%, exhibiting a maximum improvement of 24.5%. The precisions, recalls, and F-measures for all seven faults are above 90%, indicating an extremely low false positive rate and missed diagnosis rate. Moreover, the F-measure shows notable enhancements of 34.5% and 32.8% for the challenging-to-diagnose faults of RefLeak and ExOil, respectively. By integrating evidence from both types of nodes, the hybrid method effectively utilizes a wider range of information, surpassing the diagnostic capabilities of using a single type of node alone and leading to enhanced diagnostic performance.

(3) n comparison to the latest advanced methods, the hybrid method has demonstrated comparable, and in some cases, even superior diagnostic performance with higher accuracy and F-measures. Additionally, the proposed method requires shorter training and application times for the model, further highlighting its effectiveness and superiority.

(4) The application of the hybrid method on another actual chiller further validates its effectiveness, achieving a diagnostic accuracy of 98.3% and F-measures above 95% for all considered faults.

Author Contributions: Conceptualization, Z.W.; Methodology, Z.W.; Software, B.L., J.G. and X.L.; Validation, Z.W. and B.L.; Formal analysis, Z.W. and L.W.; Investigation, J.G. and Y.T.; Resources, Y.T.; Data curation, B.L.; Writing—original draft, B.L.; Writing—review & editing, Z.W., X.L. and S.Z.; Visualization, J.G.; Supervision, L.W. All authors have read and agreed to the published version of the manuscript.

Funding: The authors gratefully acknowledge the support of the National Natural Science Foundation of China (No.51806060, No.51876055), the Program for Science & Technology Innovation Talents in Universities of Henan Province (No.22HASTIT025), and the Program for Innovative Research Team (in Science and Technology) in University of Henan Province (No. 22IRTSTHN006).

Data Availability Statement: No new data were created or analyzed in this study. Data sharing is not applicable to this article.

Abbreviations

CNN	Convolutional neural network
BN	Bayesian network
DNN	Deep neural network
SVM	Support vector machine
LS-SVM	Least squares support vector machine
SA	Simulated annealing
SLENet	Sparsely local embedding CNN
SP-CNN	Data self-production algorithm with CNN
RedCdW	Reduced condenser water flow
RedEvW	Reduced evaporator water flow
RefLeak	Refrigerant leakage
RefOver	Refrigerant overcharge
CdFoul	Condenser fouling
NcG	Non-condensable gas in refrigerant
ExOil	Excess oil

References

1. Nalley, S. *Annual Energy Outlook 2021*; Energy Information Administration: Washington, DC, USA, 2021.
2. Katipamula, S.; Brambley, M.R. Methods for fault detection, diagnostics, and prognostics for building systems—A review, Part I. *HVAC&R Res.* **2005**, *11*, 3–25.
3. Browne, M.W.; Bansal, P.K. Different modelling strategies for in situ liquid chillers. *Proc. Inst. Mech. Eng. Part A J. Power Energy* **2001**, *215*, 357–374. [CrossRef]
4. Zhao, Y.; Wang, S.; Xiao, F.; Ma, Z. A simplified physical model-based fault detection and diagnosis strategy and its customized tool for centrifugal chillers. *HVAC&R Res.* **2013**, *19*, 283–294.
5. Kim, W.; Braun, J.E. Extension of a virtual refrigerant charge sensor. *Int. J. Refrig.* **2015**, *55*, 224–235. [CrossRef]
6. Kim, W.; Braun, J.E. Development and evaluation of virtual refrigerant mass flow sensors for fault detection and diagnostics. *Int. J. Refrig.* **2016**, *63*, 184–198. [CrossRef]
7. Chen, J.; Zhang, L.; Li, Y.; Shi, Y.; Gao, X.; Hu, Y. A review of computing-based automated fault detection and diagnosis of heating, ventilation and air conditioning systems. *Renew. Sustain. Energy Rev.* **2022**, *161*, 8112395. [CrossRef]
8. Yan, K.; Ma, L.; Dai, Y.; Shen, W.; Ji, Z.; Xie, D. Cost-sensitive and sequential feature selection for chiller fault detection and diagnosis. *Int. J. Refrig.* **2018**, *86*, 401–409. [CrossRef]
9. Li, G.; Yao, Q.; Fan, C.; Zhou, C.; Wu, G.; Zhou, Z.; Fang, X. An explainable one-dimensional convolutional neural networks based fault diagnosis method for building heating, ventilation and air conditioning systems. *Build. Environ.* **2021**, *203*, 108057. [CrossRef]
10. Chen, K.; Wang, Z.; Gu, X.; Wang, Z. Multicondition operation fault detection for chillers based on global density-weighted support vector data description. *Appl. Soft Comput.* **2021**, *112*, 107795. [CrossRef]
11. Liu, J.; Shi, D.; Li, G.; Xie, Y.; Li, K.; Liu, B.; Ru, Z. Data-driven and association rule mining-based fault diagnosis and action mechanism analysis for building chillers. *Energy Build.* **2020**, *216*, 109957. [CrossRef]
12. Guo, Y.; Liu, J.; Liu, C.; Zhu, J.; Lu, J.; Li, Y. Operation Pattern Recognition of the Refrigeration, Heating and Hot Water Combined Air-Conditioning System in Building Based on Clustering Method. *Processes* **2023**, *11*, 812. [CrossRef]
13. Han, H.; Cui, X.; Fan, Y.; Qing, H. Least squares support vector machine (LS-SVM)-based chiller fault diagnosis using fault indicative features. *Appl. Therm. Eng.* **2019**, *154*, 540–547. [CrossRef]
14. Han, H.; Cui, X.; Fan, Y.; Qing, H. Ensemble learning with member optimization for fault diagnosis of a building energy system. *Energy Build.* **2020**, *226*, 110351. [CrossRef]
15. Liu, X.; Li, Y.; Sun, S.; Liu, X.; Shen, J. Fault diagnosis of chillers using sparsely local embedding deep convolutional neural network. *CIESC J.* **2018**, *69*, 5155–5163.
16. Gao, J.; Han, H.; Ren, Z.; Fan, Y. Fault diagnosis for building chillers based on data self-production and deep convolutional neural network. *J. Build. Eng.* **2021**, *34*, 102043. [CrossRef]
17. Han, H.; Xu, L.; Cui, X.; Fan, Y. Novel chiller fault diagnosis using deep neural network (DNN) with simulated annealing (SA). *Int. J. Refrig.* **2021**, *121*, 269–278. [CrossRef]
18. Wang, Z.; Wang, L.; Tan, Y.; Yuan, J.; Li, X. Fault diagnosis using fused reference model and Bayesian network for building energy systems. *J. Build. Eng.* **2021**, *34*, 101957. [CrossRef]
19. Comstock, M.C.; Braun, J.E.; Bernhard, R. *Development of Analysis Tools for the Evaluation of Fault Detection and Diagnostics for Chillers*; ASHRAE Research Project 1043-RP, HL 99-20, Report #4036-3; Purdue University: West Lafayette, IN, USA, 1999.
20. Li, T.; Zhao, Y.; Zhang, C.; Luo, J.; Zhang, X. A knowledge-guided and data-driven method for building HVAC systems fault diagnosis. *Build. Environ.* **2021**, *198*, 107850. [CrossRef]
21. Wang, P.; Li, C.; Liang, R.; Yoon, S.; Mu, S.; Liu, Y. Fault detection and calibration for building energy system using Bayesian inference and sparse autoencoder: A case study in photovoltaic thermal heat pump system. *Energy Build.* **2023**, *290*, 113051. [CrossRef]
22. Li, G.; Xiong, J.; Tang, R.; Sun, S.; Wang, C. In-situ sensor calibration for building HVAC systems with limited information using general regression improved Bayesian inference. *Build. Environ.* **2023**, *234*, 110161. [CrossRef]
23. Chen, Y.; Wen, J.; Pradhan, O.; Lo, L.J.; Wu, T. Using discrete Bayesian networks for diagnosing and isolating cross-level faults in HVAC systems. *Appl. Energy* **2022**, *327*, 120050. [CrossRef]
24. Hu, M.; Chen, H.; Shen, L.; Li, G.; Guo, Y.; Li, H.; Li, J.; Hu, W. A machine learning Bayesian network for refrigerant charge faults of variable refrigerant flow air conditioning system. *Energy Build.* **2018**, *158*, 668–676. [CrossRef]
25. Wang, Z.; Wang, Z.; He, S.; Gu, X.; Yan, Z.F. Fault detection and diagnosis of chillers using Bayesian network merged distance rejection and multi-source non-sensor information. *Appl. Energy* **2017**, *188*, 200–214. [CrossRef]
26. Wang, Z.; Wang, L.; Tan, Y.; Yuan, J. Fault detection based on Bayesian network and missing data imputation for building energy systems. *Appl. Therm. Eng.* **2021**, *182*, 116051. [CrossRef]
27. Verron, S.; Tiplica, T.; Kobi, A. Fault diagnosis of industrial systems by conditional Gaussian network including a distance rejection criterion. *Eng. Appl. Artif. Intell.* **2010**, *23*, 1229–1235. [CrossRef]
28. Tan, X.; Gao, X.; Wang, Z.; Han, H.; Liu, X.; Chen, D. Learning the structure of Bayesian networks with ancestral and/or heuristic partition. *Inf. Sci.* **2022**, *574*, 719–775. [CrossRef]
29. Atoui, M.A.; Verron, S.; Kobi, A. A Bayesian network dealing with measurements and residuals for system monitoring. *Trans. Inst. Meas. Control* **2016**, *38*, 373–384. [CrossRef]

30. Kim, M.; Yoon, S.H.; Domanski, P.A.; Payne, W.V. Design of a steady-state detector for fault detection and diagnosis of a residential air conditioner. *Int. J. Refrig.* **2008**, *31*, 790–799. [CrossRef]

31. Xiao, F.; Zheng, C.; Wang, S.W. A fault detection and diagnosis strategy with enhanced sensitivity for centrifugal chillers. *Appl. Therm. Eng.* **2011**, *31*, 3963–3970. [CrossRef]

32. Zhao, Y.; Wang, S.; Xiao, F. A statistical fault detection and diagnosis method for centrifugal chillers based on exponentially-weighted moving average control charts and support vector regression. *Appl. Therm. Eng.* **2013**, *51*, 560–572. [CrossRef]

33. Wang, Z.; Wang, Z.; Gu, X.; He, S.; Yan, Z. Feature selection based on Bayesian network for chiller fault diagnosis from the perspective of field applications. *Appl. Therm. Eng.* **2018**, *129*, 674–683. [CrossRef]

34. Tran, D.A.T.; Chen, Y.; Jiang, C. Comparative investigations on reference models for fault detection and diagnosis in centrifugal chiller systems. *Energy Build.* **2016**, *133*, 246–256. [CrossRef]

35. Zhu, H.; Yang, W.; Li, S.; Pang, A. An Effective Fault Detection Method for HVAC Systems Using the LSTM-SVDD Algorithm. *Buildings* **2022**, *12*, 246. [CrossRef]

36. Zhou, Q.; Wang, S.; Xiao, F. A Novel Strategy for the Fault Detection and Diagnosis of Centrifugal Chiller Systems. *HVAC&R Res.* **2009**, *15*, 57–75.

37. Cui, J.; Wang, S. A model-based online fault detection and diagnosis strategy for centrifugal chiller systems. *Int. J. Therm. Sci.* **2005**, *44*, 986–999. [CrossRef]

Article

Research on Passive Design Strategies for Low-Carbon Substations in Different Climate Zones

Shuizhong Zhao [1], Jiangfeng Si [1], Gang Chen [1], Hong Shi [1], Yusong Lei [2], Zhaoyang Xu [3] and Liu Yang [4],*

[1] State Gird Huzhou Electric Power Supply Company State Gird ZheJiang Electric Power Supply Company, LTD, Huzhou 313000, China; zhaoshuizhong@sgepri.sgcc.com.cn (S.Z.); sijiangfeng@sgepri.sgcc.com.cn (J.S.); chengang@sgepri.sgcc.com.cn (G.C.); shihong@sgepri.sgcc.com.cn (H.S.)
[2] Huzhou Electric Power Design Institute Co., Huzhou 313000, China; leiyusong@sgepri.sgcc.com.cn
[3] State Grid Electric Power Research Institute Wuhan Efficiency Evaluation Co., Ltd., Wuhan 430077, China; xuzhaoyang@sgepri.sgcc.com.cn
[4] College of Civil Engineering and Architecture, Henan University of Technology, Zhengzhou 450001, China
* Correspondence: yl@haut.edu.cn; Tel.: +86-186-2371-2023

Abstract: In the energy-saving design of substations, the building envelope thermal parameters, window-to-wall ratio, and shape factor are three crucial influencing factors. They not only affect the building's appearance but also have an important impact on the total building energy consumption. In this paper, we applied the energy consumption simulation software DeST-c to study the influence of the above three elements on the total energy consumption of the building in a representative city with different thermal zones. The optimal envelope thermal parameters, optimal window-to-wall ratio, and optimal shape factor were derived through combination with economic analysis. Finally, the sensitivity analysis of different elements was carried out to determine the suitable passive design solutions for substations in different climate zones. It was found that the thickness of roof insulation has the greatest influence on the energy consumption of substation buildings among all envelopes. The optimal window-to-wall ratios were 0.4, 0.4~0.5, 0.3, 0.3~0.4, and 0.5 for severe cold, cold, hot summer and cold winter, hot summer and warm winter, and mild regions, respectively; and the optimal shape factors were 0.29, 0.30, 0.23, 0.31, and 0.33, respectively. The conclusions of this study can provide architects with energy-saving design strategies and suggestions for substations in different climate zones, and provide references for building energy-saving designs and air conditioning and heating equipment selection.

Keywords: low-carbon substation; climate zones; envelope thermal design; window to wall ratio; shape factor

Citation: Zhao, S.; Si, J.; Chen, G.; Shi, H.; Lei, Y.; Xu, Z.; Yang, L. Research on Passive Design Strategies for Low-Carbon Substations in Different Climate Zones. *Processes* **2023**, *11*, 1814. https://doi.org/10.3390/pr11061814

Academic Editors: Yabin Guo, Zhanwei Wang, Yunpeng Hu and João M. M. Gomes

Received: 23 April 2023
Revised: 9 June 2023
Accepted: 10 June 2023
Published: 14 June 2023

1. Introduction

Currently, global energy demand is rising dramatically, and electricity consumption has increased 3.9 times since 1973 [1,2]. In order to mitigate global warming and achieve carbon peak and carbon neutrality, significant decarbonization of the global energy system has become the key to the sustainability of the world [3]. As the main part and key node of the power grid, substations have high energy consumption and a lack of energy-saving concepts in their design, resulting in serious energy wastage [4,5]. Therefore, the energy-saving design of substations is a crucial step to reduce carbon emissions [6].

With the growth of electricity consumption in recent years, the number of substations has gradually increased, and they are widely distributed with various building facilities [7]. According to the category of buildings, substation buildings belong to industrial buildings [8,9], and their construction process must meet the requirements of both electrical processes and special buildings, and also requires comprehensive consideration of various technical requirements, such as lighting, ventilation, and fire protection [10]. Therefore, compared with ordinary industrial buildings, substation buildings are characterized by higher energy consumption, high environmental control requirements, and complex energy consumption influencing factors [11]. Meanwhile, detailed energy-saving regulations have not been made for industrial buildings, especially substation buildings [12]. In addition, their design process takes less consideration of comfort issues such as indoor dehumidification [13] and energy consumption of air conditioning systems [14] and lacks the concept of energy saving. Therefore, it is desirable to carry out passive energy-saving designs for them.

Passive buildings are developed on the basis of low-energy buildings. The so-called passive building refers to the use of materials with higher insulation performance and windows and doors with lower heat transfer coefficients, the use of optimized building construction practices to achieve efficient thermal insulation performance, and the use of clean energy to provide heating or cooling to the room, reducing or eliminating the use of actively supplied energy to make the building reach the requirements of a comfortable temperature. Passive buildings can save more than 80–90% of energy compared to conventional buildings. Passive building technology has become the most effective way to reduce building energy consumption, lower carbon emissions, and improve indoor thermal comfort [15–17]. Therefore, it has received wide attention from scholars in recent years. The main technical routes for realizing passive buildings include thermal bridge-free design and construction, rational arrangement of building orientation, design of building openings that facilitate natural ventilation, thermal insulation technology for the building envelope, building shading technology, energy-efficient windows and doors [18], building body shape, optimal window-to-wall ratios, and renewable energy use. Among them, the optimization of the envelope's thermal performance is one of the most effective methods for building energy saving. Kini et al. [19] used passive design solutions in the building envelope in warm and humid regions of India. The study showed that the envelope design improved the indoor thermal environment by 5.82 °C on a typical summer day and saved 77% of cooling energy throughout the year. Mushtaha et al. [20] conducted a passive design based on simulation methods for temperate climate zones in Palestine, including measures such as natural ventilation, external shading, and building envelope insulation. The results showed that passive design reduced building energy consumption by 59%. Jung et al. [21] studied passive design strategies for multi-story buildings in Korea. In the process, the energy, environmental impact, and economic feasibility of the dwelling could be improved by 52.7%, 39.5%, and 36.9%, respectively. Optimization of the window-to-wall ratio and shape factor of buildings are also common approaches for passive building technologies. Pajek et al. [22] investigated the effect of different passive design measures such as thermal transmittance, window-to-wall ratio, and window distribution on energy use in single-family residential buildings for a representative European climate. It was found that zero-energy buildings may be obtained in Porto by appropriate passive design. Li et al. [23] aimed to improve the indoor environment by utilizing solar energy in existing residential buildings in Lhasa, increasing the window-to-wall ratio on the south wall while reducing the depth of south-facing rooms and increasing the width of bays. Yin [24] evaluated the energy performance of passive buildings in typical cities in three climate zones in China and concluded that a smaller shape factor could lead to lower energy consumption. Zhang [25] found that energy consumption increased by 5%, 10%, and 20% when the building shape factor increased from 0.30 to 0.32, 0.34, and 0.36, respectively, through a study of existing buildings.

Passive building technology is also relatively widely used in substations. Kang et al. [26] discussed energy-saving measures for substations in severe cold regions in terms of building exterior walls, roofs, doors, and windows, respectively. It was concluded that energy-saving designs can reduce the total energy consumption of heating, ventilation, air conditioning, and lighting by 50% throughout the year. Zhang et al. [27] used a passive design concept for an extra-high voltage substation in Shijiazhuang, which led to an 85% energy saving in the building through the design of the exterior insulation system and airtightness design. Nie et al. [12] discovered that, based on the climatic conditions in Inner Mongolia, the energy-saving designs of the building's general plan, walls, windows, and roofs can achieve better building energy saving. Yang [28] studied the ventilation and air conditioning system of the Expo substation and obtained an optimized energy-saving solution by changing parameters such as air supply method, height, and effective area of air outlets. Wang et al. [29] proposed a joint centralized operation of air conditioners and exhaust fans based on the self-control strategy of the substation air conditioning system, which eventually reduced the electricity consumption of the substation. Xiong et al. [30] used passive design to enhance natural ventilation and reduce building energy demand based on field research and DeST software simulation. Zhang [31] achieved excellent economic, social, and environmental benefits by controlling the substation building shape factor and designing energy-saving strategies for exterior walls, roofs, windows, and doors. Liu et al. [32] conducted a dynamic energy consumption simulation of the Guangzhou substation by DeST-C to analyze the influence of thermal parameters of the exterior envelope on air conditioning energy consumption in the absence of natural ventilation. It was found that the building's air conditioning energy consumption decreased with increasing heat transfer coefficients of exterior walls, roofs, and exterior windows. Zhou [33] analyzed the building energy-saving design points of the Kunming substation from general layout optimization, reasonable building layout and functional zoning optimization, the energy-saving design of the envelope, and ventilation design. Zhang et al. [34] took the substation building in the Wenzhou area as the research object, analyzed the building energy consumption of the substation, and proposed specific measures for energy saving in terms of substation building monolith, shape factor, and envelope, which effectively reduced the substation's building energy consumption. Zhang et al. [35] calculated the heat dissipation of electrical equipment by collecting the internal and external temperatures of the substation building and testing the heat transfer coefficient of the enclosure system. It provided a basis for the overall design scheme of energy saving in substation buildings in severely cold regions.

Through the existing literature, it can be found that although there were some studies on substations, most of them only proposed some energy-saving methods without giving specific optimized design parameters and did not take into consideration the window-to-wall ratio and shape factors. At the same time, there were few references for the research on the energy-saving indexes of substation buildings, and there were only a few thermal performance tests of substation buildings and guidance for energy-saving design work of substation buildings. Therefore, in this study, typical cities in five climate zones, namely (a) severe cold, (b) cold, (c) hot summer and cold winter, (d) hot summer and warm winter, and (e) mild, were selected, and the optimal envelope thermal parameters, optimal window-to-wall ratio, and optimal shape factors for each climate zone were derived by controlling variables and using DeST-c simulation. The regression equations and errors of total building energy consumption versus window-to-wall ratio and total energy consumption versus shape factors were also obtained. Finally, combined with sensitivity analysis, the optimal design strategies for each climate zone were obtained. The study provides a reference basis for the design scheme of substation buildings and provides data support for the construction of 220 kV substations in different climate zones.

This paper is structured as follows: Section 2 first analyzed the key impact parameters of the substation building and determined the economic evaluation index. Then simulation tools, climate zones, and thermal disturbance parameters were selected to build a physical model of the substation; Section 3's energy simulation yielded results for a simple analysis; Section 4 derived the optimal design solutions for the thermal characteristics of the envelope, the window-to-wall ratio, and shape factor in each region; Section 5 concluded the study.

2. Methodology

In this study, using the energy consumption simulation software DeST-c applied to commercial buildings, the technical route of the study was shown in Figure 1. First, on-site research was conducted on several substations to establish a typical substation building model by using thermal disturbance parameters, thermal parameters, and building information. Then, using the control variable method, the total energy consumption of the substation building was derived by varying the envelope heat transfer coefficient, window-to-wall ratio, and shape factor, respectively, in the simulation. Then, the optimal window-to-wall ratio and optimal shape factor were obtained based on the energy consumption level, and the optimal envelope thermal parameters also needed to be combined with the Benefits-to-Cost Ratio (R_{BC}). Finally, a sensitivity analysis was performed to determine the optimal optimization strategy for substations in different climate zones.

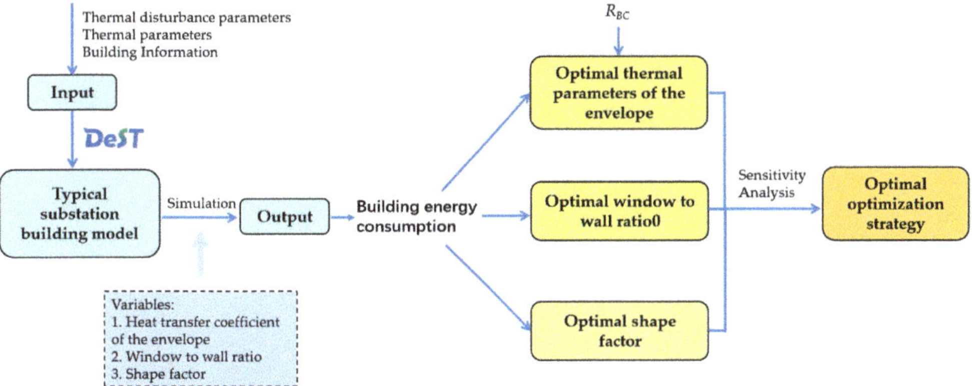

Figure 1. Technology Roadmap.

2.1. Analysis of Key Impact Parameters of Passive Energy Saving in Substation Buildings

First, a theoretical analysis of the heat transfer process of the substation building was carried out to obtain the key influencing factors of energy consumption in the substation building. As shown in Figure 2, in addition to the energy consumption of the production equipment itself, the main energy consumption of the substation building was the energy consumption of the air conditioning system needed to maintain the normal operation of the production equipment and the thermal comfort of the personnel on duty.

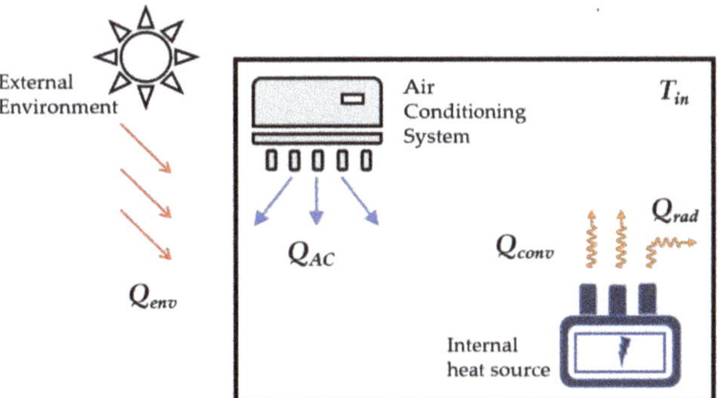

Figure 2. Schematic diagram of heat transfer in substation building. Note: T_{in} is the indoor air temperature, K; Q_{env} is the heat transfer from the envelope to the room, kW; Q_{conv} is the convection heat dissipation from the indoor heat source, kW; Q_{rad} is the radiation heat dissipation from the indoor heat source, kW; Q_{AC} is the heat supplied to the room by the air conditioning system, kW.

According to the heat transfer from the envelope of the substation and the internal heat source, the indoor heat balance equation of the substation was listed as Equation (1). Among them, the heat dissipation from indoor heat sources was divided into two parts, convection and radiation. However, only the convection part could directly heat the indoor air, while the radiation part was heating other solid surfaces first, and then transmitting to the air by convection. Therefore, only the convection part of the indoor heat source could be reflected in Equation (1).

$$\rho_{in} V_{in} c_{in} \frac{\partial T_{in}}{\partial t} = \sum Q_{env} + \sum Q_{conv} + \sum Q_{AC} \quad (1)$$

where, ρ_{in} is the indoor air density, kg/m^3; V_{in} is the indoor air volume, m^3; c_{in} is the specific heat of air, kJ/kg·K; t is the time analyzed, s.

As can be seen, for substation buildings, their production equipment and staff were necessary for their normal working configuration, so the heat dissipation $\sum Q_{conv}$ from indoor heat sources cannot be regulated. Therefore, if we want to reduce the energy consumption of the substation, i.e., reduce the extra heat added or removed $\sum Q_{AC}$, we must try to control the heat transfer $\sum Q_{env}$ of the building envelope.

The heat transfer of the building envelope was further analyzed to establish a set of dynamic heat transfer equations.

Outer envelopes such as exterior walls and roofs were usually analyzed according to one-dimensional unsteady heat transfer. For a less heat-indolent envelope such as windows, the heat storage effect was usually ignored in the heat transfer analysis, and they were analyzed according to the steady-state heat transfer process. The heat transfer diagram and heat balance equation are shown in Table 1.

Table 1. Heat transfer in the substation's building envelope.

Envelope	Heavy Envelope	Light Envelope
Heat transfer diagram	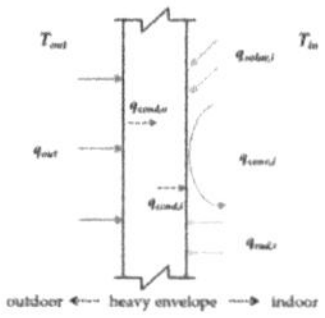	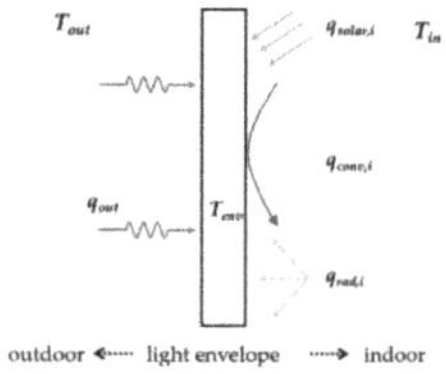
Heat balance relationship	outer surface: $q_{out} = q_{cond,o}$ (2) inner surface: $q_{cond,i} + q_{solar,i} + q_{rad,i} = q_{conv,i}$ (3) Unsteady thermal conduction: $\frac{\partial T_{env}}{\partial t} = a_{env} \frac{\partial^2 T_{env}}{\partial x_{env}^2}, 0 \le x_{env} \le \sigma_{env}, t > 0$ (4)	$q_{out} + q_{solar,i} + q_{rad,i} = q_{conv,i}$ (6)
Expanding equations	$-\lambda_{env,o} \left(\frac{\partial T_{env}}{\partial x_{env}} \right)_{x_{env}=0} = \alpha_{env} (T_{out} - T_{env,o})$ $-\lambda_{env,i} \left(\frac{\partial T_{env}}{\partial x_{env}} \right)_{x_{env}=\sigma_{env}} + \gamma_{env} Q_{solar,i}/A_{env} + \alpha_{env,rad,i}(T_{env,i} - T_{ave,i}) + \kappa_{env} Q_{rad}/A_{env} = \alpha_{env,i}(T_{env,i} - T_{in})$ (5)	$\alpha_{env,o}(T_{out} - T_{env}) + \gamma_{env} Q_{solar,i}/A_{env} + \alpha_{env,rad,i}(T_{env} - T_{ave,i}) + \kappa_{env} Q_{rad}/A_{env} = \alpha_{env,i}(T_{env} - T_{in})$ (7)
Notes	q_{out} denotes the total heat gain of the outer surface, W/m^2; $q_{cond,o}/q_{cond,i}$ denotes the heat gains from thermal conduction of the outer/inner surface, W/m$^{2;}$ $q_{solar,i}$ indicates the heat from sunlight entering the room with windows on the inner surface, W/m$^{2;}$ $q_{rad,i}$ denotes the heat obtained from the inner surface by radiation from an internal heat source, W/m^2; $q_{conv,i}$ denotes the convective heat exchange between the inner surface and the room air, W/m^2; $T_{env,o}/T_{env,i}$ indicates the outer/inner surface temperature of the envelope, K; $T_{ave,i}$ denotes the average temperature of the envelope, K; a_{env} denotes the heat diffusion coefficient, m^2/s; σ_{env} denotes the thickness of this envelope, m; $\lambda_{env,o}/\lambda_{env,i}$ denotes the thermal conductivity of the outer/inner surface, W/m·K; γ_{env} denotes the proportion of solar radiation received by the inner surface to the total incident radiation; $\alpha_{env,rad,i}$ denotes the radiation heat transfer coefficient of the inner surface, W/m^2·K; κ_{env} denotes is the ratio of the radiant heat transfer from indoor heat sources received by the inner surface to the total radiant heat dissipation from indoor heat sources. See Nomenclature for detailed explanation.	

Through the above theoretical analysis, it could be found that the heat transfer of the outer envelope, driven by the temperature difference between indoors and outdoors, was mainly determined by the area and thermal parameters of the envelope. For functional buildings such as substations, their internal use space was usually required for production equipment and functions, which could hardly be changed. Therefore, the building shape factor (the ratio between the area of the outer surface of the building in direct contact with the outdoor atmosphere and its enclosed volume) became one of the main influences on the heat transfer of the building. The windows were less heat-indolent and easily transformed into cold and hot bridges for heat transfer. The overall window-to-wall ratio of the building envelope determined the percentage of window area under the same total area of the envelope. In summary, "envelope thermal parameters", "window-to-wall ratio", and "shape factor" were selected as key parameters for passive optimization in this study. The impact of the changes in the above parameters on the energy consumption of substation buildings was studied.

2.2. Definition of the Economic Evaluation Index

By changing the thickness of the insulation layer of the envelope, the thermal performance of the building will be improved and the energy consumption of the building will be reduced accordingly. However, after the energy consumption is reduced to a certain degree, if the thickness of the insulation layer continues to be changed blindly, the speed of energy consumption reduction will be slowed down significantly, and the initial investment in the building will continue to increase [36]. Therefore, in the building design period, we should not blindly increase the thickness of the insulation layer to achieve the purpose of saving energy. Instead, it should be combined with the economic index to seek the optimal design parameters to achieve the purpose of saving energy and money.

When the thickness of the insulation layer of the initial model envelope is changed, the initial investment of the building will increase, i.e., incremental costs will be generated. The reduction in energy consumption will bring corresponding incremental economic benefits.

When the thickness of the insulation layer of the initial model envelope is changed, the initial investment of the building will increase, i.e., incremental costs will be generated, and the reduction of energy consumption will bring corresponding incremental economic benefits. Incremental costs and economic benefits are related in all phases of the building life cycle, and the Benefits-to-Cost Ratio (R_{BC}) can be used as an economic evaluation index. R_{BC} is the ratio of the total present value of benefits to the total present value of costs of the project over the calculation period and can be calculated according to Equation (8) [37].

$$R_{BC} = \frac{\sum_{t=0}^{N} B_t(1+d)^{-t}}{\sum_{t=0}^{N} C_t(1+d)^{-t}} = \frac{LCB}{LCC} \tag{8}$$

$$LCC = \sum_{t=0}^{N} \frac{C_t}{(1+d)^t} \tag{9}$$

$$LCB = \sum_{t=0}^{N} \frac{B(1+e)^t}{(1+d)^t} \tag{10}$$

where, N—the calculation period, taken as $N = 50$, i.e., the life cycle of the technical solution; t—year; B—savings in energy costs, i.e., economic benefits, B_t—economic benefits in the year t; C—cost, C_t— the cost in the year t. $t = 0$, C_0 represents the initial costs; $t = N$, C_N represents the dismantling costs; $0 < t < N$, C_t represents the operation and maintenance costs; d—social discount rate, taken as $d = 6\%$; LCB—Life Cycle Benefits; LCC—Life Cycle Costs; e— the average annual growth rate of energy prices, take $e = 0.04$.

When the R_{BC} is greater than or equal to 1, the total present value of the benefits of the project is greater than the total present value of its costs. Therefore, it is reasonable to invest in the project. On the contrary, if R_{BC} is less than 1, then the economic efficiency of investing in the project is not at an acceptable level. The larger the value of R_{BC}, the higher the economic efficiency that the project can obtain at a certain present value of cost, in other words, the project is more profitable and more worthy of investment.

2.3. Simulation Study

(1)   Selection of Simulation Tools

The study of energy consumption impact factors of substations was one of the important elements of this paper's research, i.e., the analyzed impact of different optimization strategies on the change in building energy consumption. Through a large number of actual engineering tests, the building energy simulation software can quickly solve the simulation results of energy consumption brought by different variables. At the same time, it could calculate the actual situation of building energy consumption more accurately by analyzing the change in building energy consumption with each impact factor that had a high reference value. DeST is the mainstream energy consumption simulation software in China. Its official website is https://www.dest.net.cn/ (acessed on 22 April 2023). DeST simulates the year-round time-by-time temperatures and humidity in each room of a building in a specified area, taking into account solar radiation, multi-room relationships, inter-radiation between wall surfaces, time-by-time internal disturbances, atmosphere, and earth. The meteorological data required for the simulations are generated by Medpha, which calculates day-by-day meteorological data based on 30 years of measured meteorological data using the time-series analysis method of stochastic processes according to the principles of the time-by-time meteorological model, and further uses the ARMA model and Markov transfer matrix to perform time-by-time calculations on the basis of the day-by-day data, taking into account the interactions and time-series relationships of various meteorological parameters [38]. The time series relationship of various meteorological parameters is considered, and various topographic features of the country are corrected. It can truly reflect the climatic conditions in different regions under different months and has a high degree of simulation. In addition, the software relies on the AutoCAD platform and is relatively easy to use and highly visualized. The reliability of the simulation calculation results is confirmed by the theoretical calibration of the software and the simulation calibration with the internationally recognized building energy simulation software [39]. It was the focus of this paper to analyze the energy consumption impact factors of low-carbon substations based on DeST and propose passive design strategies.

(2)   Selection of simulated climate zones

China has a vast territory, and the "Thermal Design Code for Civil Buildings" [40] (GB50176-2016) divides the country into five climatic zones: severe cold, cold, hot summer and cold winter, hot summer and warm winter, and mild. Through literature combing, it was found that the meteorological parameters in different climate zones had significant differences. Therefore, the energy-saving optimization strategies of their substations were also significantly different. In this paper, representative cities located in five different climate zones were selected: Harbin (located in the severe cold region), Beijing (located in the cold region), Wuhan (located in the hot summer and cold winter region), Guangzhou (located in the hot summer and warm winter region), and Kunming (located in the mild region). The geographical location of each city and the average daily dry bulb temperature throughout the year were shown in Figure 3.

(3)   Establishment of the physical model of the substation

The initial model was established through on-site research of several substations, and its floor plan and axonometric drawing are shown in Figures 4 and 5. The model was a 220 kV internal guarded and unmanned substation with a construction area of 760 m^2 and a floor height of 3.6 m. The main functional rooms were the equipment room, pump room, battery room, office, data room, guard room, and bathroom, etc. The building information is shown in Table 2.

In this regard, the shape factor is the ratio of the exterior area of the building in contact with the outdoor atmosphere to the volume it encloses. The exterior surface area does not include the area of the floor and the interior wall of the unheated stairwell and the door of the house. The window-to-wall ratio is the ratio of the total area of exterior windows (including transparent curtain walls) of a certain orientation to the total area of walls (including window area) of the same orientation. The building scheme was basically similar to the completed outdoor-type substation, and it had a certain reference and representativeness for the study of energy-saving designs of substation buildings.

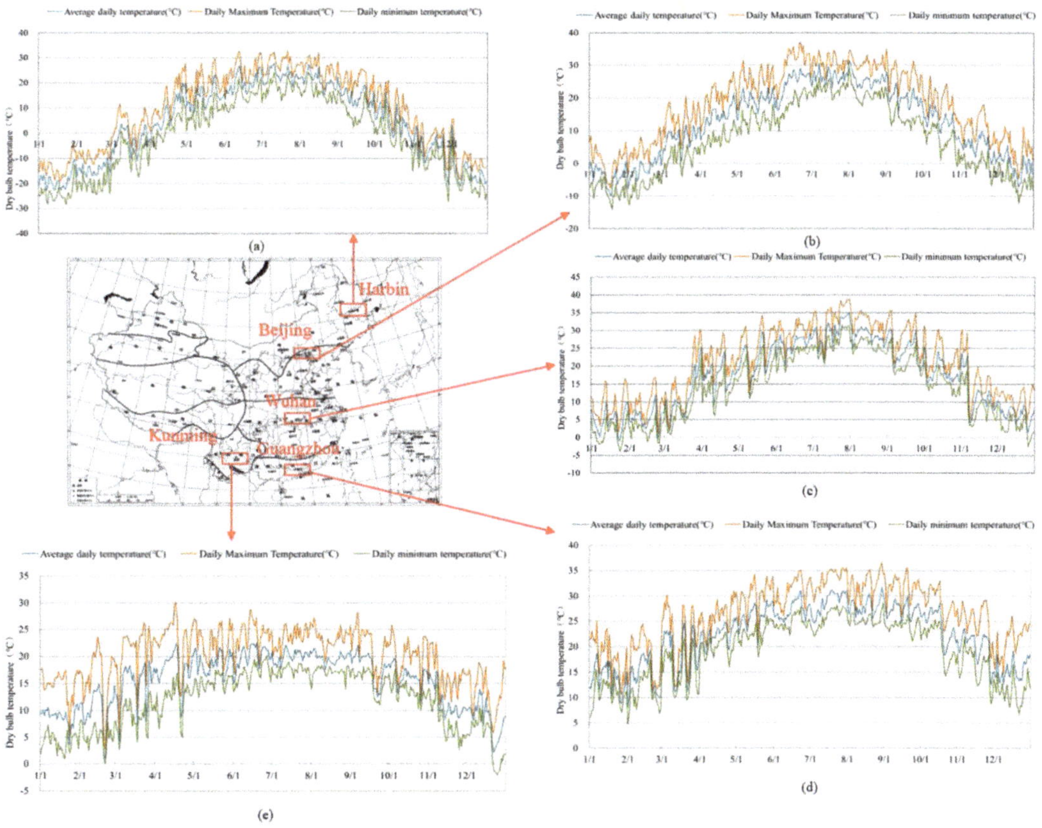

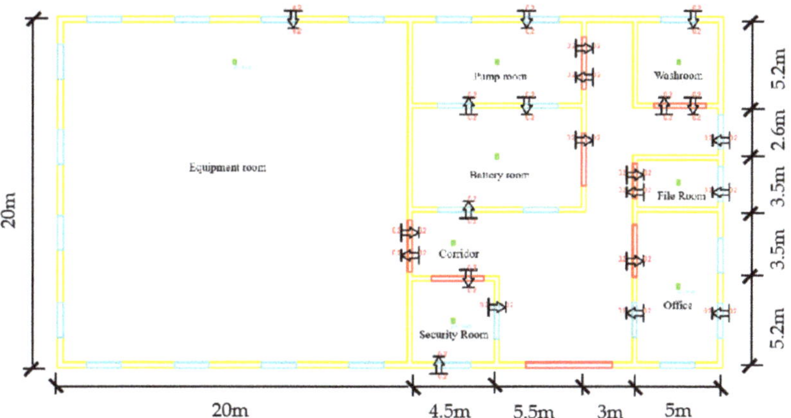

Figure 3. Geographical locations of different cities and the variation in daily average dry bulb temperatures. ((**a**) Harbin, (**b**) Beijing, (**c**) Wuhan, (**d**) Guangzhou, (**e**) Kunming).

Figure 4. Substation model plan.

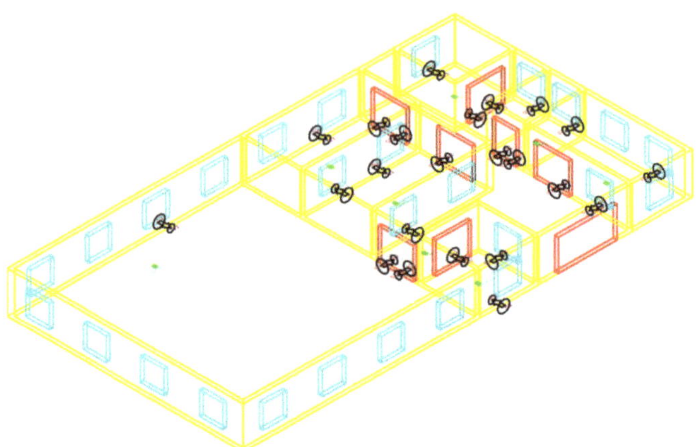

Figure 5. Axonometric drawing of the substation model.

Table 2. Initial model building information table.

Building Height/m	Building Area/m^2	Shape Factor	Window to Wall Ratio (East)	Window to Wall Ratio (South)	Window to Wall Ratio (West)	Window to Wall Ratio (North)
3.6	760	0.33	0.26	0.38	0.26	0.39

The construction of the envelope of the low-carbon substation was summarized through on-site research. Meanwhile, the material and thermal parameters of the model were set with the actual situation. As shown in Table 3.

(4) Determining the boundary conditions and working conditions of the model

The energy consumption simulation of substation buildings differed from that of other types of buildings, mainly because the internal disturbance parameters of each room were different. Through on-site research and combined with the "Technical Regulations for the Design of 220–750 kV Substation" [41] (DL/T 5218-2012), rooms in substations were classified into three categories according to the frequency of personnel activities: areas of frequent personnel activities (such as offices and guard rooms), areas of general personnel activities (such as data rooms, battery rooms, and bathrooms), and areas of electrical equipment (such as pump rooms and equipment rooms). Different types of rooms had different thermal disturbance settings. The thermal disturbance settings were as follows:

Personnel thermal disturbance: maximum number of people 1, minimum number of people 0, heat per capita 66 W, moisture production per capita 0.102, minimum fresh air per capita 30 m^3/hr. Thermal disturbance allocation mode: default personnel thermal disturbance. Indicator type: square meter indicator.

Light thermal disturbance: maximum power 18 W, minimum power 0 W, electric heat conversion efficiency 0.9. Thermal disturbance allocation mode: default light thermal disturbance. Indicator type: square meter indicator.

Equipment thermal disturbance 1: maximum power 13 W, minimum power 0 W. Thermal disturbance allocation mode: default equipment thermal disturbance. Indicator type: square meter indicator. (Applicable to the areas of frequent personnel activities and general personnel activities).

Equipment thermal disturbance 2: maximum power 100 W, minimum power 0 W. Thermal disturbance allocation mode: default equipment thermal disturbance. Indicator type: square meter indicator. (Applicable to area of electrical equipment).

Air conditioning thermal disturbance: the upper limit of the air conditioning start temperature was 26 °C, and the lower limit of the air conditioning start temperature was

18 °C. (Only the area of frequent personnel activities and the area of electrical equipment are set up with air conditioners). The air conditioner turned on when the indoor temperature reached the air conditioner start temperature.

Table 3. Initial model thermal parameters.

Envelope	Area (m^2)	Structure	Heat Transfer Coefficient (W/m$^2 \cdot$K)
Exterior Walls	322.8	240 mm mortar clay + 10 mm pure gypsum board + 60 mm polystyrene foam + 8 mm pure gypsum board	0.564
Exterior Windows	84	3 mm flat glass + 20 mm air laminated + 3 mm flat glass	2.8
Roof	760	20 mm cement mortar + 200 mm reinforced concrete + 46 mm polystyrene + 20 mm cement mortar	0.595
Floor	760	20 mm cement mortar + 200 mm porous concrete	1.02
Exterior Door	10	25 mm pine and spruce	3.5
Interior Wall	200	20 mm cement mortar + 180 mm vitrified concrete + 20 mm cement mortar	1.515
Interior Door	40	25 mm pine and spruce	3.5
Interior Window	24	3 mm flat glass + 20 mm air laminated + 3 mm flat glass	2.8

Through inquiries and surveys of station personnel, a table of parameters for each thermal disturbance at work and rest in accordance with the substation was summarized. According to the set series of internal disturbance parameters, a complete set of frameworks for substation energy consumption simulation was formed by entering personnel thermal disturbance, light thermal disturbance, and equipment thermal disturbance, respectively, into the parameter interface. Each small square in the heat disturbance diagram represented an hourly set value, vertically by week, from the first week of January until the last week of December; horizontally by hour, from 0:00 on Monday until 23:00 on Sunday. Different colors in the figure indicated different work or usage frequencies, including the rate of fixture turn-on, equipment utilization, and occupancy of people in the room. Specific legends were marked in the figures. The larger the number, the higher the frequency of work or use. The thermal disturbance settings of the model were shown in Figures 6–9.

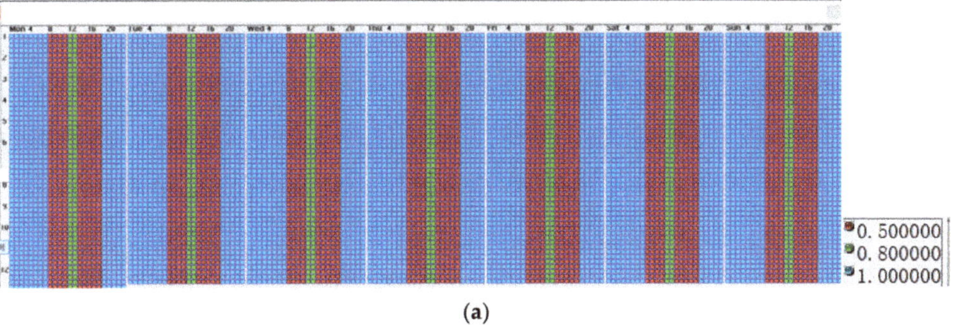

(a)

Figure 6. *Cont.*

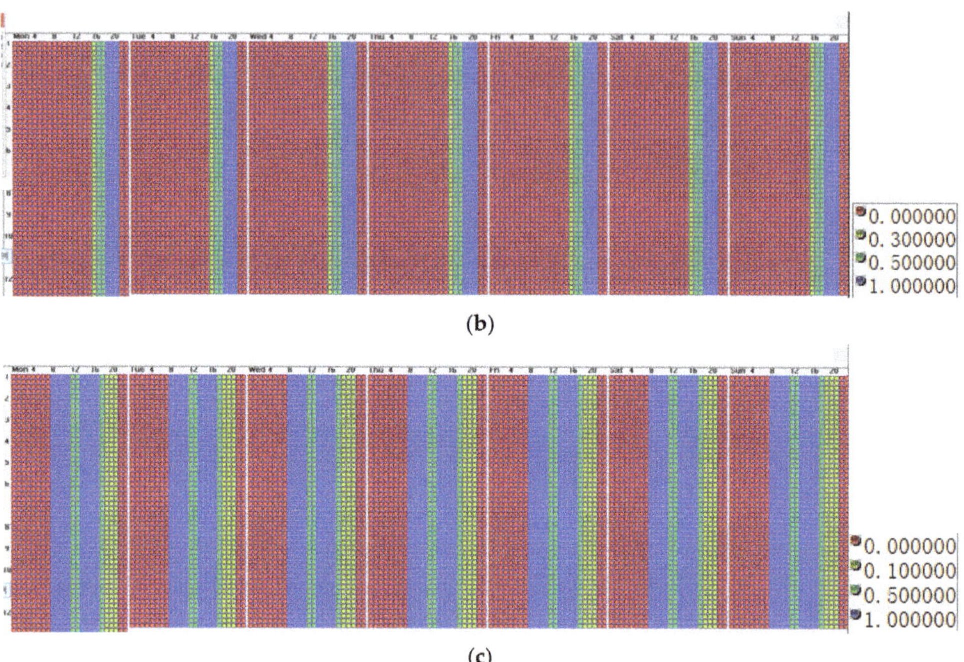

(b)

(c)

Figure 6. Thermal disturbance settings in areas of frequent personnel activities: (**a**) Personnel thermal disturbance, (**b**) Light thermal disturbance, (**c**) Equipment thermal disturbance.

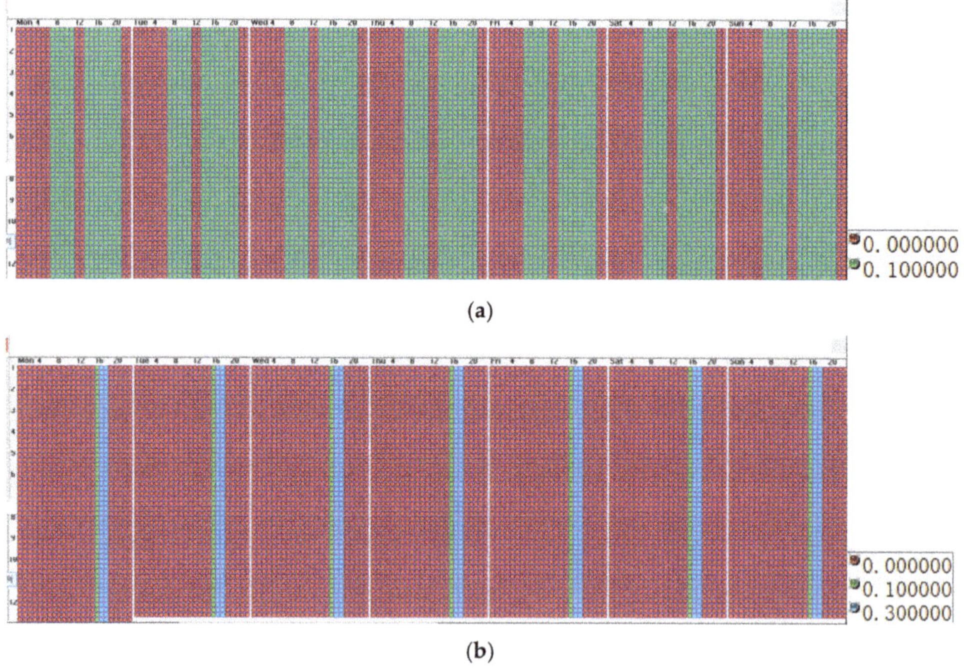

(a)

(b)

Figure 7. *Cont.*

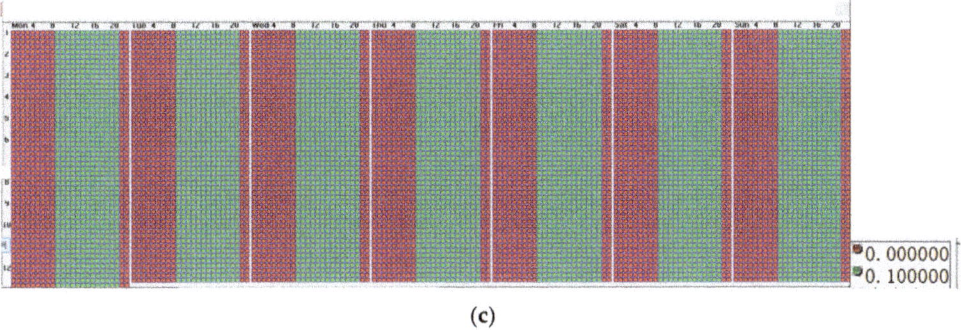

(c)

Figure 7. Thermal disturbance settings in areas of general personnel activities: (**a**) Personnel thermal disturbance, (**b**) Light thermal disturbance, (**c**) Equipment thermal disturbance.

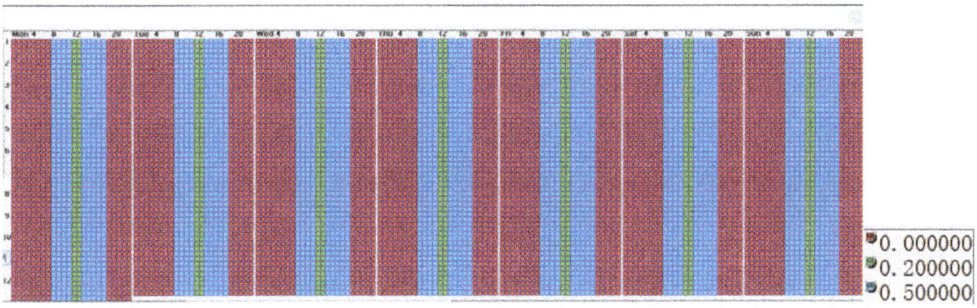

(a)

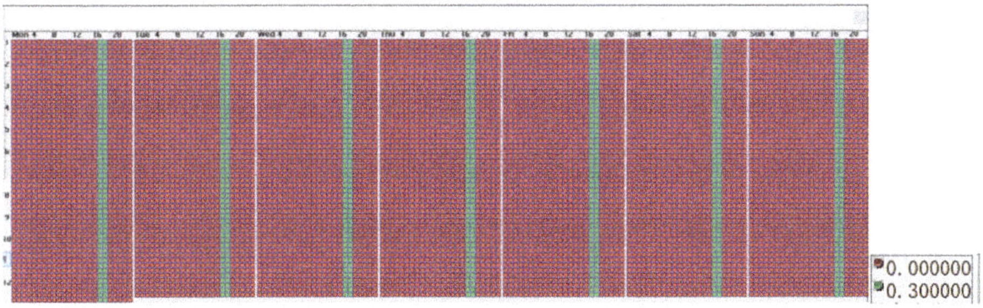

(b)

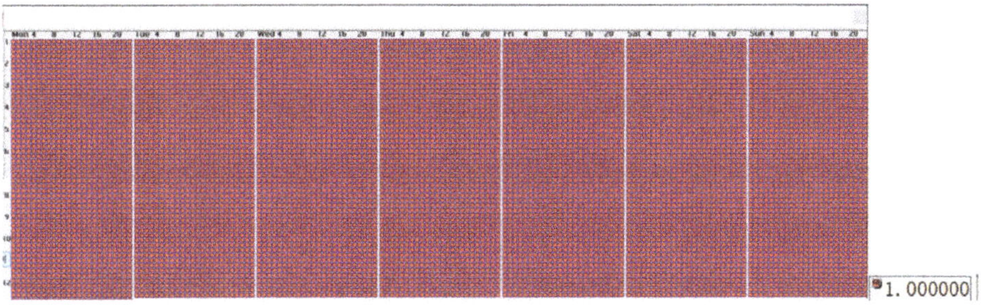

(c)

Figure 8. Thermal disturbance settings in areas of electrical equipment: (**a**) Personnel thermal disturbance, (**b**) Light thermal disturbance, (**c**) Equipment thermal disturbance.

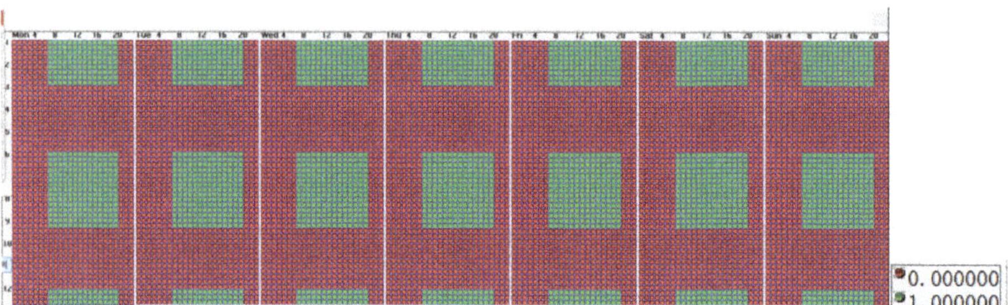

Figure 9. Air conditioning thermal disturbance in three areas.

At the same time, we set up steady-state ventilation in the model, and the number of air changes was taken as 0.4 times/h for buildings with room volumes of 2501~3000 m^3, with reference to the "Unified Standard for Energy-Saving Design of Industrial Buildings" [42] (GB 51245-2017).

3. Results

3.1. Optimal Design Solution for Envelope Thermal Parameters

The building envelope was composed of three main parts, namely the exterior walls, roofs, and exterior windows. For the exterior walls and roofs, physical models were designed for twelve working conditions with the thickness of the insulation layer increasing from 0 mm to 110 mm in steps of 10 mm, and the respective heat transfer coefficients are shown in Figure 10. For the exterior windows, physical models were designed for ten working conditions, and the respective heat transfer coefficients are shown in Table 4. In this design phase, the values of the heat transfer coefficients of the envelope were fixed, and changes caused by changes in operating conditions and the outdoor environment were not considered.

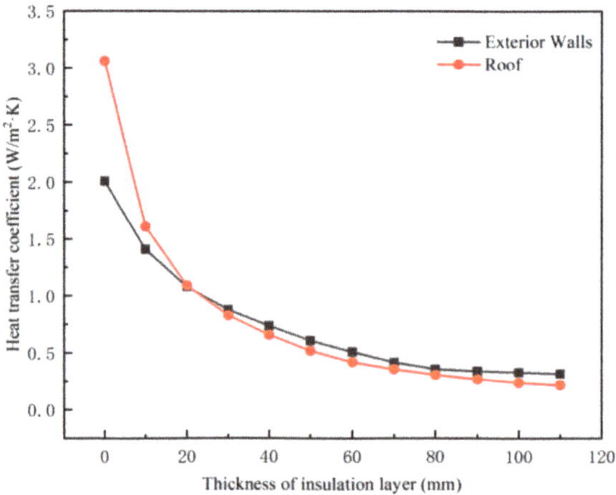

Figure 10. Heat transfer coefficients of exterior walls and roofs with different insulation layer thicknesses.

The heat transfer coefficients of one envelope part were changed individually while maintaining other parameters, and the change in building energy consumption and benefit-cost rate with the thermal properties of the envelope were obtained through computerized energy simulation, so as to seek the optimal design strategies for substations in different climate zones.

In the simulation, the heat transfer coefficients in Table 4 were entered separately in the software, and the effects of outdoor wind speed and wind direction were reflected in the weather file. Through simulation, it was found that the energy consumption of buildings in the severe cold, cold, and mild regions varied greatly, while the energy consumption of the remaining two regions were less than that in the above three regions, but the most economical and energy-saving thermal parameters of the envelope can still be obtained from them. Therefore, the most economical and energy-saving thermal parameters of substations in different climate zones were analyzed in three regions: severe cold, cold, and mild regions.

Table 4. Heat transfer coefficients of different exterior windows.

Exterior Windows	Heat Transfer Coefficient (W/m^2·K)
Old standard exterior window 3.50–0.80	3.50
Standard exterior window 3.2	3.20
Standard exterior window 3.00–0.80	3.00
Standard exterior window	2.80
6 (low-e) + 9 + 6	2.40
3 (low-e) + 0.1 + 3	2.20
6 (low-e) + 9 (Argon) + 6	2.00
6 + 6 air + Pet(low-e) + 6 air + 6	1.70
3 (low-e) + 0.1 + 3 + 9 + 6	1.40
6 + 6 air + Pet (low-e) + 6 air + Pet (low-e) + 6 air + 6	1.00

(1)　Optimal values of envelope thermal parameters in severe cold regions

As shown in Figure 11, the total energy consumption and R_{BC} of the three types of envelopes had similar trends with the change of heat transfer coefficient. For exterior walls and roofs, when the insulation layer was thin, an appropriate increase in the thickness of the insulation layer can significantly reduce building energy consumption. However, when the thickness of exterior wall insulation reached 90~100 mm and roof insulation reached about 100 mm, the building's energy consumption was no longer significantly reduced, which meant that the energy-saving effect was limited when the thickness of the insulation layer was increased. For exterior windows, the building's energy consumption decreased linearly with the reduction of the heat transfer coefficient.

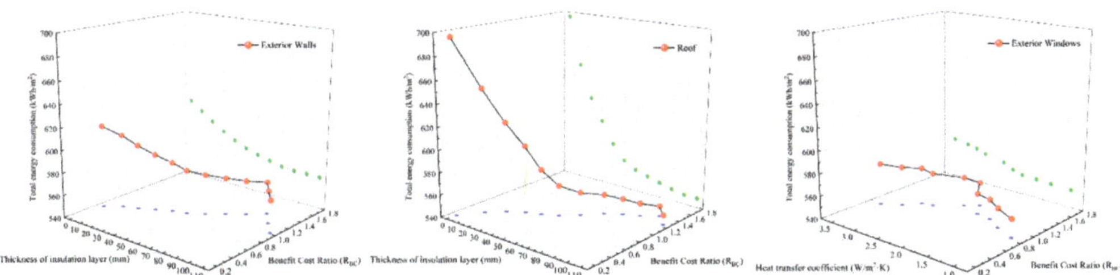

Figure 11. Variations in building energy consumption and benefit-cost rate with thermal properties of exterior walls, roofs, and exterior windows in severe cold regions.

From the economic point of view, the high energy cost brought by an insulation layer that is too thin and the surge of initial investment brought by an insulation layer that is too thick led to a significant difference in the benefit-cost rate of different insulation layers as well. From Figure 11, as the thickness of the insulation layer increased, the heat transfer coefficient decreased continuously, and the R_{BC} of all three showed a trend of rising first and then falling. When the thickness of exterior wall insulation reached 90 mm, the roof insulation thickness reached 100 mm, and the heat transfer coefficient of exterior windows reached 2.2 W/m^2·K, R_{BC} reached the maximum values of 1.37, 1.40, and 1.28, respectively, i.e., the envelope in such cases was the most economical.

Using the two indicators of comprehensive energy consumption and benefit-cost rate, it can be found that the most economical and energy-saving insulation layer for exterior walls in severe cold regions was 90 mm, when the heat transfer coefficient was 0.34 W/m^2·K; the most economical and energy-saving roof insulation layer thickness was 100 mm, when the heat transfer coefficient was 0.24 W/m^2·K. When the heat transfer coefficient of exterior windows was about 2.32 W/m^2·K, the building had the best energy-saving effect and the lowest cost.

(2) Optimal values of envelope thermal parameters in cold regions

As shown in Figure 12, the total energy consumption and R_{BC} of the three envelopes in cold regions also had similar trends compared to those in severe cold regions. For exterior walls and roofs, when the insulation layer was thin, there was a significant reduction in building energy consumption as the thickness of insulation increased. However, when the thickness of the exterior wall insulation and roof insulation reached 70~80 mm, the trend of reducing building energy consumption slowed down significantly. When the thickness of the insulation layer was greater than this limit, there was no significant change in building energy consumption. For exterior windows, the building energy consumption decreased by different degrees with the reduction of the heat transfer coefficient. When the heat transfer coefficient of exterior windows reached 2.4 W/m^2·K, the building energy consumption did not change much with the heat transfer coefficient.

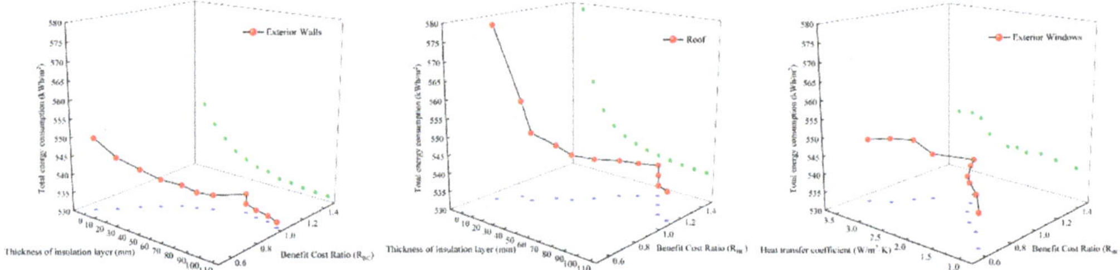

Figure 12. Variations in building energy consumption and benefit-cost rate with thermal properties of exterior walls, roofs, and exterior windows in cold regions.

From the viewpoint of economy, R_{BC} also showed different trends as the heat transfer coefficient decreased with the increase of insulation thickness. When the thickness of exterior wall insulation reached 70 mm, roof insulation reached 80 mm, and the heat transfer coefficient of exterior windows reached 2.4 W/m^2·K, R_{BC} reached the maximum values of 1.184, 1.383, and 1.189, respectively, which meant that the envelope in such cases was the most economical.

The two indicators of comprehensive energy consumption and benefit-cost rate can be found when the thickness of the exterior wall insulation in cold areas was 70 mm and the heat transfer coefficient was 0.42 W/m^2·K, which can achieve the purpose of economic energy saving at this time, while the optimal insulation thickness of the roof was 80 mm. In addition, when the heat transfer coefficient of exterior windows was maintained at 2.4 W/m^2·K, it can also have an adequate energy-saving effect while ensuring economic benefits.

(3) Optimal values of envelope thermal parameters in mild regions

The simulation results for the mild regions are shown in Figure 13. From the figure, it can be seen that the change pattern in the mild regions was obviously different from that in the severe cold and cold regions, which was mainly reflected by the building energy consumption increasing instead of decreasing with the increase in insulation thickness. This was mainly due to the fact that cooling energy consumption in mild regions accounted for a relatively large share of the total energy consumption, and the operation of equipment in substations generated a large amount of heat load, which was not conducive to heat exhaust to the outdoors when the thickness of insulation increased, leading to an increase in total energy consumption. This change law was similar in hot summer and cold winter and hot summer and warm winter regions.

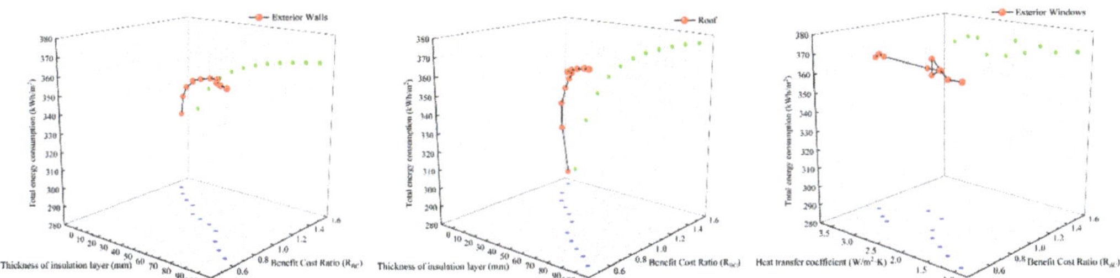

Figure 13. Variations in building energy consumption and benefit-cost rate with thermal properties of exterior walls, roofs, and exterior windows in mild regions.

As shown in Figure 13, the variation of total energy consumption and R_{BC} with the heat transfer coefficient of the envelope in the mild region was significantly different from the two regions of severe cold and cold. For exterior walls and roofs, the thicker the thickness of the insulation layer, the higher the building energy consumption would be. Therefore, in the design of exterior walls and roofs in mild regions, insulation materials with thinner and larger heat transfer coefficients should be used to promote heat exchange between indoor and outdoor areas in summer and reduce building energy consumption. Meanwhile, although the change in the heat transfer coefficient of exterior windows would cause certain fluctuations in building energy consumption, in general, the heat transfer coefficient of exterior windows had no significant effect on building energy consumption.

From the economic point of view, as the thickness of the exterior wall and roof insulation layer increased, both energy consumption cost and initial investment cost gradually increased, resulting in its R_{BC} gradually decreasing. Therefore, the most energy-saving and economical thickness threshold range of the exterior wall and roof insulation layer in mild areas was 0~10 mm, showing that the thickness of the insulation layer of the envelope should be reduced as much as possible. In addition, it can be seen from Figure 13 that when the heat transfer coefficient of exterior windows was 3.5 W/m²·K, the R_{BC} was the largest, and the building energy consumption at this time was also maintained at a low level. Therefore, the most economical and energy-saving heat transfer coefficient of exterior windows in mild regions should be kept at about 3.5 W/m²·K.

(4)　Optimal threshold values for thermal parameters of envelopes in different climatic zones

Through simulation, it was found that the optimal thermal parameters of the envelope varied from one climate zone to another due to the variability of climate characteristics in each region, and the results are shown in Table 5.

In the thermal design of the envelope of low-carbon substations, the most economical and energy-saving form of envelope should be selected according to local conditions.

3.2. Optimal Design Solution for the Window-to-Wall Ratio

As an important building design parameter, the size of the window-to-wall ratio not only affects the façade effect of the building but also has a significant impact on the energy consumption of the building. China is a vast country, and the building energy consumption in different climatic zones is very different, so the window-to-wall ratio design is also different in the building design of each climatic zone. This section sought the best window-to-wall design for substations in different climatic zones by simulating the building energy consumption of typical cities in five climatic zones with different window-to-wall ratios.

Table 5. Optimal values of envelopes' thermal parameters in different climatic zones.

Climate Zones	Exterior Walls		Roofs		Exterior Windows
	Thickness of Insulation Layer (mm)	Heat Transfer Coefficient (W/m²·K)	Thickness of Insulation Layer (mm)	Heat Transfer Coefficient (W/m²·K)	Heat Transfer Coefficient (W/m²·K)
Severe cold	90	0.34	100	0.24	2.2
cold	70	0.42	80	0.31	2.4
Hot summer and cold winter	60	0.51	60	0.42	2.6
Hot summer and warm winter	30	0.88	50	0.52	3.0
Mild	10	1.41	10	1.61	3.5

In the field study, the window-to-wall ratios of several substations were calculated and found to fluctuate from 0.2 to 0.8 for existing substations. Therefore, in this section, a set of window-to-wall ratios was designed and the areas of windows in each direction were calculated based on the dimensions of a typical substation, as shown in Table 6. The building window-to-wall ratio was changed in the global settings of the DeST-c software and used as a variable to simulate energy consumption, while the rest of the parameters were kept the same, and the simulation results are shown in Figure 14.

Table 6. List of design values of building window-to-wall ratios.

Window Area \ Window to Wall Ratio	0.1	0.2	0.3	0.4	0.5
East window (m²)	7.95	15.91	23.86	31.81	39.77
South Window (m²)	11.93	23.86	35.79	47.72	59.65
West Window (m²)	7.95	15.91	23.86	31.81	39.77
North Window (m²)	13.92	27.84	41.76	55.68	69.60

Window Area \ Window to Wall Ratio	0.6	0.7	0.8	0.9	1.0
East window (m²)	47.72	55.68	63.63	71.58	72.00
South Window (m²)	71.59	83.52	95.45	107.38	136.80
West Window (m²)	47.72	55.68	63.63	71.58	72.00
North Window (m²)	83.52	83.52	111.36	125.28	136.80

From Figure 14, it could be seen that there was a significant effect on reducing building energy consumption by controlling the window-to-wall ratio. With the increase in the window-to-wall ratio, the building energy consumption in different climate zones showed a similar linear growth trend, but the growth rate changed slightly differently. For the severe cold regions of Harbin, the building energy consumption only increased by 5.7 kWh/m² when the window-to-wall ratio increased from 0.1 to 0.4, while the increase was much larger than 5.7 kWh/m² when the window-to-wall ratio was greater than 0.4. Therefore, the optimal window-to-wall ratio for the severe cold regions should be maintained at about 0.4. For the cold regions where Beijing is the representative city, when the window-to-wall ratio was between 0.4 and 0.5, the increase in building energy consumption was smaller, so the optimal window-to-wall ratio for this climate zone should be maintained between 0.4 and 0.5. Similarly, for hot summer and cold winter, hot summer and warm winter, and mild regions, the minimum increase in building energy consumption occurred at window-to-wall ratios of 0.3, 0.35, and 0.5, respectively. Therefore, the optimal window-to-wall ratio should be maintained at about 0.3 for hot summer and cold winter regions, between 0.3 and 0.4 for hot summer and warm winter regions, and 0.5 for mild regions.

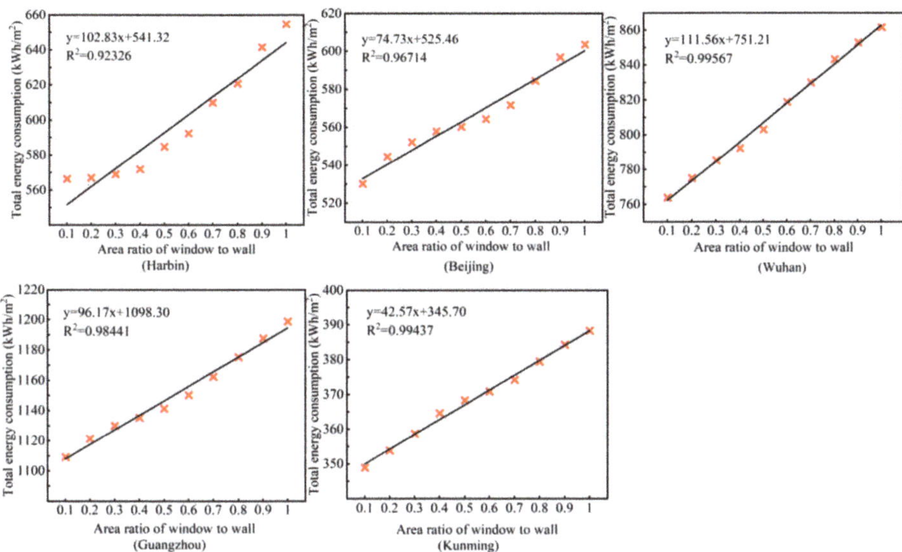

Figure 14. Variations in total building energy consumption with window-to-wall ratios in different climate zones.

3.3. Optimal Design Solutions for Shape Factors

The building shape factor is an important factor influencing the energy consumption of a building, and as the building shape factor changes, so does the energy consumption of the building. The aim of this part of the simulation was to analyze the changing pattern of building energy consumption as the building shape factor changed, and to analyze the range of shape factors that were most energy-saving for substations in different climate zones.

After field research and data queries on outdoor-type substations in severe cold regions, it was concluded that the current shape factor of the substation was about 0.35 to 0.40. Based on existing research data, this experiment was carried out for outdoor-type substations in different climatic zones by varying the height of the model's story so that the shape factor increased in steps of 0.01 between 0.33 and 0.43, while the rest of the parameters remained the same. In addition, the simulation results were fitted to a polynomial and the trends and optimum values were analyzed according to the fitting results in order to find the optimum values of the substation shape factors for different climatic zones. The simulation results are shown in Figure 15.

Based on the fitting results in Figure 15, it can be found that the fitting results for all five regions were quadratic, so the optimum shape factors for different climatic zones could be investigated based on the fitting results combined with the method of finding the optimum value of the quadratic equation. By solving the five equations above, it was found that the shape factors were taken differently for different climatic zones in order to pursue the minimum value of building energy consumption. For the severe cold regions represented by Harbin, the minimum building energy consumption was achieved when the shape factor was 0.29; For the cold regions represented by Beijing, the minimum building energy consumption was achieved when the shape factor was 0.30; For the hot summer and cold winter regions represented by Wuhan, the minimum building energy consumption was obtained when the shape factor was 0.23; For the hot summer and warm winter regions represented by Guangzhou, the minimum building energy consumption was achieved when the shape factor was 0.31, while for the mild regions represented by Kunming, the minimum building energy consumption was achieved when the shape factor was 0.33. It can be seen that the optimum shape factor varied from one climatic zone to another. Therefore, the building design should be adapted to the geographical location of the building and different building forms should be used to maximize energy saving.

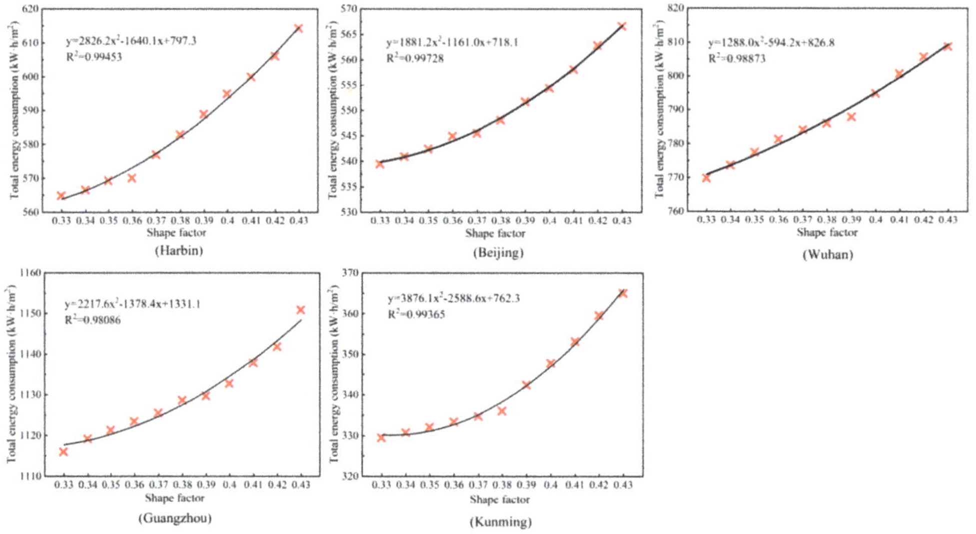

Figure 15. Variations in total building energy consumption with shape factor in different climate zones.

4. Discussion

The above study of passive design strategies for substations in different climatic zones found that the three passive strategies of changing the thermal parameters of the envelope, building window-to-wall ratio, and shape factor had significant differences in the degree of impact on building energy consumption. In order to further study the main influencing factors of building energy consumption in different climate zones, this section conducted a sensitivity analysis on the energy-saving effect of the above three passive strategies, so as to propose more accurate optimization suggestions for the relevant climate zones.

In terms of envelope design, the thickness of the roof insulation had the greatest impact on the total energy consumption of the building, the thickness of the exterior wall insulation had the second greatest impact, and the heat transfer coefficient of the exterior windows had the least impact. This was mainly due to the fact that the equipment inside the substation operated uninterruptedly throughout the day, which generated a large amount of heat. At this time, due to the building height and indoor-outdoor temperature difference, the chimney effect would be generated inside the building, which made a large amount of heat gather at the top of the building space, resulting in a large temperature difference between the inside and outside of the roof. Therefore, once the heat transfer coefficient of the roof changed slightly, it would lead to a corresponding change in the total energy consumption of the building. In other words, the energy-saving design of the roof should be focused on the thermal design of the envelope of substations in different climate zones.

In terms of building window-to-wall ratios, it can be seen from Table 5 that the sensitivity of different climate zones for different window-to-wall ratios varied. Among them, the window-to-wall ratio had the greatest impact on building energy consumption in hot summer and cold winter and severe cold regions, followed by hot summer and warm winter and cold regions, and the least impact on building energy consumption in mild regions. Therefore, the energy saving design in hot summer and cold winter regions and severe cold regions should focus on the setting of the best window-to-wall ratio, while in mild regions the building window-to-wall ratio should be reduced as much as possible to achieve the best energy-saving effect.

The effect of shape factors on building energy consumption in different regions also varied. When the shape factor fluctuated between 0.33 and 0.43, the variation of building energy consumption in severe cold regions was 49.4 kWh/m^2, in cold regions was 27.1 kWh/m^2, in hot summer and cold winter regions was 38.9 kWh/m^2, in hot summer and warm winter regions was 34.9 kWh/m^2, and in mild regions was 35.5 kWh/m^2. Therefore, the design process of substations in severe cold regions should focus on the setting of the optimal shape factor.

Overall, the results of this study were a guided optimization scheme applicable to both new and existing substation buildings. Of course, for existing buildings, this could only be achieved by retrofitting techniques, for which the choice is limited. Furthermore, the window-to-wall ratio in this study was a comprehensive indicator for different climate zones, and different rooms can further select different window-to-wall ratios according to their functions. At the same time, considering the actual construction and application, the insulation thickness was the same in the design, and a different insulation thickness was not considered according to the room type.

In the simulation process, this study only considered the effects of three single factors, namely the thermal physical properties of the envelope, window-to-wall ratio, and shape factor, on building energy consumption, while ignoring the coupled effects of each factor on building energy consumption. The synergistic influence of the above three factors on building energy consumption is very meaningful and worth studying. Future research will focus on different combinations of the above three factors in order to arrive at an integrated and comprehensive design strategy suitable for substations in different climate zones. In addition, combining natural and mechanical ventilation can further reduce building energy consumption, and the impact of different ventilation amounts can be considered for future studies.

5. Conclusions

In this paper, the most economical and energy-saving thermal parameters of envelopes in different climates were obtained by simulating the thermal parameters of envelopes, building window-to-wall ratios, and shape factors in typical cities in different climatic zones, and the optimal window-to-wall ratios and optimal shape factors in different climatic zones were obtained by data analysis. Through the analysis of the data, the following conclusions are summarized:

(1) The thermal design strategy of the envelopes is different in different climate zones, and the most economical and energy-saving envelopes' thermal parameters also differ. Through simulation, it was found that in order to maximize energy savings and minimize budget investment in substation buildings, each region should choose the most suitable thermal parameters of envelopes according to local conditions. Among the envelopes, the thickness of the roof insulation had the greatest influence on the energy consumption of substation buildings, the thickness of the exterior wall insulation was the second, and the heat transfer coefficient of exterior windows had the least influence on the energy consumption of buildings.

(2) Low-carbon substations selected the optimal window-to-wall ratio to effectively reduce the comprehensive energy consumption of the building, and the optimal window-to-wall ratio for substations in different climate zones was different. The optimal window to wall ratio should be maintained at around 0.4 in severe cold regions, 0.4~0.5 in cold regions, 0.3 in hot summer and cold winter regions, 0.3~0.4 in hot summer and warm winter regions, and 0.5 in mild regions. In particular, in the hot summer and cold winter regions and severe cold regions, special attention should be paid to the selection of the optimal window-to-wall ratio.

(3) The geometry of the building should be designed to maximize the energy-saving effect of the building. The optimal shape factor for severe cold regions was 0.29, 0.30 in cold regions, 0.23 in hot summer and cold winter regions, 0.31 in hot summer and

warm winter regions, and 0.33 in mild regions. Among them, the shape factor had the greatest influence on building energy consumption in cold regions.

(4) Energy efficiency optimization strategies for substations vary greatly from one climate zone to another. In terms of the thermal design of the enclosure, colder regions should use thicker insulation materials to reduce the heat exchange between the interior and exterior. On the contrary, in hot and humid regions, the thickness of the insulation material should be reduced as much as possible to maximize the energy saving of the building. In terms of the building's physical design, the window area in cold areas should not be too large to prevent heat outflow, while in hot and humid areas, a larger window area should be controlled. In addition, the design of geometric dimensions should also be adapted to local conditions, and the building form factor should be reasonably controlled.

Author Contributions: Conceptualization, S.Z. and G.C.; Methodology, H.S.; Software, S.Z.; Validation, J.S., Z.X. and Y.L.; Formal analysis, J.S.; Investigation, L.Y.; Resources, S.Z.; Data curation, J.S.; Writing—original draft, S.Z.; Writing—review and editing, G.C.; Visualization, Y.L.; Supervision, H.S. and L.Y.; Project administration, S.Z.; Funding acquisition, Z.X. All authors have read and agreed to the published version of the manuscript.

Funding: This research was funded by the Collective Enterprise Science and Technology Project (CY840800JS20230018) of State Gird Huzhou Electric Power supply company State Gird ZheJiang Electric Power supply Company, LTD, Huzhou 313000, China.

Data Availability Statement: Data are available on request due to restrictions. The data presented in this paper are available on request from the corresponding author.

Acknowledgments: This article was supported by State Gird Huzhou Electric Power supply company State Gird ZheJiang Electric Power supply Company, LTD, Huzhou 313000, China.

Conflicts of Interest: The authors declare no conflict of interest.

Abbreviations

ρ_{in}	Indoor air density	kg/m^3
V_{in}	Indoor air volume	m^3
c_{in}	Specific heat of air	$kJ/kg{\cdot}K$
T_{in}	Indoor air temperature	K
t	Time analyzed	s
$\sum Q_{env}$	The total heat transfer from the envelope to the room	kW
$\sum Q_{conv}$	The total convection heat dissipation from the indoor heat source	kW
$\sum Q_{AC}$	The total heat supplied to the room by the air conditioning system	kW
$\alpha_{env,i}$	Convective heat transfer coefficient of the inner surface of the outer envelope	$W/m^2{\cdot}K$
$T_{env,i}$	Temperature of the inner surface of the outer envelope	K
A_{env}	Surface area of the envelope	m^2
q_{out}	The sum of the total heat gain of the outer surface of the outer envelope	W/m^2
$q_{cond,o}$	The sum of all heat gains from thermal conduction of the outer surface of the outer envelope	W/m^2
$q_{cond,i}$	The sum of all heat gain from thermal conduction of the inner surface of the outer envelope	W/m^2
$q_{solar,i}$	The sum of all the heat from sunlight entering the room with windows on the inner surface of the outer envelope	W/m^2
$q_{rad,i}$	The heat obtained from the inner surface by radiation from an internal heat source	W/m^2
$q_{conv,i}$	The convective heat exchange between the inner surface and the room air	W/m^2
T_{env}	The temperature of the outer envelope at time t and distance x_{env}	K

a_{env}	Heat diffusion coefficient inside the outer envelope	m^2/s
x_{env}	Location of the outer envelope analyzed at time t	m
σ_{env}	Thickness of the outer envelope	m
$\lambda_{env,o}$	Thermal conductivity of the outer surface of the outer envelope	W/m·K
$\lambda_{env,i}$	Thermal conductivity of the inner surface of the outer envelope	W/m·K
γ_{env}	The proportion of solar radiation received by the inner surface of the outer envelope to the total incident radiation	
$\alpha_{env,rad,i}$	Radiant heat transfer coefficient of the inner surface of the outer envelope	W/m^2·K
$T_{ave,i}$	Average temperature of inner surfaces other than the analyzed surface	K
κ_{env}	The ratio of the radiant heat transfer from indoor heat sources received by the inner surface of the outer envelope to the total radiant heat dissipation from indoor heat sources	
T_{out}	Outside air temperature	K
$T_{env,o}$	Outer surface temperature of the outer envelope	K
$T_{env,i}$	Inner surface temperature of the outer envelope	K
α_{env}	Convective heat transfer coefficient of the outer surface of the outer envelope	W/m^2·K
$Q_{solar,i}$	Total indoor solar heat radiation	kW
Q_{rad}	Total indoor heat source radiation heat dissipation	kW
N	Calculation period	
t	Year	
B	Economic benefits	
B_t	Economic benefits in year t	
C_0	Initial costs	
C_N	Dismantling costs	
C_t	Operation and maintenance costs	
d	Social discount rate	
e	Average annual growth rate of energy prices	
R_{BC}	Benefits to Cost Ratio	
$DeST\text{-}C$	Designer's Simulation Toolkit-Commercial	
LCB	Life Cycle Benefits	
LCC	Life Cycle Cost	

References

1. IEA. *Key World Energy Statistics 2016*; IEA: Paris, France, 2016. [CrossRef]
2. Holmatov, B.; Hoekstra, A.Y.; Krol, M.S. Land, water and carbon footprints of circular bioenergy production systems. *Renew. Sustain. Energy Rev.* **2019**, *111*, 224–235. [CrossRef]
3. Elshkaki, A. The implications of material and energy efficiencies for the climate change mitigation potential of global energy transition scenarios. *Energy* **2023**, *267*, 126596. [CrossRef]
4. Guo, J.H. Introduction to the application of energy saving technology in substations. *Shanxi Archit.* **2010**, *36*, 248–249. [CrossRef]
5. Li, S.Q. Construction and Empirical Study of Energy Saving Assessment System for Substation Construction Projects. Master's Thesis, North China Electric Power University, Baoding, China, 2012.
6. Feng, Z.; Zhang, G.; Wu, J.; Han, W.; Fang, S.; Yan, X.; Huang, S. Research on corona characteristics and type selection of fittings in extra high voltage alternating current substation. *Energy Rep.* **2022**, *8*, 176–186. [CrossRef]
7. Zhang, X.; Liu, L.Q.; Wu, M.L.; Wang, W.; Chen, G.; Yan, W.; Cao, M.; Liang, H.S.; Zhang, C.S.; Chen, B.; et al. Research and application of energy efficiency assessment and energy saving retrofit technology for substations. *Power Ind.* **2019**, *470*, 40.
8. China Electricity Council Technical and Economic Consulting Center for Electricity Construction. *Substation Building Engineering*; China Electric Power Press: Beijing, China, 2008.
9. Du, C. Research on Substation Building Design. Master's Thesis, Shenyang Jianzhu University, Shenyang, China, 2020. [CrossRef]
10. Wang, Y.F. Research on Energy Saving Design of Substation Buildings in Severe Cold Regions Based on Computer Simulation. Master's Thesis, Harbin Institute of Technology, Harbin, China, 2013.
11. Chen, Z.; Sun, D.W. Analysis of factors affecting energy consumption of buildings in China. *Ind. Technol. Forum* **2008**, *6*, 150–151.
12. Nie, J.C.; Zhang, Z.; Zhang, X.Y.; SaRen, G.W.; Luo, D.M. Energy saving design of substation buildings in Inner Mongolia. *Inn. Mong. Electr. Power Technol.* **2018**, *36*, 43–50.
13. Ybyraiymkul, D.; Chen, Q.; Burhan, M.; Akhtar, F.H.; AlRowais, R.; Shahzad, M.W.; Ja, M.K.; Ng, K.C. Innovative solid desiccant dehumidification using distributed microwaves. *Sci. Rep.* **2023**, *13*, 7386. [CrossRef] [PubMed]
14. Krarti, M.; Ybyraiymkul, D.; Ja, M.K.; Burhan, M.; Chen, Q.; Shahzad, M.; Ng, K. Energy performance of hybrid evaporative-vapor compression air conditioning systems for Saudi residential building stocks. *J. Build. Eng.* **2023**, *69*, 106344. [CrossRef]
15. Wolfgang, F.; Guo, L.; Sheng, S.C.; Jiang, H.J. Passive house in China: Health, Comfort, affordability and sustainability. *Eco-City Green Build* **2016**, *3*, 31–37.

16. Liu, H.Z.; Ding, H.T.; Li, T.Y.; Guo, Y.Y.; Zhao, G.P. Research on building energy consumption status of Civil buildings and development trend in China. *Build. Technol.* **2018**, *8*, 10–11. [CrossRef]
17. Lin, Y.L.; Zhao, L.Q.; Yang, W.; Hao, X.L.; Li, C.Q. A review on research and development of passive building in China. *J. Build. Eng.* **2021**, *42*, 102509. [CrossRef]
18. Brito-Coimbra, S.; Aelenei, D.; Gloria Gomes, M.; Moret Rodrigues, A. Building Façade Retrofit with Solar Passive Technologies: A Literature Review. *Energies* **2021**, *14*, 1774. [CrossRef]
19. Kini, P.G.; Garg, N.K.; Kamath, K. An assessment of the impact of passive design variations of the building envelope using thermal discomfort index and energy savings in warm and humid climate. *Energy Rep.* **2022**, *8*, 616–624. [CrossRef]
20. Mushtaha, E.; Salameh, T.; Kharrufa, S.; Mori, T.; Aldawoud, A.; Hamad, R.; Nemer, T. The impact of passive design strategies on cooling loads of buildings in temperate climate. *Case Stud. Therm. Eng.* **2021**, *28*, 101588. [CrossRef]
21. Jung, Y.J.; Heo, Y.; Lee, H. Multi-objective optimization of the multi-story residential building with passive design strategy in South Korea. *Build. Environ.* **2021**, *203*, 108061. [CrossRef]
22. Pajek, L.; Kosir, M. Strategy for achieving long-term energy efficiency of European single-family buildings through passive climate adaptation. *Appl. Energy* **2021**, *297*, 117116. [CrossRef]
23. Li, E.; Liu, J.P.; Yang, L. Research on the passive design optimization of direct solar gain house for residential buildings in Lhasa. *Ind. Constr.* **2012**, *42*, 27–32. [CrossRef]
24. Yin, M.Z. Adaptability Analysis of Passive House in the Northern Part of China. Master's Thesis, Shandong Jianzhu University, Jinan, China, 2016.
25. Zhang, S.L. Research on Methods of Passive House Design for Congregate Housing in Cold Region. Master's Thesis, Hebei University of Technology, Tianjin, China, 2015.
26. Kang, J.; Yin, X. Energy saving design of 500 KV substation building. *Jilin Electr. Power* **2013**, *41*, 27–28. [CrossRef]
27. Zhang, G.L.; Zhao, S.Y.; Wang, N.; Zhao, B.J. Research on the application of passive building technology in the extra-high voltage substation buildings. *Constr. Technol.* **2018**, *37*, 18–26. [CrossRef]
28. Yang, X.P. Research on Ventilation and Air Conditioning System of Underground Substation and Energy Saving Optimization. Master's Thesis, Donghua University, Shanghai, China, 2011.
29. Wang, W.B.; Hua, B. Research on energy saving of automatic control of air conditioning system in substation. *East China Electr. Power* **2009**, *37*, 777–780.
30. Xiong, T.J.; Li, G.P.; Liu, Y.; Li, S.J. Research on ways to achieve near-zero energy substation buildings. *Build. Energy Effic.* **2021**, *12*, 39–44.
31. Zhang, C. Building optimization innovation and energy saving and environmental protection. *Build. Energy Sav.* **2014**, *51*, 239–240.
32. Liu, Y.F.; Zhang, Y.F.; Cai, Z.H.; Zhang, J.; Liu, Q.N. Influence of thermal parameters of outer envelope on energy consumption of substation buildings. *Build. Energy Sav.* **2020**, *2*, 29–33.
33. Zhou, S.R. Energy saving design of substation buildings in mild regions. *Sci. Technol. Innov. Her.* **2015**, *28*, 93–94. [CrossRef]
34. Lin, Y.L.; Zhang, F. Energy saving design of substation buildings in hot summer and cold winter regions. *Shanxi Archit.* **2009**, *35*, 238–240.
35. Zhang, S.J.; Zhang, J.B.; Du, H.B. Energy saving testing of substation buildings in severe cold regions. *Heilongjiang Sci.* **2014**, *5*, 70–71.
36. Han, X.L.; Chen, B.; Zhu, J.Y. Economic Analysis and Case Study of Building Thermal Design Optimization. *Build. Sci.* **2014**, *30*, 65–71. [CrossRef]
37. Ministry of Construction, National Development and Reform Commission. *Methods and Parameters for Economic Evaluation of Construction Projects*, 3rd ed.; China Planning Press: Beijing, China, 2006.
38. Xie, X.N.; Song, F.T.; Yan, D.; Jiang, Y. Building environment design simulation software DeST(2): Dynamic thermal process of buildings. *Heat. Vent. Air Cond.* **2004**, *34*, 35–47.
39. Zhu, D.D.; Yan, D.; Wang, C. Comparison of Building Energy Simulation Programs: DeST, EnergyPlus and DOE-2. *Build. Sci.* **2012**, *28*, 213–222. [CrossRef]
40. *GB 50176-2016*; Thermal Design Code for Civil Buildings. China Construction Industry Press: Beijing, China, 2016.
41. *DL/T 5218-2012*; Technical Regulations for the Design of 220 kV~750 kV Substations. China Plan Press: Beijing, China, 2012.
42. *GB 51245-2017*; Unified Standard for Energy-Saving Design of Industrial Buildings. China Plan Press: Beijing, China, 2017.

 *processes*

 **MDPI**

Article

Operation Pattern Recognition of the Refrigeration, Heating and Hot Water Combined Air-Conditioning System in Building Based on Clustering Method

Yabin Guo [1], Jiangyan Liu [2], Changhai Liu [1], Jiayin Zhu [1], Jifu Lu [1] and Yuduo Li [1,*]

1   School of Water Conservancy and Civil Engineering, Zhengzhou University, Zhengzhou 450001, China
2   Key Laboratory of Low-Grade Energy Utilization Technologies and Systems, Chongqing University, Chongqing 400044, China
*   Correspondence: yd_li@gs.zzu.edu.cn

Abstract: Air-conditioning system operation pattern recognition plays an important role in the fault diagnosis and energy saving of the building. Most machine learning methods need labeled data to train the model. However, the difficulty of obtaining labeled data is much greater than that of unlabeled data. Therefore, unsupervised clustering models are proposed to study the operation pattern recognition of the refrigeration, heating and hot water combined air-conditioning (RHHAC) system. Clustering methods selected in this study include K-means, Gaussian mixture model clustering (GMMC) and spectral clustering. Further, correlation analysis is used to eliminate the redundant characteristic variables of the clustering model. The operating data of the RHHAC system are used to evaluate the performance of proposed clustering models. The results show that clustering models, after removing redundant variables by correlation analysis, can also identify the defrosting operation mode. Moreover, for the GMMC model, the running time is reduced from 27.80 s to 10.04 s when the clustering number is 5. The clustering performance of the original feature set model is the best when the number of clusters of the spectral clustering model is two and three. The clustering hit rate is 98.99%, the clustering error rate is 0.58% and the accuracy is 99.42%.

Keywords: air conditioning system; pattern recognition; clustering; correlation analysis; defrosting operation mode

Citation: Guo, Y.; Liu, J.; Liu, C.; Zhu, J.; Lu, J.; Li, Y. Operation Pattern Recognition of the Refrigeration, Heating and Hot Water Combined Air-Conditioning System in Building Based on Clustering Method. *Processes* 2023, 11, 812. https://doi.org/10.3390/pr11030812

Academic Editor: Jean-Pierre Corriou

Received: 8 February 2023
Revised: 1 March 2023
Accepted: 3 March 2023
Published: 8 March 2023

1. Introduction

According to the report of the International Energy Agency, buildings currently account for 36% of global carbon dioxide emissions [1]. China has also put forward carbon peak and carbon neutral targets, so energy conservation and emission reduction in various energy industries will become increasingly important [2,3]. Building cooling, heating and hot water supply account for a large proportion of building energy [4,5]. Therefore, the combined air-conditioning system that can achieve cooling, heating and hot water production will help achieve further energy-saving and emission-reduction goals. Analyzing the operation data of a combined air-conditioning system and identifying the operation mode through data mining will further help to improve the energy-saving potential.

At present, many researchers have carried out research on the refrigeration, heating and hot water combined air-conditioning (RHHAC) system. Gong et al. [6] studied the combined chiller system, which can provide a sanitary hot water supply and air conditioning simultaneously. Zhao et al. [7] studied the heating performance of an air-source heat pump with water tank for thermal energy storage. Byrne et al. [8,9] carried out the experimental study of an air-source heat pump for simultaneous heating and cooling. The research mainly consists of two parts, including basic concepts and performance verification and dynamic behavior and the two-phase thermosiphon defrosting technique. In addition, the operating characteristics of the combined air-conditioning system under different operating conditions have also been studied, and especially the start-up characteristics have been analyzed. In the hot-water mode, the exhaust pressure of the unit reached 3.7 Mpa [10]. However, there is less research on the further in-depth analysis of the operation data of the RHHAC system. In particular, data-mining methods have been used more and more widely in the field of air conditioning [11–13], so it is necessary to carry out data-mining research on the RHHAC system. The clustering method is an unsupervised machine learning method [14–16] that can analyze the data without labeling data and obtain valuable information [17]. Since labeled data are more difficult to obtain than unlabeled data, the application of clustering methods will have advantages [18]. At present, the widely used clustering methods mainly include K-means clustering [19–22], Gaussian mixture model clustering (GMMC) [23–25] and spectral clustering (SC) [26–28]. Xia et al. [29] proposed the cluster model based on the K-means method to find out the difference between the climate and the building thermal environment of three regions. David et al. [30] proposed a new climate classification method based on K-means, and the towns incorrectly classified were reduced between 0.128 and 7.702%. Nurseda et al. [31] combined the K-means method and the association rule mining method to study the balance of solar power generation. Islam et al. [32] proposed an enhanced brain tumor detection scheme based on the K-means method and principal component analysis. A framework using clustering methods for modeling time-varying operations in complex energy systems was developed. Different clustering methods in the domain of the objective function of two example operational optimization problems were compared [33]. Zhao et al. [34] developed the Gaussian mixture model to preprocess historical data to obtain steady-state measurements under various operating conditions. Further, the Gaussian mixture model was used to optimize the energy management of heterogeneous building neighborhoods [35]. Shen et al. [36] used the principal analysis and the Gaussian mixture model to establish a building type clustering model, which is computationally efficient and a more accurate reflection of the local urban microclimate. Wang et al. [37] proposed an optimal scheduling strategy for an electricity–hydrogen–gas–heat integrated energy system based on the spectral clustering method, and used the spectral clustering method to describe the uncertainty of the system. Guo et al. [38] proposed multi-view spectral clustering combined with simultaneous consensus graph learning and discretization. From the above research, it can be concluded that the data-mining method can carry out knowledge discovery from the operation data of the air-conditioning system. The data-mining method has important potential for air-conditioning system operation pattern recognition. However, there is no research on applying the clustering method to data analysis of the RHHAC system. Therefore, this study focuses on the application of the clustering method in the operation pattern recognition of the RHHAC system.

In addition, different feature variable sets also have an important impact on the complexity and running time of the model [39]. The correlation analysis approach can eliminate the redundant characteristic variables in the model [40–42]. Therefore, the correlation analysis method is used to eliminate the redundant characteristic variables in this study.

The main contribution of this paper is to establish cluster models of a RHHAC system using various clustering methods for system operation pattern recognition. The clustering performance of the model with different clustering methods, different numbers of clusters and different feature variable sets is analyzed. This method can be used for data mining and energy-saving pattern recognition of air-conditioning systems, and plays an important role in the fault identification and energy saving of air-conditioning systems.

The paper is organized as follows. Section 2 outlines the pattern recognition approach based on clustering. Section 3 describes the experiment and data introduction of the RHHAC system. Section 4 introduces the results and discussion of the clustering model in detail. Finally, the paper is concluded in Section 5.

2. Pattern Recognition Approach Based on Clustering

The pattern recognition approach proposed in this study is based on the unsupervised clustering algorithm, as shown in Figure 1. The unsupervised clustering method can identify valuable operating modes under the condition of unlabeled data. This section introduces the principles of three unsupervised clustering methods and the redundant variable elimination approach based on correlation analysis. Further, quantitative evaluation indexes are proposed to evaluate the performance of different clustering models.

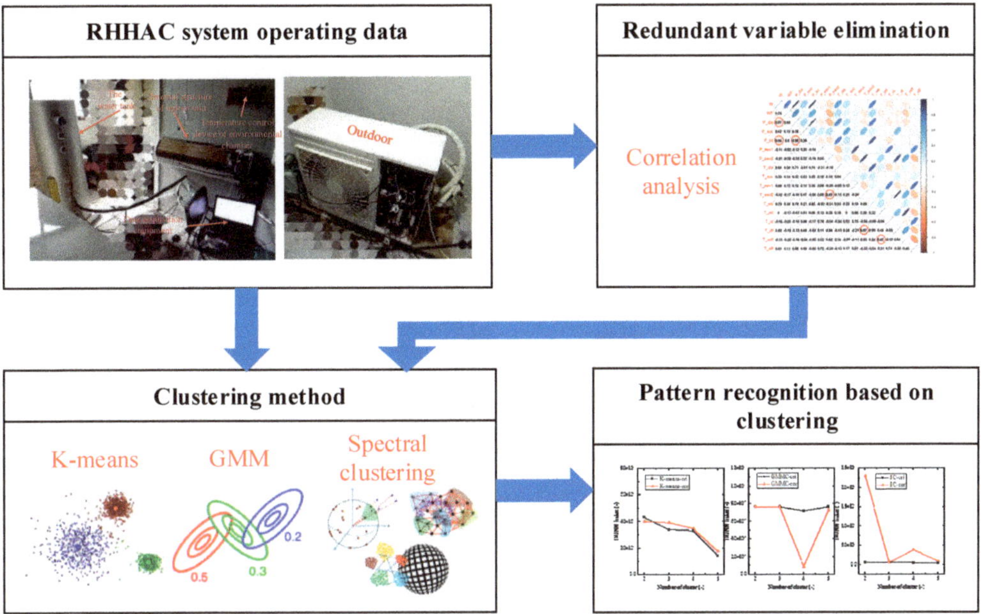

Figure 1. Pattern recognition approach flow chart based on clustering.

2.1. Redundant Variable Elimination Strategy Based on Correlation Analysis

The correlation coefficient reflects the degree of linear correlation among different variables. A positive correlation coefficient indicates a positive correlation among variables. A negative correlation coefficient indicates a negative correlation among variables. The larger the absolute value of the correlation coefficient, the stronger the correlation among the variables, and vice versa, the weaker the correlation. The Pearson simple correlation coefficient method is used in this study, which describes the correlation between two variables of different scales.

The correlation coefficient between the two variable data is defined as follows:

$$r_{xy} = \frac{\sum\limits_{i=1}^{n} (x_i - \overline{x})(y_i - \overline{y})}{\sqrt{\sum\limits_{i=1}^{n} (x_i - \overline{x})^2}\sqrt{\sum\limits_{i=1}^{n} (y_i - \overline{y})^2}} \tag{1}$$

In the above formula, r_{xy} is the value of the correlation coefficient, and the meaning of different values is shown in Figure 2:

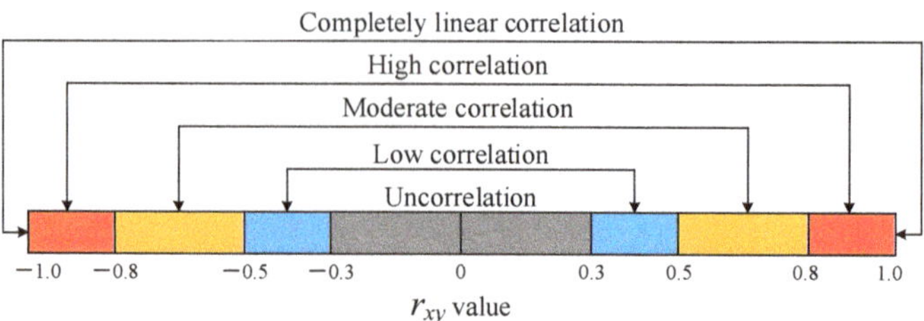

Figure 2. Correlation coefficient graph.

Variables with high correlation contain redundant information, and too many feature variables may increase the complexity of the model. Therefore, the correlation analysis method is selected to eliminate redundant variables, thereby reducing feature variables. When using the correlation analysis method to select feature variables, the choice of threshold is very important for the performance of eliminating redundant variables. If the threshold is too large, the performance of eliminating redundant variables will be poor. On the contrary, there will be a risk of loss of useful information. When the absolute value of r_{xy} is larger than 0.8, there is a high correlation between the two variables. However, in order to avoid excessive removal of feature variables and loss of effective information, when the absolute value of the correlation between the 2 feature variables is larger than 0.9, 1 of the 2 feature variables will be removed to achieve the purpose of reducing redundant variables.

2.2. Clustering Method

Cluster analysis is an important statistical analysis method for studying classification problems, and it is also an important algorithm in data mining. There are many implementation forms of cluster analysis, usually by calculating the sample distance in a multi-dimensional space. This research mainly uses the K-means method, Gaussian mixture model clustering method and spectral clustering method.

2.2.1. K-Means Approach

The K-means algorithm is a dynamic clustering algorithm, which belongs to the category of dynamic grouping. The basic idea is to randomly select K objects as the center of the initial K sets for a database containing N data objects. Then, the center distances of other samples in each set are calculated, and the set closest to the center sample is found. The average method is used to calculate the new cluster center after adjustment. If there is no change in the centers of two adjacent clusters, the sample clustering has been completed.

Calculation steps:

1. Select the number of clusters K.
2. Select K samples $C_1, C_2, \ldots, C_k$ as the initial cluster centers.
3. Calculate the distance d from other samples to the cluster center point, as shown in the formula:

$$d = \sqrt{\sum_{i=1}^{n} (x_i - C_{ji})^2} \quad C_j \in C \quad 1 < j < k \tag{2}$$

In the formula, x is a certain sample, and C_j is the center of a certain cluster.

4. According to the principle of being closest to the center point, all samples are classified into K categories.
5. Then, calculate the centroid of the cluster and use it as the new cluster center.
6. Repeat steps (3)–(5), and iterate until the cluster centers no longer change.

2.2.2. Gaussian Mixture Model Clustering Method

Gaussian mixture model clustering is a clustering method that uses multiple Gaussian models to represent data distribution. The clustering principle is as follows:

Assuming that the operating data sample of the combined air-conditioning system is $x_i(i = 1, 2, \ldots N)$, the Gaussian mixture model is shown in Equation (3):

$$P(x) = \sum_{k=1}^{K} \pi_k f\left(x_k \middle| \mu_k, \sum_k\right) \tag{3}$$

In the Gaussian mixture model, π, μ and $\sum$ need to be estimated, and Equation (3) can be transformed into Equation (4):

$$P\left(x \middle| \pi, \mu, \sum\right) = \sum_{k=1}^{K} \pi_k f\left(x_k \middle| \mu_k, \sum_k\right) \tag{4}$$

The above three parameters can be estimated by the Expectation-Maximization (EM) method. The specific steps are as follows:

1. Set the initial values of π, μ and $\sum$.
2. Calculate the posterior probability $p(Z_{nk})$:

$$p(Z_{nk}) = \frac{\pi_k f(x_n | \mu_k, \Sigma_k)}{\sum\limits_{j=1}^{K} \pi_j f\left(x_n \middle| \mu_j, \Sigma_j\right)} \tag{5}$$

3. Calculate the posterior probability $p(Z_{nk})$:

$$\mu_k = \frac{1}{f_k} \sum_{n=1}^{N} p(Z_{nk}) x_n \tag{6}$$

4. Find the maximum likelihood value of Σ_k:

$$\sum_k = \frac{1}{f_k} \sum_{n=1}^{N} p(Z_{nk})(x_n - \mu_k)(x_n - \mu_k)^T \tag{7}$$

5. Calculate the maximum likelihood function of π_k:

$$\pi_k = \frac{f_k}{f} \tag{8}$$

6. Perform iterative calculations on steps (2)–(5) until convergence.

2.2.3. Spectral Clustering Method

Spectral clustering (SC) is a clustering algorithm based on spectrogram theory. The main idea is to treat sample point data as points in space and connect the points with lines. The weight value is lower when the distance between two points is far, and vice versa. Then, the graph composed of all sample points is segmented. Furthermore, the weights in the sub-pictures are as high as possible after the segmentation process, and the weights among different sub-pictures are as low as possible, so as to achieve the clustering performance. The specific steps are shown in Figure 3.

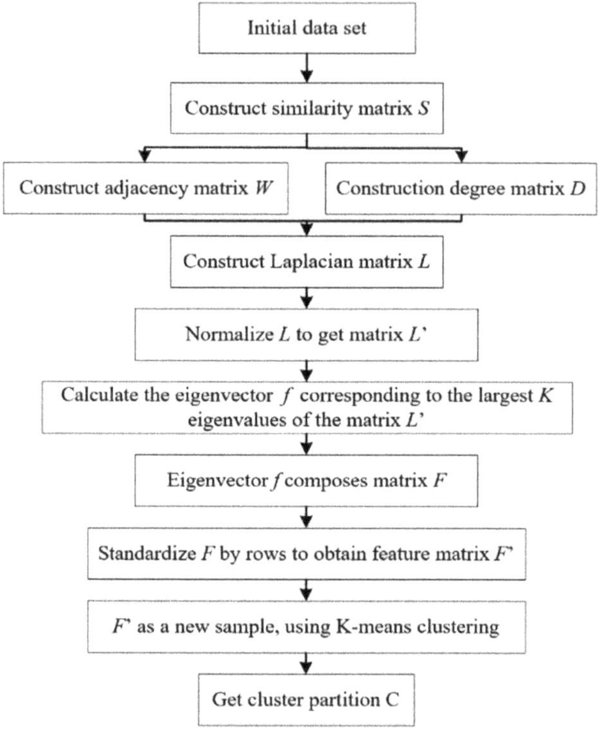

Figure 3. Schematic diagram of spectral clustering.

2.3. The Evaluation Index of Pattern Recognition Performance

In order to be able to evaluate the performance of clustering quantitatively, this study mainly uses two types of evaluation indicators, the DUNN index and the clustering accuracy index.

2.3.1. The DUNN Index

The DUNN index is widely used in the evaluation of clustering performance. This index is the shortest distance between any two samples in different clusters, divided by the maximum distance between two samples in any one cluster, and the calculation method is shown in Equation (9). The larger the DUNN index, the better the clustering performance.

$$\text{DUNN} = \frac{\min\limits_{0<m\neq n<K}\left\{\min\limits_{\substack{\forall x_i\in\Omega_m \\ \forall x_j\in\Omega_n}}\left\{\left\|x_i - x_j\right\|\right\}\right\}}{\max\limits_{0<l\leq K}\max\limits_{\forall x_{i'},\forall x_{j'}\in\Omega_l}\left\{\left\|x_{i'} - x_{j'}\right\|\right\}} \tag{9}$$

2.3.2. Clustering Accuracy

In order to analyze the clustering performance more pertinently and accurately, this study proposes some new indicators to evaluate the clustering performance, mainly including the clustering hit rate, clustering error rate and accuracy. In the clustering result, the true category is Ψ_m, the expected cluster category is Ω_m, the number of correct samples in the expected cluster category is NT, and the number of incorrect samples in the expected cluster category is NF. The cluster hit rate C_h reflects the ratio of the number of clustered correct samples in the expected cluster category, divided by the number of samples in the true category, and the calculation method is shown in Equation (10). In order to avoid a large number of wrong samples in the expected clustering category, the clustering accuracy index is also introduced. The accuracy C_p is the number of correct samples of the expected clustering category, divided by the number of samples of the expected clustering category. The calculation method is shown in Equation (11). In addition, the clustering error rate F represents the number of error samples in the expected cluster category, divided by the number of samples in the expected cluster category. The calculation method is shown in Equation (12). The higher the clustering hit rate and accuracy, the better. The lower the clustering error rate, the better.

$$C_h = \frac{NT}{N(\Psi_m)} \tag{10}$$

$$C_p = \frac{NT}{N(\Omega_m)} \tag{11}$$

$$F = \frac{NF}{N(\Omega_m)} \tag{12}$$

3. Experiment and Data Introduction

3.1. Introduction of Experimental Subjects

A RHHAC system is a system with refrigeration, heating, hot water and multiple complex operation modes. For example, a RHHAC system can produce hot water while cooling. Therefore, the structure of the RHHAC system has been improved compared with the ordinary air-conditioning system. The RHHAC system, mainly through two four-way reversing valves and two electronic expansion valves to realize the transformation of the refrigerant flow path, realizes the switching of various working conditions, such as cooling, cooling and hot water at the same time, in heating and hot-water modes. Figure 4 shows the schematic diagram of the RHHAC system structure and the refrigerant flow path layout under cooling mode conditions. The compressor type of the experimental system is QXA-C18B030, and the rated power is 1.5 kW. The type of the electronic expansion valve is DPF (TS1) 1.3C-01. The water tank capacity is 150 L. The RHHAC system implementation site diagram is shown in Figure 5. The RHHAC experimental system mainly includes an indoor unit, outdoor unit, heating water tank, environmental room and data acquisition device. The biggest feature of the RHHAC system is that it can meet the needs of users for hot water while realizing the cooling and heating functions.

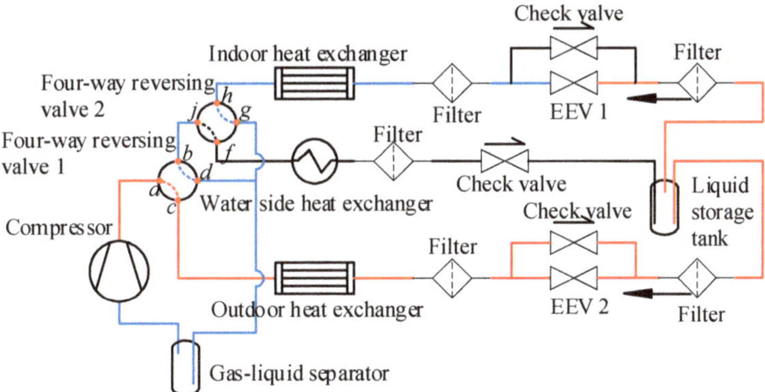

Figure 4. Schematic diagram of the RHHAC system.

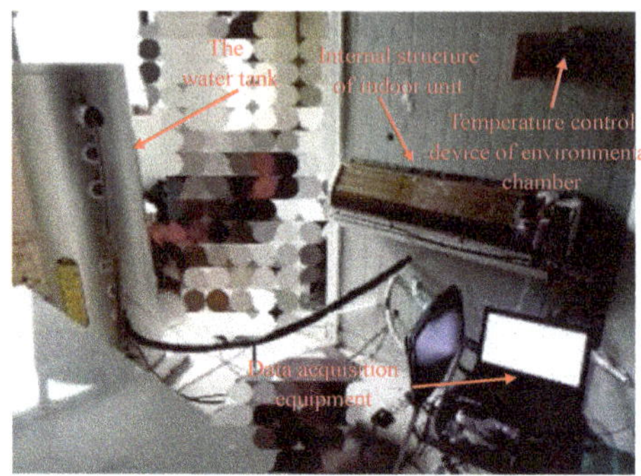

Figure 5. The RHHAC system implementation site diagram.

3.2. Data Collection and Processing

The experimental condition is that the outdoor temperature is 7 °C and the indoor temperature is 20 °C. Further, the initial temperature of the water tank is 30 °C., and the operating mode of the RHHAC system is the hot-water mode. The experiment collects a total of 50 variables in temperature, pressure and various control parameters. After removing invalid variables and some control parameters, 17 variables are retained, and the details of the selected variables are listed in Table 1. All sensors are installed by the unit itself. Table 2 lists the information of the temperature sensor and the pressure sensor. The data collection time interval is 1 s, and the collected data is transmitted to the computer through the host computer and stored. The experiment lasted 3.49 h, and 12,551 samples were collected. Each sample contains 17 selected variables.

Table 1. Details of collected experimental data.

No.	Variables	Unit	Maximum	Minimum	Average Value
1	Input power (W)	W	2072.60	352.40	1517.96
2	Power factor (WF)	-	0.97	0.47	0.94
3	Discharge pressure (P_{dis})	Mpa	3.69	0.47	2.31
4	Suction pressure (P_{suc})	Mpa	1.23	0.13	0.45
5	Intermediate pressure (P_{int})	Mpa	3.64	0.86	2.26
6	Pressure after electronic expansion valve 1 throttling (P_{eev1})	Mpa	3.18	0.19	0.67
7	Pressure after electronic expansion valve 2 throttling (P_{eev2})	Mpa	1.71	0.21	0.47
8	Discharge temperature (T_{dis})	°C	104.50	50.20	74.35
9	Suction temperature (T_{suc})	°C	20.80	−30.70	−0.48
10	Temperature after electronic expansion valve 1 (T_{eev1})	°C	48.30	−28.10	5.56
11	Temperature after electronic expansion valve 2 (T_{eev2})	°C	25.20	−22.50	−8.80
12	Water tank temperature (T_{wb})	°C	54.70	24.90	34.03
13	Outdoor temperature (T_{od})	°C	11.10	−0.10	3.67
14	Indoor temperature (T_{id})	°C	22.80	19.00	20.30
15	Defrost temperature (T_{df})	°C	17.10	−15.00	−5.23
16	Outdoor fan outlet temperature (T_{odf})	°C	21.43	0.02	3.74
17	Indoor fan outlet temperature (T_{idf})	°C	32.68	0.21	20.57

Table 2. Information and accuracy of sensors.

No.	Sensors	Type	Brand	Range	Accuracy
1	Temperature	Thermal resistance	UNIOHM	−30~120 °C	±1%
2	Pressure	Strain mode	Huadian	0~10 Mpa	±0.2%

4. Results and Discussion

4.1. Correlation Analysis Results

Figure 6 shows the results of the correlation analysis of the RHHAC system operating data. The lower left part of the figure is the specific correlation coefficient value, and the upper right part represents the correlation between variables in the form of a graph. The depth of the color represents the value of the correlation coefficient, and the deflection direction of the ellipse indicates that the variables are positively correlated or negatively correlated. By analyzing the correlation between the various variables, there are a total of 6 sets of variables with a correlation of more than 0.90. They mainly include W–P_{dis}, W–P_{int}, P_{int}–P_{dis}, T_{eev2}–P_{eev2}, T_{df}–T_{eev2} and T_{odf}–T_{od}. There is also overlap between these variable sets. Based on the consideration of eliminating more variables and better reflecting the operation of the system, the five variables of P_{dis}, P_{int}, P_{eev2}, T_{df} and T_{odf} are eliminated. The feature variable set after the correlation eliminates redundant variables (cor-feature) contains 12 variables, namely, W, WF, P_{suc}, P_{eev1}, T_{dis}, T_{suc}, T_{eev1}, T_{eev2}, T_{wb}, T_{od}, T_{id} and T_{idf}.

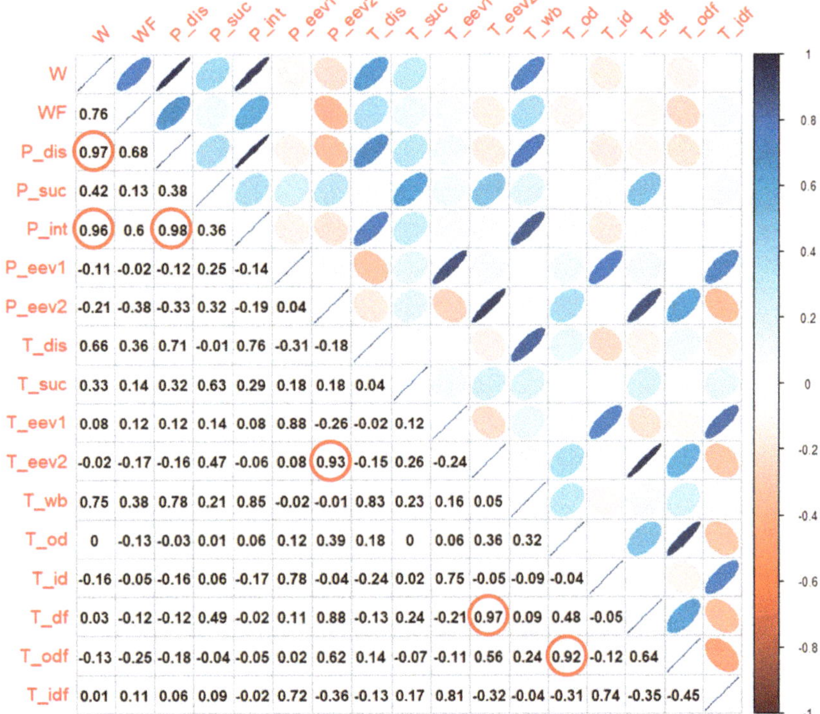

Figure 6. Result of correlation analysis of the RHHAC system operating data.

4.2. Clustering Result Analysis

Firstly, the original feature variable set (ori-feature) is used for cluster analysis. A total of 17 variables are included; detailed variable information is listed in Table 1.

The collected data are first used for K-means clustering analysis. The number of clusters is selected as two and three, and the results are shown in Figure 7. In order to present the clustering performance of high-dimensional data in a two-dimensional graph, the principal component result is used as the horizontal and vertical axis. The contribution of the abscissa is 30%, and the contribution of the ordinate is 25.5%. When clustering into two categories, it can be seen that the two categories are more distinct. Only in the middle three regions, the two categories have an intersection, and the data are divided into two types, the upper left and the lower right. When the data are clustered into three categories, the boundaries of the three categories are also obvious, and there is not too much crossover between each category. There is not much overlap between the various categories. Through the comparative analysis of the two clustering results, it can be seen that cluster 1 is consistent. Cluster 2 and cluster 3, when the model clustered into three categories, correspond to cluster 2 when the model clustered into two categories.

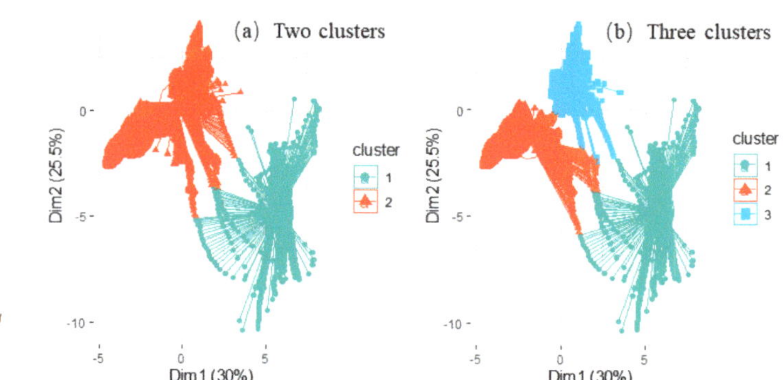

Figure 7. K-means clustering performance when numbers of clusters are two and three.

In order to further analyze the clustering results, each cluster is analyzed separately. The parameters that can better characterize the operation of the RHHAC system are selected, which mainly include the compressor discharge temperature, compressor suction temperature and defrost temperature. Figure 8 shows the results when the clustering is two categories. The figure shows the distribution of the operation data of two categories. Figure 9 shows the data distribution of cluster 1 when the number of clusters is 3. The other two categories have no clear knowledge information, so they are not displayed. First of all, it can be seen from Figure 8 that the data distributions of the two clusters have obvious differences. Cluster 1 compressor suction and discharge temperature fluctuate sharply, while cluster 2 compressor suction and discharge temperature change smoothly. The difference in defrosting temperature is even more obvious. The defrost temperature of cluster 1 changes regularly, which is worthy of further analysis. Comparing with the result of Figure 9, the changes in the two cases are similar. In addition, the changes in the suction and discharge temperature in cluster 1 of Figures 8 and 9 are also similar, which also illustrates that these two clusters are the same cluster and have the regular change period. Combining with the operating characteristics of the RHHAC system, it is found that cluster 1 is the defrosting process of the RHHAC system. It can also be seen from the figure that this cluster contains a total of four defrosting processes.

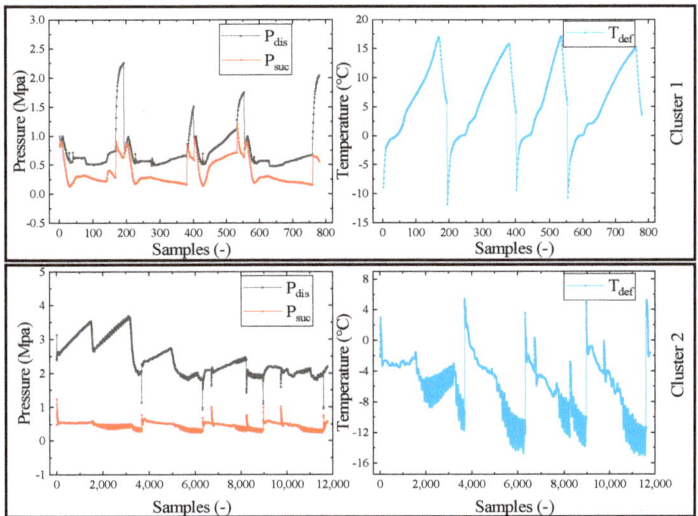

Figure 8. The K-means clustering result when the number of clusters is 2.

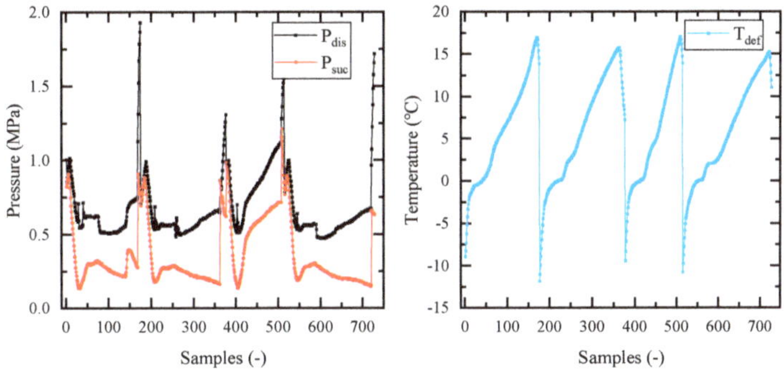

Figure 9. The K-means clustering result cluster 1 when the number of clusters is three.

4.3. Clustering Performance of Different Clustering Models

In order to carry out a more in-depth study on the clustering of the RHHAC system operating mode, this study uses three clustering methods to cluster the data of the RHHAC system, namely, K-means clustering, Gaussian mixture model clustering and spectral clustering. The cluster model data set includes the original feature variable set (ori) and the feature variable set after the correlation analysis removes redundant variables (cor). The number of clusters includes two, three, four and five, a total of four clustering situations.

4.3.1. K-Means Clustering Results

In order to analyze the clustering performance more intuitively, a clustering category diagram is used to show the clustering results of the RHHAC system data with different numbers of clusters. Figure 10 shows the comparison of the clustering results of the K-means algorithm. By analyzing the operating status of the system, the defrosting operation mode is artificially marked, as shown in the figure. The black line is the artificial marking state, the state 0.5 is the defrosting state, and the state -0.5 is the normal running state. The defrosting status marks in the results of other models are consistent with this. It can be seen from the figure that when the original feature variable set is used, there is a corresponding defrost category for different cluster numbers, which are cluster1, cluster1, cluster1 and cluster5. The model can effectively cluster the defrost categories and has a good consistency. For the cor-feature set model, it is found that when the number of clusters is two and three, the clustering performance is poor, and the defrost category cannot be identified. When the number of clusters is four and five, the clustering performance is improved, and the defrosting operation modes can be identified, which are cluster 1 and cluster 4, respectively. These results show that the clustering method can identify the mode of defrosting operation. When the number of clusters is small, the clustering performance of the K-means model with the ori-feature set is better than the cor-feature set model.

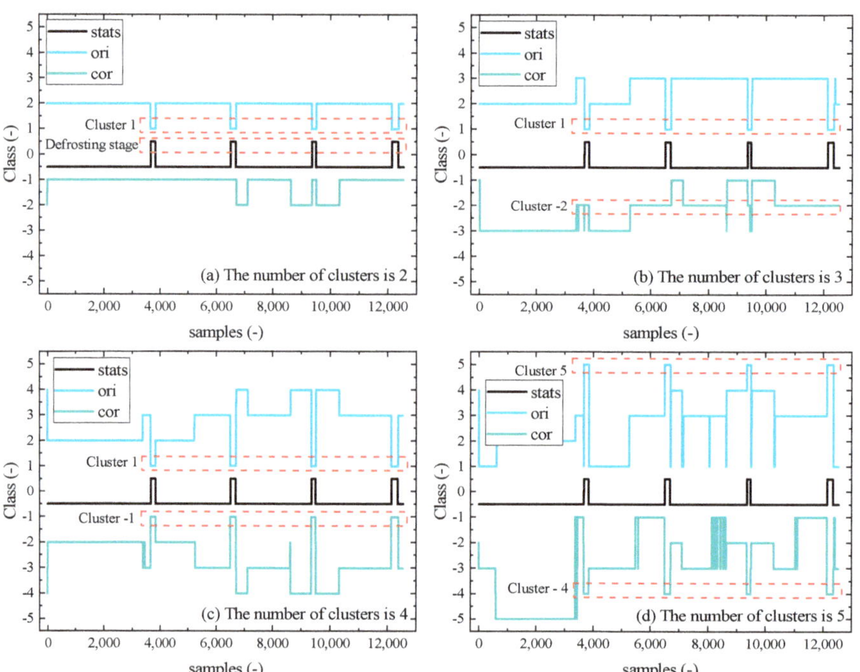

Figure 10. Comparison of the K-means clustering model results. (The "stats" represents the defrosting operation mode of the system. The "ori" represents clustering using original data sets. The "cor" represents the data set after removing redundant variables for clustering.).

4.3.2. Gaussian Mixture Model Clustering Results

Figure 11 shows the clustering results of the GMMC model. First, the model results of the ori-feature set are analyzed. When the number of clusters is 2, it can be found from the figure that cluster 2 clustered by GMMC is more consistent with the defrost category. Yet, it is obviously different from the result of the K-means algorithm, that is, the range of cluster 2 clustered by the GMMC algorithm exceeds the actual defrost category. From another perspective, the GMMC algorithm recognizes part of the normal operating state data as the defrosting category. Comparing the situation with other cluster numbers, there is a category corresponding to the defrost mode in the results of these models. The distribution of these categories is also relatively similar; these categories are cluster 2, cluster 3, cluster 3 and cluster 3. Then, the clustering performance of the cor-feature set is compared and analyzed. When the number of clusters is different, the clustering performance of the model is consistent with the results of the ori-feature set model. The corresponding clusters of defrosting mode obtained by clustering are cluster 2, cluster 3, cluster 3 and cluster 3. From these results, it can be concluded that for the GMMC model, removing redundant variables through correlation analysis will not affect the clustering performance of the model.

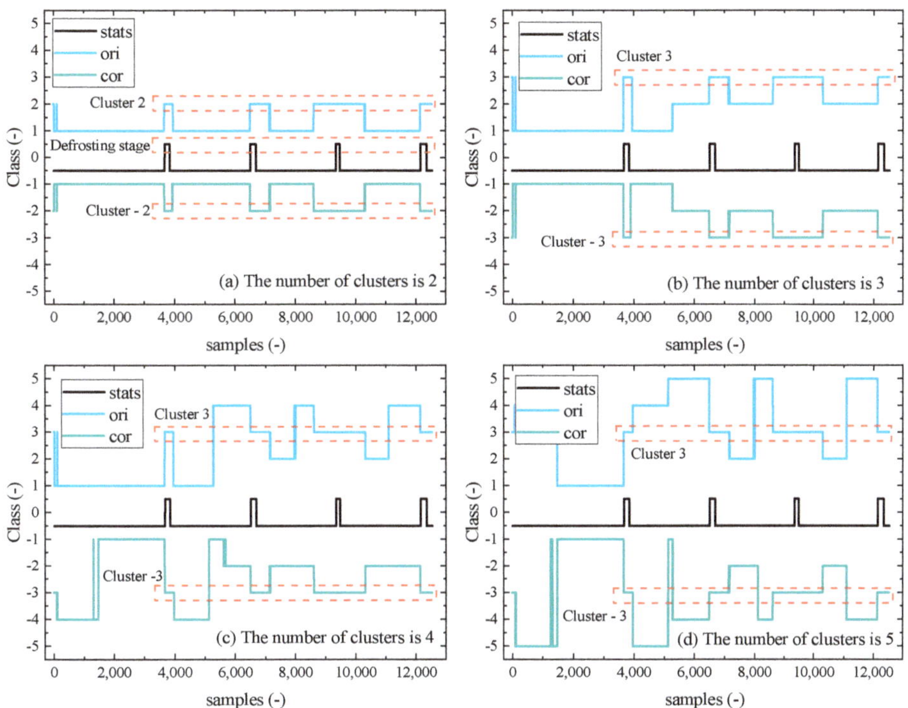

Figure 11. Comparison of the GMM clustering model results (The "stats" represents the defrosting operation mode of the system. The "ori" represents clustering using original data sets. The "cor" represents the data set after removing redundant variables for clustering.).

4.3.3. Spectral Clustering Results

Figure 12 shows the clustering results of the spectral clustering algorithm. For the clustering model established by the ori-feature set, when the number of clusters is two and three, cluster 2 and cluster 1 in the clustering result are consistent with the actual defrost category. It can be seen from the figure that no excessive operating data are identified as the defrost category. Yet, when the number of analysis clusters is four and five, category 2 and category 3 in the clustering result correspond to the actual defrost category. It can be seen from the figure that although defrosting can be identified, the clustering results contain more running data. Yet, when the number of clusters is four and five, cluster 2 and cluster 3 in the clustering result correspond to the actual defrost category. It can be seen from the figure that although the defrost category can be identified, the clustering result contains a lot of normal running data. For the cor-feature set model, when the number of clusters is three and four, the model can identify the defrosting operation mode accurately. Yet, when the other two cluster numbers are selected, the cluster performance is poor. Therefore, combining the clustering performance of the two feature sets, the best number of clusters is three.

4.4. Comparative Analysis of Clustering Performance

In order to analyze the clustering performance of each algorithm in different numbers of clusters more specifically, the DUNN index is used to evaluate the results of different numbers of clusters, as shown in Figure 13. For the K-means algorithm, as the number of clusters increases, the DUNN index gradually decreases, while for the GMMC and SC algorithms, the DUNN index fluctuates. In addition, for the K-means algorithm and the SC algorithm, the DUNN index of the cor-feature set model is basically higher than that of the ori-feature set model. Yet, for the GMMC algorithm, the DUNN index of the cor-feature set model is lower than that of the ori-feature set model. The overall analysis found that the DUNN index of the model is higher when the number of clusters is small, which corresponds to the above clustering results. Yet, for the clustering models of different methods, the DUNN index cannot accurately evaluate the clustering performance of the model.

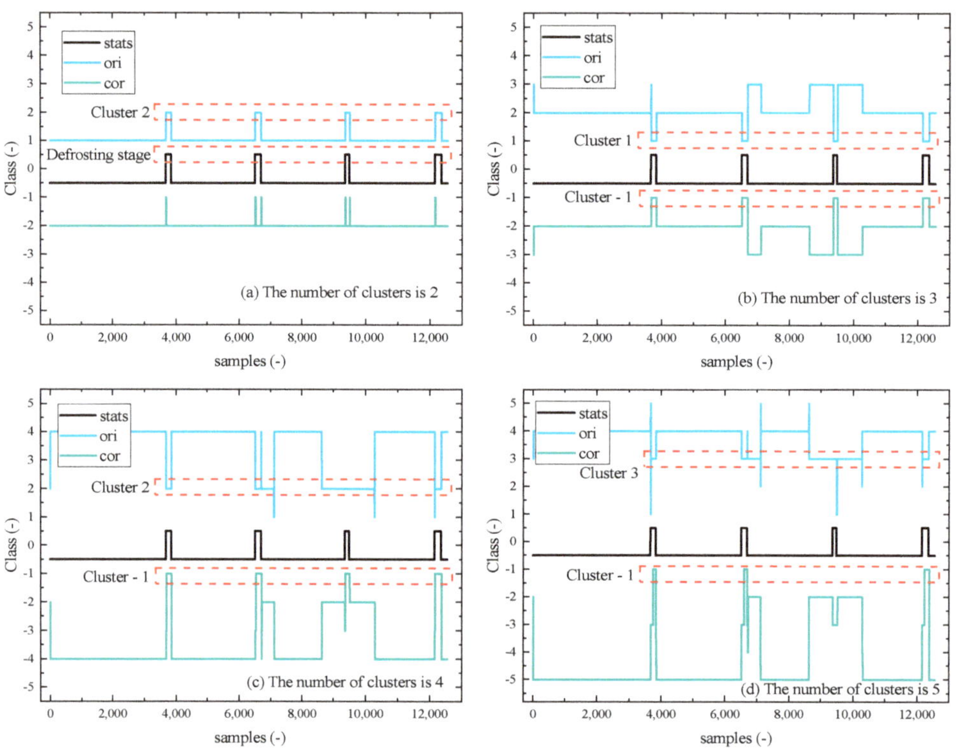

Figure 12. Comparison of the SC clustering model results (The "stats" represents the defrosting operation mode of the system. The "ori" represents clustering using original data sets. The "cor" represents the data set after removing redundant variables for clustering.).

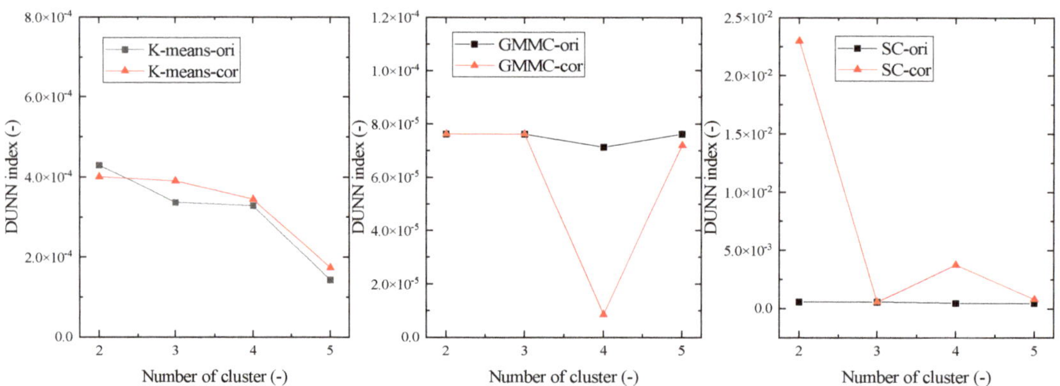

Figure 13. The DUNN index value under different numbers of clusters.

Table 3 lists the clustering accuracy rate of different clustering methods with different numbers of clusters. It can be seen from the overall results that the hit rate of clustering is relatively high. Among them, for the GMMC model with different numbers of clusters, the hit rate reached 100%, that is, the categories obtained by the cluster method completely included the defrosting operation mode data. Yet, from another aspect, the non-defrosting categories included in the clustering results are also worthy of attention. For the GMMC algorithm also, although the hit rate is higher, the error rate is also higher. When the number of clusters is 5, the clustering error rate reaches 77.94%, that is, the clustering result contains a large amount of running data. Further, this study also uses the clustering accuracy evaluation index, and it can be concluded that the model clustering accuracy is not high for the GMMC algorithm. Therefore, it is effective and accurate to use the three indicators of hit rate, error rate and accuracy to evaluate the clustering performance of the model.

Table 3. Performance of different clustering models.

Number of Clusters			2 Clusters	3 Clusters	4 Clusters	5 Clusters
K-means	ori	True	98.99%	98.99%	98.99%	98.99%
		False	11.79%	10.65%	11.11%	7.90%
		Precision	88.21%	89.35%	88.89%	92.10%
	cor	True	98.13%	93.53%	**98.99%**	98.99%
		False	93.58%	88.60%	**10.07%**	8.27%
		Precision	6.42%	11.40%	**89.93%**	91.73%
GMMC	ori	True	100%	100%	100%	100%
		False	77.33%	77.28%	77.38%	77.94%
		Precision	22.67%	22.72%	22.62%	22.06%
	cor	True	**100%**	**100%**	**100%**	**100%**
		False	**77.06%**	**76.83%**	**77.96%**	**77.92%**
		Precision	**22.94%**	**23.17%**	**22.04%**	**22.08%**
SC	ori	True	**98.99%**	**98.99%**	99.14%	98.99%
		False	**0.58%**	**0.58%**	73.70%	73.70%
		Precision	**99.42%**	**99.42%**	26.30%	26.30%
	cor	True	7.48%	98.99%	91.51%	46.47%
		False	10.34%	1.43%	7.02%	4.44%
		Precision	89.66%	98.57%	92.98%	95.56%

Combining the three evaluation indexes, we compare and analyze the performance of the ori-feature set and cor-feature set model under different methods. First, for the K-means method, the clustering performance of the cor-feature set model is better when the number of clusters is four. Then, for the GMMC method, the clustering performance of the cor-feature set model and the ori-feature set model is not significantly different. Finally, for the SC method, the clustering performance of the ori-feature set model is better. This result also shows that the clustering model still has a good clustering effect and greatly reduces the complexity of the model after eliminating redundant variables through correlation analysis.

Based on the above analysis, it is concluded that the clustering performance of the ori-feature set model is the best when the number of clusters of the SC algorithm model is two and three. The hit rate is 98.99%, the error rate is 0.58%, and the accuracy is 99.42%. When the number of clusters of the K-means model is five, the clustering performance of the ori-feature set is the second. For the GMMC model, its clustering performance is worse than the other two clustering methods.

4.5. The Clustering Model Running Time Analysis

Figure 14 shows the running time results of the model when the number of clusters is different. All models are run on a desktop computer with a CPU of Intel Core I7-6700 3.4 GHz, two memories of 8 G and a Windows 10 64-bit operating system. For the K-means algorithm, the running time of the model gradually increases as the number of clusters increases. For the GMMC method, when the ori-feature set is used, the running time of the model increases as the number of clusters increases. Yet, when the cor-feature set is used, the running time of the model is relatively stable. Compared with the ori-feature set model, the running time is greatly reduced. For the SC method, as the number of clusters increases, the running time of the model tends to decrease. There is little difference in the running time of models with different feature sets. When comparing and analyzing different clustering algorithms, the running time of the K-means algorithm is the shortest, which is less than 1 s. The running time of the SC algorithm is relatively long, more than 12 h, and the running time far exceeds the K-means algorithm and the GMMC algorithm. From the perspective of different feature sets, for the GMMC method, the running time of the cor-feature set model is greatly reduced. Especially when the number of clusters is 5, the running time of the model is reduced from 27.80 s to 10.04 s, a reduction of 63.88%. However, the running time is slightly increased under some other models.

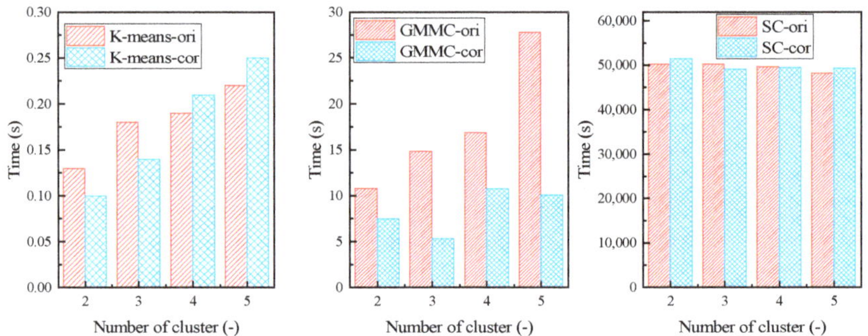

Figure 14. Comparison of the model running times with different numbers of clusters.

5. Conclusions

In this study, data mining is carried out based on the operating data of the RHHAC system. The following conclusions are drawn:

1. The operation data of the RHHAC system are analyzed by the clustering method. Different clustering methods can identify the defrosting operation mode.
2. Correlation analysis can eliminate redundant feature variables in the model. The cor-feature set model can also identify the defrosting operation mode under the condition of fewer feature variables. Some models are even better than the ori-feature set model. The running time of the model is also improved with different feature sets. Especially for the GMMC method when the number of clusters is 5, the model running time is reduced from 27.80 s to 10.04 s, which is a reduction of 63.88%.
3. The DUNN index cannot evaluate the clustering performance of the model very accurately. Analyzing the clustering performance evaluation indexes proposed in this study, the clustering performance of the ori-feature set model is the best when the number of clusters of the SC algorithm is two and three. The clustering hit rate is 98.99%, the clustering error rate is 0.58%, and the accuracy is 99.42%. When the number of clusters of the K-means model is five, the clustering performance of the ori-feature set is the second. For the GMMC model, its clustering performance is worse than the other two clustering methods.

The clustering model established in this study can effectively identify the operation mode of the RHHAC system. In the future, clustering will be further used to carry out relevant research on identifying energy-saving modes and fault modes of air-conditioning systems.

Author Contributions: Methodology, Y.G.; resources, C.L.; data curation, J.Z.; writing—original draft preparation, Y.G.; writing—review and editing, J.L. (Jiangyan Liu); visualization, C.L.; supervision, Y.L.; funding acquisition, J.L. (Jifu Lu) and Y.L. All authors have read and agreed to the published version of the manuscript.

Funding: This research was funded by the China Construction Seventh Bureau Technology Research Project (No. CSCEC7b-2015-Z-24) and the Science and Technology Department of Henan Province (No. 222102320051 and No. 222102320113).

Institutional Review Board Statement: Not applicable.

Informed Consent Statement: Not applicable.

Data Availability Statement: The data presented in this study are available on request from the corresponding author.

Conflicts of Interest: The authors declare no conflict of interest.

Nomenclature

C_h	cluster hit rate
C_j	center of a certain cluster
C_p	clustering accuracy
cor	feature variable set after the correlation analysis removes redundant variables
f	eigenvector
F	clustering error rate
K	number of clusters
ori	original feature variable set
P_{dis}	discharge pressure
P_{suc}	suction pressure
P_{int}	intermediate pressure
P_{eev1}	pressure after electronic expansion valve 1 throttling
P_{eev2}	pressure after electronic expansion valve 2 throttling
r_{xy}	value of the correlation coefficient
Ψ_m	true category
T_{dis}	discharge temperature
T_{suc}	suction temperature
T_{eev1}	temperature after electronic expansion valve 1
T_{eev2}	temperature after electronic expansion valve 2
T_{wb}	water tank temperature
T_{od}	outdoor temperature
T_{id}	indoor temperature
T_{df}	defrost temperature
T_{odf}	outdoor fan outlet temperature
T_{idf}	indoor fan outlet temperature
	Greek symbols
π	weighting coefficient
μ	the mean
Σ	covariance matrix
Ω_m	expected cluster category

References

1. Dai, B.; Tong, Y.; Hu, Q.; Chen, Z. Characteristics of thermal stratification and its effects on HVAC energy consumption for an atrium building in south China. *Energy* **2022**, *249*, 123425. [CrossRef]
2. Jin, B. Research on performance evaluation of green supply chain of automobile enterprises under the background of carbon peak and carbon neutralization. *Energy Rep.* **2021**, *7*, 594–604. [CrossRef]

3.   Fang, K.; Li, C.; Tang, Y.; He, J.; Song, J. China's pathways to peak carbon emissions: New insights from various industrial sectors. *Appl. Energy* **2022**, *306*, 118039. [CrossRef]

4.   Rogers, A.; Guo, F.; Rasmussen, B. A review of fault detection and diagnosis methods for residential air conditioning systems. *Build. Environ.* **2019**, *161*, 106236. [CrossRef]

5.   Fan, C.; Liu, Y.; Liu, X.; Sun, Y.; Wang, J. A study on semi-supervised learning in enhancing performance of AHU unseen fault detection with limited labeled data. *Sustain. Cities Soc.* **2021**, *70*, 102874. [CrossRef]

6.   Gong, G.; Chen, F.; Su, H.; Zhou, J. Thermodynamic simulation of condensation heat recovery characteristics of a single stage centrifugal chiller in a hotel. *Appl. Energy* **2012**, *91*, 326–333. [CrossRef]

7.   Zhao, X.; Long, E.; Zhang, Y.; Liu, Q.; Jin, Z.; Liang, F. Experimental Study on Heating Performance of Air—Source Heat Pump with Water Tank for Thermal Energy Storage. *Procedia Eng.* **2017**, *205*, 2055–2062. [CrossRef]

8.   Byrne, P.; Miriel, J.; Lenat, Y. Experimental study of an air-source heat pump for simultaneous heating and cooling—Part 2: Dynamic behaviour and two-phase thermosiphon defrosting technique. *Appl. Energy* **2011**, *88*, 3072–3078. [CrossRef]

9.   Byrne, P.; Miriel, J.; Lenat, Y. Experimental study of an air-source heat pump for simultaneous heating and cooling—Part 1: Basic concepts and performance verification. *Appl. Energy* **2011**, *88*, 1841–1847. [CrossRef]

10.   Pei, X.; Guo, Y.; Guo, Y.; Li, R.; Yang, J.; Mao, M. Research on the characteristics of the refrigeration, heating and hot water combined air-conditioning system. *Int. J. Refrig.* **2021**, *130*, 150–160. [CrossRef]

11.   Zhou, Z.; Chen, H.; Li, G.; Zhong, H.; Zhang, M.; Wu, J. Data-driven fault diagnosis for residential variable refrigerant flow system on imbalanced data environments. *Int. J. Refrig.* **2021**, *125*, 34–43. [CrossRef]

12.   Guo, Y.; Chen, H. Fault diagnosis of VRF air-conditioning system based on improved Gaussian mixture model with PCA approach. *Int. J. Refrig.* **2020**, *118*, 1–11. [CrossRef]

13.   Bai, X.; Zhang, M.; Jin, Z.; You, Y.; Liang, C. Fault detection and diagnosis for chiller based on feature-recognition model and Kernel Discriminant Analysis. *Sustain. Cities Soc.* **2022**, *79*, 103708. [CrossRef]

14.   Li, G.; Hu, Y. Improved sensor fault detection, diagnosis and estimation for screw chillers using density-based clustering and principal component analysis. *Energy Build.* **2018**, *173*, 502–515. [CrossRef]

15.   An, J.; Yan, D.; Hong, T. Clustering and statistical analyses of air-conditioning intensity and use patterns in residential buildings. *Energy Build.* **2018**, *174*, 214–227. [CrossRef]

16.   Nweye, K.; Nagy, Z. MARTINI: Smart meter driven estimation of HVAC schedules and energy savings based on Wi-Fi sensing and clustering. *Appl. Energy* **2022**, *316*, 118980. [CrossRef]

17.   Afaifia, M.; Djiar, K.A.; Bich-Ngoc, N.; Teller, J. An energy consumption model for the Algerian residential building's stock, based on a triangular approach: Geographic Information System (GIS), regression analysis and hierarchical cluster analysis. *Sustain. Cities Soc.* **2021**, *74*, 103191. [CrossRef]

18.   Gilanifar, M.; Wang, H.; Cordova, J.; Ozguven, E.E.; Strasser, T.I.; Arghandeh, R. Fault classification in power distribution systems based on limited labeled data using multi-task latent structure learning. *Sustain. Cities Soc.* **2021**, *73*, 103094. [CrossRef]

19.   Molokomme, D.; Chabalala, C.; Bokoro, P. Enhancement of Advanced Metering Infrastructure Performance Using Unsupervised K-Means Clustering Algorithm. *Energies* **2021**, *14*, 2732. [CrossRef]

20.   Yu, W.; Zhao, F.; Yang, W.; Xu, H. Integrated analysis of CFD simulation data with K-means clustering algorithm for soot formation under varied combustion conditions. *Appl. Therm. Eng.* **2019**, *153*, 299–305. [CrossRef]

21.   Wang, K.; Qi, X.; Liu, H.; Song, J. Deep belief network based k-means cluster approach for short-term wind power forecasting. *Energy* **2018**, *165*, 840–852. [CrossRef]

22.   Jiménez Torres, M.; Bienvenido-Huertas, D.; May Tzuc, O.; Bassam, A.; Ricalde Castellanos, L.J.; Flota-Bañuelos, M. Assessment of climate change's impact on energy demand in Mexican buildings: Projection in sin-gle-family houses based on Representative Concentration Pathways. *Energy Sustain. Dev.* **2023**, *72*, 185–201. [CrossRef]

23.   Jin, H.; Shi, L.; Chen, X.; Qian, B.; Yang, B.; Jin, H. Probabilistic wind power forecasting using selective ensemble of finite mixture Gaussian process regression models. *Renew. Energy* **2021**, *174*, 1–18. [CrossRef]

24.   Qureshi, M.; Ghiaus, C.; Ahmad, N. A blind event-based learning algorithm for non-intrusive load disaggregation. *Int. J. Electr. Power Energy Syst.* **2021**, *129*, 106834. [CrossRef]

25.   Yi, D.H.; Kim, D.W.; Park, C.S. Prior selection method using likelihood confidence region and Dirichlet process Gaussian mixture model for Bayesian inference of building energy models. *Energy Build.* **2020**, *224*, 110293. [CrossRef]

26.   Gao, S.; Hu, B.; Xie, K.; Niu, T.; Li, C.; Yan, J. Spectral clustering based demand-oriented representative days selection method for power system expansion planning. *Int. J. Electr. Power Energy Syst.* **2021**, *125*, 106560. [CrossRef]

27.   Pacella, M.; Papadia, G. Fault Diagnosis by Multisensor Data: A Data-Driven Approach Based on Spectral Clustering and Pairwise Constraints. *Sensors* **2020**, *20*, 7065. [CrossRef]

28.   Li, P.-H.; Pye, S.; Keppo, I. Using clustering algorithms to characterise uncertain long-term decarbonisation pathways. *Appl. Energy* **2020**, *268*, 114947. [CrossRef]

29.   Xia, B.; Han, J.; Zhao, J.; Liang, K. Technological adaptation zone of passive evaporative cooling of China, based on a clustering analysis. *Sustain. Cities Soc.* **2021**, *66*, 102564. [CrossRef]

30.   Bienvenido-Huertas, D.; Marín-García, D.; Carretero-Ayuso, M.J.; Rodríguez-Jiménez, C.E. Climate classification for new and restored buildings in Andalusia: Analysing the current regulation and a new approach based on k-means. *J. Build. Eng.* **2021**, *43*, 102829. [CrossRef]

31. Yürüşen, N.Y.; Uzunoğlu, B.; Talayero, A.P.; Estopiñán, A.L. Apriori and K-Means algorithms of machine learning for spatio-temporal solar generation balancing. *Renew. Energy* **2021**, *175*, 702–717. [CrossRef]
32. Islam, M.K.; Ali, M.S.; Miah, M.S.; Rahman, M.M.; Alam, M.S.; Hossain, M.A. Brain tumor detection in MR image using superpixels, principal component analysis and template based K-means clustering algorithm. *Mach. Learn. Appl.* **2021**, *5*, 100044. [CrossRef]
33. Teichgraeber, H.; Brandt, A.R. Clustering methods to find representative periods for the optimization of energy systems: An initial framework and comparison. *Appl. Energy* **2019**, *239*, 1283–1293. [CrossRef]
34. Zhao, T.; Li, J.; Wang, P.; Yoon, S.; Wang, J. Improvement of virtual in-situ calibration in air handling unit using data preprocessing based on Gaussian mixture model. *Energy Build.* **2022**, *256*, 111735. [CrossRef]
35. Shafiullah, D.; Vergara, P.P.; Haque, A.; Nguyen, P.; Pemen, A. Gaussian Mixture Based Uncertainty Modeling to Optimize Energy Management of Heterogeneous Building Neighborhoods: A Case Study of a Dutch University Medical Campus. *Energy Build.* **2020**, *224*, 110150. [CrossRef]
36. Shen, P.; Liu, J.; Wang, M. Fast generation of microclimate weather data for building simulation under heat island using map capturing and clustering technique. *Sustain. Cities Soc.* **2021**, *71*, 102954. [CrossRef]
37. Wang, Z.; Hu, J.; Liu, B. Stochastic optimal dispatching strategy of electricity-hydrogen-gas-heat integrated energy system based on improved spectral clustering method. *Int. J. Electr. Power Energy Syst.* **2021**, *126*, 106495. [CrossRef]
38. Zhong, G.; Shu, T.; Huang, G.; Yan, X. Multi-view spectral clustering by simultaneous consensus graph learning and discretization. *Knowl.-Based Syst.* **2022**, *235*, 107632. [CrossRef]
39. Hu, M.; Ge, D.; Telford, R.; Stephen, B.; Wallom, D.C. Classification and characterization of intra-day load curves of PV and non-PV households using interpretable feature extraction and feature-based clustering. *Sustain. Cities Soc.* **2021**, *75*, 103380. [CrossRef]
40. Wang, J.; Li, G.; Chen, H.; Liu, J.; Guo, Y.; Hu, Y.; Li, J. Liquid floodback detection for scroll compressor in a VRF system under heating mode. *Appl. Therm. Eng.* **2017**, *114*, 921–930. [CrossRef]
41. Guo, Y.; Li, G.; Chen, H.; Wang, J.; Guo, M.; Sun, S.; Hu, W. Optimized neural network-based fault diagnosis strategy for VRF system in heating mode using data mining. *Appl. Therm. Eng.* **2017**, *125*, 1402–1413. [CrossRef]
42. Zhang, Q.; Xu, D.; Zhou, D.; Yang, Y.; Rogora, A. Associations between urban thermal environment and physical indicators based on meteorological data in Foshan City. *Sustain. Cities Soc.* **2020**, *60*, 102288. [CrossRef]

Article

Peak Load Shifting Control for a Rural Home Hotel Cluster Based on Power Load Characteristic Analysis

Weilin Li [1,2], Yonghui Liang [2,*], Jianli Wang [1], Zhenhe Lin [3], Rufei Li [2] and Yu Tang [1]

1 Huadian Zhengzhou Mechanical Design Institute Co., Ltd., Zhengzhou 450001, China
2 School of Hydraulic and Civil Engineering, Zhengzhou University, Zhengzhou 450000, China
3 School of Mechanical and Storage and Transportation Engineering, China University of Petroleum (Beijing), Beijing 102249, China
* Correspondence: 202022212014189@gs.zzu.edu.cn; Tel.: +86-186-3907-2408

Abstract: The large-scale rural home hotel clusters have brought huge pressure to the rural power grid. However, the load of rural home hotels not only has the inherent characteristics of rural residential buildings but is also greatly impacted by the occupancy rate, which is very different from conventional buildings. Therefore, the existing peak shifting strategies are difficult to apply to rural home hotels. In view of the above problems, this study took a typical visitor village in Zhejiang Province as the research object, which had more than 470 rural home hotels. First, through a basic information survey and power load data collection, the characteristics of its power load for heating, cooling and transition months were studied, and a "No Visitors Day" model was proposed, which was split to obtain the seasonal load curve for air conditioning. Then, combined with the characteristics of the air conditioning power load and the natural conditions of the rural house, a cluster control peak-load-shifting system using phase change energy storage was proposed, and the system control logic was determined and established. Finally, the collected power load data was brought into the model for actual case analysis to verify its feasibility and the effect of peak-load shifting. The results showed that due to the influence of the number of tourists, the electricity loads on weekends and holidays were higher, especially the electricity load of air conditioning equipment in the heating and cooling seasons. An actual case was simulated to verify the peak-shifting effect of the proposed regulation strategy; it was found that the maximum peak load of the cluster was reduced by 61.6%, and the peak–valley difference was 28.6% of that before peak shifting.

Keywords: load data analysis; demand side management (DSM); rural home hotels; cluster control strategy; PCM storage

Citation: Li, W.; Liang, Y.; Wang, J.; Lin, Z.; Li, R.; Tang, Y. Peak Load Shifting Control for a Rural Home Hotel Cluster Based on Power Load Characteristic Analysis. *Processes* **2023**, *11*, 682. https://doi.org/10.3390/pr11030682

Academic Editor: Wei Sun

Received: 18 January 2023
Revised: 9 February 2023
Accepted: 21 February 2023
Published: 23 February 2023

1. Introduction

Developing rural home hotels is an important measure to promote rural economic development in China. However, the expansion speed of the rural power grid is having trouble matching the rapid development of the rural home hotel industry. Rural roads are relatively backward [1], and thus, the transportation of construction material is difficult and the construction costs are high [2]; furthermore, equipment construction and upgrade periods are long [3], thus leading to insufficient power supply capacity [4]. Especially for the rapidly developing rural home hotels industry, the existing power grid cannot meet the increasing electricity load. Therefore, the peak-load shifting strategy of the demand side plays a crucial role in balancing the difference between supply and demand of the rural power grid [5,6].

Rural home hotels are usually clustered together, and thus, load fluctuations and load peaks caused by the change in the number of visitors are prominent. Furthermore, due to the large visitor number during winter and summer holidays, indoor thermal comfort is very important [7], and thus, the air conditioning load is the main cause of power supply tension in summer [8] and the problem is also prominent in winter for non-centrally heated villages. Therefore, as a suitable technology for shifting the peak load of air conditioning, thermal energy storage is expected to reduce the pressure on a rural power grid and improve the stability of a rural power grid.

1.1. Thermal Energy Storage Technology

Thermal energy storage technology [9–11] uses thermal storage materials as media to store and release thermal energy to solve the problem caused by the mismatch between thermal energy supply and demand in time, space or intensity. Thermal storage technologies are divided into three main categories: sensible, latent and thermochemical thermal storage [12]. Among them, phase change energy storage is a promising technology, which takes advantage of the phase change process of new types of phase change materials (PCMs). PCMs are usually divided into four major types: liquid–gas [13], solid–liquid [14], solid–gas [15] and solid–solid [16]. PCMs have the advantages of high applicable energy storage density, wide temperature range, high crystallization rate and high thermal conductivity [17–19]. At present, phase change energy storage has been widely used in energy storage and construction [20], such as air conditioning systems in buildings [21].

The utilization of PCMs in building peak-load shifting received extensive attention. Sun et al. [22] proposed a phase change thermal storage electric heater that could meet the peak-shifting demand throughout the day with 8 h of thermal storage at night and 3 h of thermal storage during the daytime. Wang et al. [23] proposed a PCM tank for clean heating projects of intertemporal heating, which could transfer at least a 1309.2 kW peak load of an office building. Liu et al. [24] analyzed the effect of the addition of a phase change thermal storage device on a solar heating system with a valley filling rate of 66.67% and a peak shaving rate of 11.9%. Riahi et al. [25] proposed a phase change energy storage vapor compression cooling system for power peak-load shifting and concluded that when the volume of the PCM was increased from 38 L to 309 L, the peak load reduction could be increased from 12.7% to 18.7%. De et al. [26] proposed a PCM-based cold storage device connected to a chiller that could store 25 kWh of energy with a charge time of 2.5 h and a discharge time of 1.6 h. Hu et al. [27] investigated the temperature variation of three different sizes of PCM storage during charging and discharging, and found that the energy cost can be reduced by 7% compared with an HVAC system without PCM storage. Koželj et al. [28] compared the conventional sensible thermal energy storage tank with the mixed latent thermal energy storage tank. The experimental results showed that a 15% PCM in the water storage tank provided 70% more thermal storage than the conventional water storage tank with only water inside.

1.2. Peak-Shifting Strategy of Rural Home Hotels

Existing studies [29–32] verified the peak-load-shifting effect by using phase change energy storage in buildings, but most of them focused on single buildings and large-scale commercial buildings, such as office buildings and shopping malls. Rural home hotels are usually converted from rural residential buildings, and thus, they have some similar characteristics to residential buildings, such as being small-scale individual buildings and the large randomness of occupant activities. Therefore, the peak-load-shifting strategy for home hotels will differ significantly from that of large-scale commercial buildings. In particular, the load of a single rural home hotel is small, but the spatial distribution of multiple buildings is concentrated. The small-load users can jointly provide a larger regulation capacity [33]. However, the current peak shifting strategies on building clusters are mostly about incentive strategies for demand response [29,30], and the research objects are also mostly large buildings [31], independent hotel buildings [32] and industrial parks [34]; therefore, they cannot be directly applied to a home hotel cluster.

Unlike most buildings, rural home hotels are mostly located in remote areas, among mountains and forests, and in remote towns [35,36], with backward rural energy structures and high retrofitting potential. Wei et al. [37] pointed out that rural home hotels could be retrofitted in many aspects, including floor plan design, energy-saving renovation of the envelope structure and the use of clean energy. They further concluded that retrofitted home hotels were significantly improved in terms of natural indoor lighting and thermal comfort, producing energy savings of more than 50%. Zhu et al. [38] modeled three types of rural tourism buildings, namely, two-in-one courtyard, triad courtyard and quadrangle courtyard, and optimized their pre-design to determine the design solution with the highest thermal comfort. Gutierrez Rodriguez et al. [39] studied the validation of three dynamic capabilities in nature tourism through the observation of dynamic capabilities absorption, adaptation and innovation in SMEs and tourism clusters composed of these companies. D'Agostino [40] applied the cost optimization method to a building in the Mediterranean region, evaluated and compared different energy measures, and found the cost-optimal solution for the existing structure. Rural tourism individuals are more suitable for management by combining them into clusters. The above studies about energy use in rural home hotels focused on the energy-saving renovations of existing buildings and the optimal design of new buildings. However, limited attention was paid to the difference between the electricity demand of home hotels and the grid supply.

1.3. Paper Contributions and Organization

The renovation period of a rural power grid is much longer than that of towns and cities, and the speed of rural power grid expansion is having trouble matching the rapid development of the rural home hotels industry. However, so far, relevant research mostly focused on the research of peak shifting strategy for a single building and the research on home hotels is still limited. The existing peak-load-shifting technologies are mismatched with the characteristics of home hotels.

Therefore, this study took a visitor home hotel village as the object, and the main contributions can be stated as follows:

(1) Based on the data collection and field research, the electricity load characteristics of a home hotel village were comprehensively analyzed;

(2) The adjacent rural home hotels were composed into a cluster to realize the electricity load regulation of the home hotel cluster through mutual energy storage;

(3) A novel phase change energy storage load regulation system was proposed for a rural home hotel cluster by combining the advantages of phase change thermal storage technology and the load characteristics of rural home hotels.

Through a case study of a building cluster in this village, the peak-load-shifting effect of this system was obtained. The study results can be applied to rural home hotels to reduce the peak load and improve the grid stability of visitor villages.

The structure of this paper is as follows: the second part provides the characteristics of the energy consumption behavior of the home hotel village, which was obtained via on-site research. The third part gives the analysis results of the load characteristics of the home hotel buildings with the electricity consumption data of the home hotel village. The fourth part introduces the proposed cluster control strategy, including the phase change energy storage system and control logic. The last part describes the use of a home hotel cluster of the village as an example to simulate and analyze the peak-load-shifting effect of the cluster control strategy. The flow chart of the detailed methodology is shown in Figure 1.

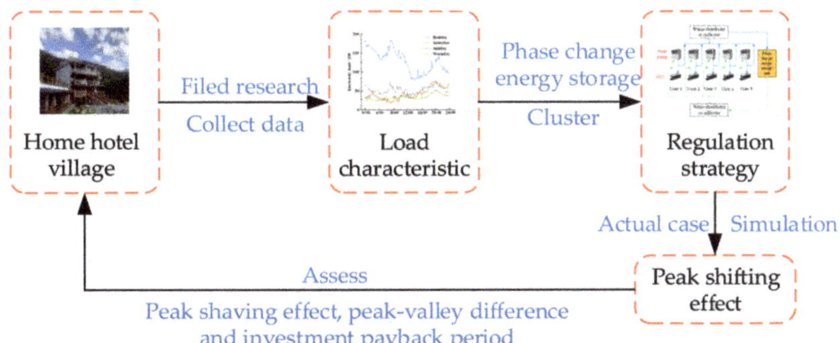

Figure 1. Flow chart of each research step.

2. Investigation of the Characteristics of Electricity Consumption Behavior of Rural Home Hotels

A home hotel village in Zhejiang Province was taken as the research object, which had a stable development, good development trend and a certain scale of home hotels; therefore, its electricity load characteristics were typical and representative. The energy consumption behavior and main energy-using equipment of the home hotel village were obtained via an on-site visit and questionnaire survey. The questionnaire regarding energy consumption behavior for home hotels and the main energy-using equipment is shown in Table 1.

Table 1. Questionnaire on the energy consumption behavior of home hotels.

NO.	Question	Answer
1	Floor area	_m^2
2	Maximum accommodation number per day	_Person/day
3	Peak season	
4	Monthly electricity bill for the last year	_$
5	Heating method	Central heating/air conditioning heating/gas heating/coal-fired heating
6	Cooling method	air conditioning cooling/other_____
7	Hot water method	Electricity hot water/gas hot water/other hot water______
8	Distributed energy and quantity	1. Energy name: Quantity and power: 2. Energy name: Quantity and power: 3. Energy name: Quantity and power:
9	Main equipment	Air source heat pump _kW Split Type Air Conditioner _kW TV _kW Refrigerator _kW Washing machine _kW Hot water kettle _kW Water heater _kW Electric stove _kW
10	Other large appliances	Please specify _________ _kW

The survey showed that the village was a typical rural tourism resort, with more than 470 home hotels in the village. The peak number of visitors was more than 200,000 during the Spring Festival and National Day holidays. The village was located in an administrative county with eight power supply stations, of which the home hotel village accounted for one-third of the electricity consumption of the entire county. To meet the peak electricity demand of the home hotel village, the grid corporation adopted a capacity increase strategy. The home hotel village had 65 supply and distribution transformers and 6 grid lines. However, the power supply was still insufficient during the peak demand period. The increase in capacity also brought a large surplus and uneconomic grid capacity during normal times. The home hotel village now focuses on safely keeping electricity and solving problems that may arise in the business of home hotels.

Combined with the questionnaire survey results, the characteristics of the energy consumption behavior of the home hotel villages were further analyzed.

(1)　Basic information about the home hotel buildings

Basic information about the building determined the maximum reception level of the village and individual home hotels, which is one of the key parameters that affect the maximum electricity demand. According to the questionnaire result, most of the home hotels were converted from rural residential buildings. The building area (Figure 2) was mostly in the range of 300~600 m^2. The maximum number of receptions was in the range of 15~40. The per capita building area was 12~25 m^2.

(2)　Configuration of energy-using equipment of the home hotels

The main energy-using equipment in the home hotels was similar to that in rural residential buildings, including air conditioning equipment, domestic hot water systems, TV, and hot water kettles. Among them, the air conditioning system was all cooling and heating dual-condition air conditioning. The system form was mostly an air source heat pump and fan coil system and a split-type air conditioner. Domestic hot water was heated using approximately 80% electric water heaters and the remaining 20% using air source heat pumps and solar water heaters (Figure 3).

(3)　Occupancy rate characteristics

The occupancy rate of the home hotels in the village was mainly influenced by the day type. The holidays, especially Spring Festival and National Day, were the "big peak season", with occupancy rates approaching 100%. Weekends, from Saturday to Sunday noon, were the "small peak season", with occupancy rates around 60% and visitors always returned home on Sunday afternoon. Weekdays were the "slack season" with an average occupancy rate of approximately 10%.

In summary, a home hotel building is a hotel building converted from a common residential building. Therefore, it maintains the dual characteristics of residential and hotel buildings in terms of energy consumption. On the one hand, the fluctuation of air conditioning load due to seasonal changes can still be the main reason for the seasonal peak and valley load difference. On the other hand, since the number of visitors is much higher than the number of villagers, the daily load difference caused by the changes in visitor number and schedule will be more prominent. Therefore, the load characteristics of the home hotel village need to be quantitatively analyzed based on the electricity consumption data.

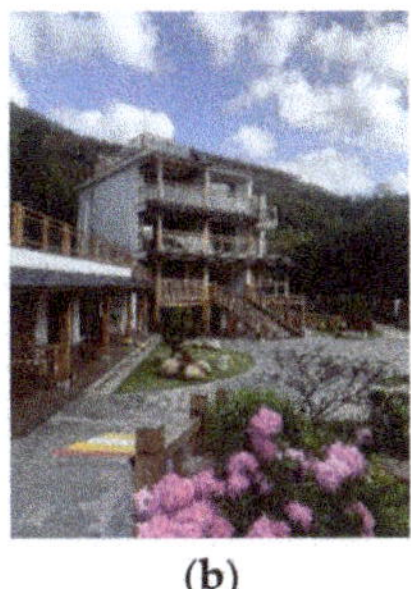

(**a**)   (**b**)

Figure 2. Photos of a site investigation of a typical home hotel: (**a**) exterior of the home hotel; (**b**) interior of the home hotel.

(**a**)   (**b**)   (**c**)

Figure 3. Photos of home hotel energy-using equipment: (**a**) air source heat pump as a water heater; (**b**) electric water heater; (**c**) air source heat pump for air conditioning.

3. Electricity Load Characteristics Analysis of Rural Home Hotels

3.1. Monthly Load Characteristics

In order to correct for the deviation in the total monthly electricity consumption caused by the inconsistency in the number of days in each month, the average daily electricity consumption in each month was compared and analyzed (Figure 4.)

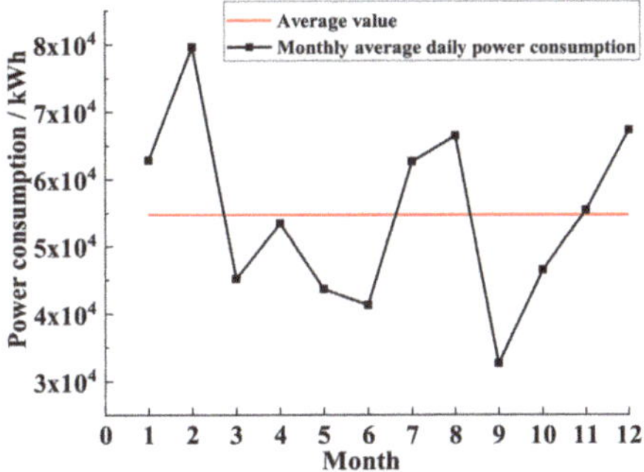

Figure 4. Monthly average daily power consumption of the home hotel village.

As can be seen from Figure 5, the electricity consumption of this rural home hotel village was obviously seasonal. This village was located in Zhejiang Province, without central heating facilities, and it needed air conditioning for heating, which led to the most prominent electricity consumption in winter. Therefore, the month type could be defined according to the average daily electricity consumption level of each month. The classification was as follows:

(1) Heating months, which included the winter in this district and the low-temperature months of late autumn and early spring, and their average daily electricity consumption was 10% higher than the annual average daily electricity consumption.

(2) Cooling months, which included the high-temperature months in summer of this district, and their daily average electricity consumption was 10% higher than the annual average daily electricity consumption.

(3) Transition months, which included other months that were not defined as heating or cooling months.

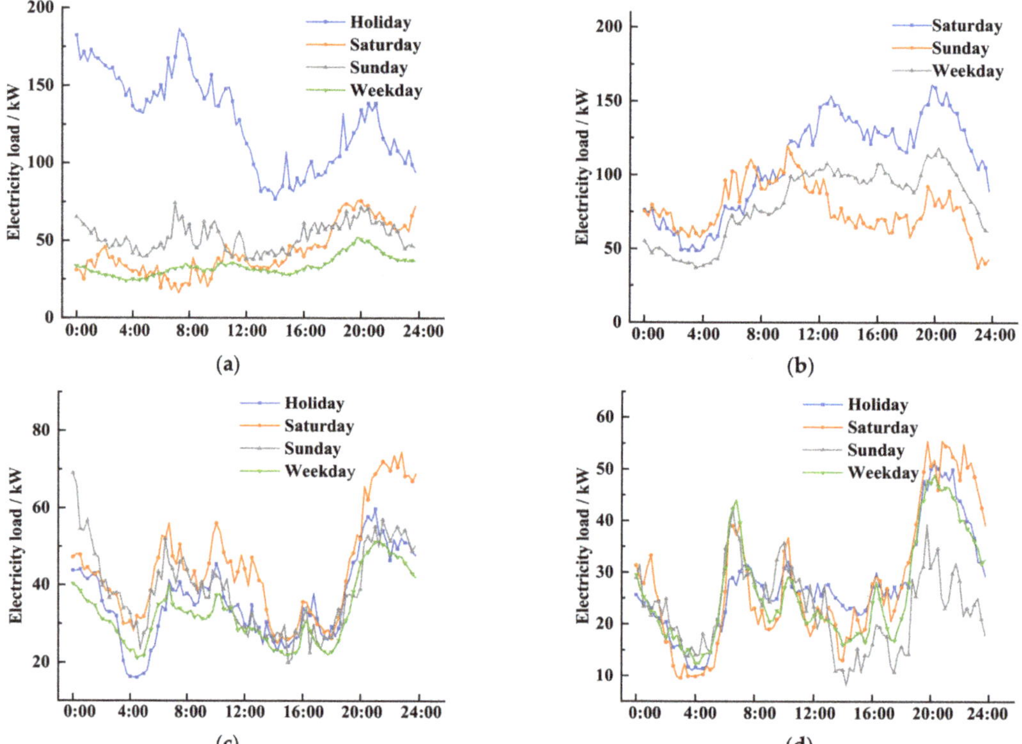

Figure 5. Typical electricity load curves of different day types: (**a**) heating month; (**b**) cooling month; (**c**) transition month (first half of the year); (**d**)transition month (second half of the year).

As shown in Table 2, based on the above division method, January, February and December were defined as heating months and July and August were defined as cooling months, which were when the average daily electricity consumption was around 60,000 kWh. Other months, namely, March, April, May, June, September, October and November, were transition months.

Then, according to the number of visitors, the day types were further divided into four types of each month type, i.e., weekday, holiday, Saturday and Sunday.

Finally, according to the behavior schedule of the occupants, the daily load could be divided into four periods: daybreak (0:00–8:00), morning (8:00–12:00), afternoon (12:00–18:00) and night (18:00–24:00).

Table 2. Study time division of electricity load characteristics.

Month Type	Month	Day Type	Time Type
Heating month	January, February and December	Working day Weekend (Saturday and Sunday) Holiday	Daybreak (0:00–8:00) Morning (8:00–12:00) Afternoon (12:00–18:00) Night (18:00–24:00)
Cooling month	July and August	Working day Weekend (Saturday and Sunday) Holiday	Daybreak (0:00–8:00) Morning (8:00–12:00) Afternoon (12:00–18:00) Night (18:00–24:00)
Transition month	March, April, May, June, September, October and November	Working day Weekend (Saturday and Sunday) Holiday	Daybreak (0:00–8:00) Morning (8:00–12:00) Afternoon (12:00–18:00) Night (18:00–24:00)

3.2. Daily Load Characteristics

According to the analysis results in Section 3.1, there were obvious seasonal differences in the electricity load of the rural home hotel village. Furthermore, according to the survey results, the occupancy rate of this village was affected by the day type, holidays, weekends and weekdays. Since visitors mostly returned home on Sunday afternoon, the electricity load curves on Saturday and Sunday were obviously different.

3.2.1. Daily Load Characteristics in Heating Months

Among the heating months (Table 2), February had all four types of days, which facilitated the comparison of different day types. Therefore, taking February as the representative heating month, the daily electricity load curves of four types of days, namely, weekday, Saturday, Sunday and Spring Festival holiday, were analyzed (Figure 5a). The data of the week closest to the Spring Festival holiday was selected for analysis regarding the weekday, Saturday and Sunday day types. The workday data was the average of the workday load for that week.

Through the comparative analysis of different day types in Figure 5a, it was found that the peak load during the Spring Festival holiday was 4.5 times the load of other day types for the same time, indicating that the number of visitors had a great influence on the electricity load.

Therefore, holidays were the main object of peak-load shifting for the heating months. Further analysis of the daily load of the holidays showed that the load of the home hotel village started to increase at 4:00 a.m. At 7:15, the load reached the peak value of 186.74 kW and then began to decline, reaching the valley point of 76.68 kW at 14:00. A small peak of 138.89 kW occurred at 21:00.

3.2.2. Daily Load Characteristics in Cooling Months

The main cooling supply months of this home hotel village were July and August (Table 2), which do not include legal holidays, and the average daily electricity consumptions were similar. By taking August as an example of a cooling month, the daily load characteristics of the three non-holiday types of days, namely, weekday, Saturday and Sunday, were analyzed (Figure 5b). The data of the week in the middle of the cooling month was selected for the analysis of the weekday, Saturday and Sunday day types. The workday data was the average of the workday load for that week.

Through the comparative analysis of different day types in Figure 5b, an obvious positive correlation between the electricity load and the number of visitors was found. On Saturday afternoon, the electricity load reached a peak when the visitor number was the largest. On Sunday afternoon, when visitors gradually departed, the electricity load reached the minimum. The load was at the intermediate level on weekdays, when the visitor number was moderate.

Therefore, Saturday was the main object of peak-load shifting for the cooling months. Further analysis of the daily electricity load of Saturday showed that the load of the home hotel village started to increase at 4:00 in the morning and reached the peak value of 153.87 kW at 12:45. At 20:00, the load began to decline, and at 3:45 the next morning, the electricity load reached its valley point of 48.48 kW (Figure 5b).

3.2.3. Daily Load Characteristics in Transition Month

Regarding the transition months (Table 2), April and October were taken as representative transition months for the first and second half of the year, respectively. The daily electricity load curves of four types of days were analyzed (Figure 5c (April) and Figure 5d (October)). The data of the week closest to the holiday was selected for the analysis of weekday, Saturday and Sunday day types. The workday data was the average of the workday load for that week.

Although according to the survey results, the number of visitors on the holiday, weekend and weekday day types were significantly different, their electrical load did not show significant differences. As visitors went out during the day and returned to the hotel at night, a small peak electricity consumption occurred from 6:00 to 9:00 and from 20:00 to 23:00. Meanwhile, the peak load of the home hotel village was 74.52 kW in April and 55.54 kW in October, which were much smaller than the peak loads in the heating and cooling month. Therefore, the peak load caused by visitors was mainly the load of air conditioning. The daily-life load was relatively stable and low and did not change significantly due to the variation in the number of visitors. That is, the number of visitors in transition months had little impact on the electricity load.

Based on the analysis of the representative daily loads of each month type, the peak-to-valley difference in the transition months was much smaller than that of the heating months and cooling months. Therefore, the peak-load-shifting strategy should be focused on the heating and cooling months. In addition, both the highest daily load and the largest peak-to-valley difference occurred in the heating season. Therefore, taking the heating season as an example, this study further proposed the peak-load-shifting strategy on the demand side.

3.3. Load Characteristics Analysis for Air Conditioning

Through the analysis of the load characteristics of the home hotel village and the occupancy rate of home hotels, it was found that although there were obvious differences in the number of visitors on different day types in the transition seasons, the electricity load difference between different day types was not obvious due to the low utilization rate of air conditioning.

However, in the heating and cooling seasons, besides other electrical equipment, the visitors mostly used the air conditioning system to maintain thermal comfort. The loads of different day types were obviously different due to the different visitor numbers. Therefore, the peak power consumption of the home hotel village was mainly caused by the air conditioning system. The air conditioning power consumption can be transferred in the form of thermal storage, which is convenient for stabilizing the grid load. Through the following data analysis, the air conditioning electricity load was separated from the total power consumption data to further analyze the air conditioning electricity load.

3.3.1. Electricity Load of a "No Visitors Day"

The electricity load without an air conditioning load was defined as a "daily-life load" in this study. The daily-life load was mainly caused by the activities of dining, washing, leisure and entertainment, which were less affected by climatic conditions and depended on the number of people.

Therefore, the daily-life electricity load only caused by the villagers would be relatively stable throughout the whole year, which was less affected by climatic conditions, the number of visitors and other factors.

Furthermore, it can be considered that there were few visitors on the day with the lowest electricity load of each month, and the electricity consumption was mainly caused by the villagers, which was called "No Visitors Day". Comparing the lowest daily loads of the heating and transition months, an interesting phenomenon was found. Both of the load curves remained at a low level, below 30 kW all day, and the general trends were similar (Figure 6). Therefore, it can be inferred that the air conditioning electricity consumption caused by villagers was negligible. Thus, it can be considered that on "No Visitors Days" the load only included the daily-life load of villages.

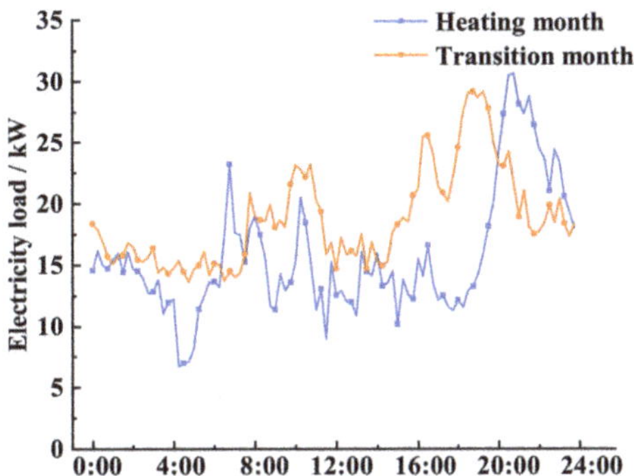

Figure 6. Minimum daily load curves in a heating month and a transition month.

3.3.2. Electricity Load of Air Conditioning

The electricity load of air conditioning can be obtained by removing the daily-life load from the total load. Based on the definition of a daily-life load, it can be simplified as a single-valued function of the number of people. Therefore, the total daily-life load caused by villagers and visitors can be represented by n times of the daily-life load only caused by villagers (the load of a no-visitor day), as represented in Equation (1). Based on the above logic, the air conditioning electricity load can be obtained by disaggregating the daily-life load from the total load (Equation (2)).

$$P_{SH} = n \cdot P_{SH \cdot J}, \tag{1}$$

$$P_{AC} = P - P_{SH}, \tag{2}$$

where P_{SH} is the total daily-life load of visitors and residents, kW; n is the coefficient of the visitors; $P_{SH \cdot J}$ is the load of a no-visitor day, kW; P_{AC} is the electricity load of the air conditioning, kW; and P is the total electricity load, kW.

By combining the electricity load curves of the holiday with a no-visitor day, the load curve of the air conditioning in heating months can be obtained using Equations (1) and (2), as shown in Figure 7. The electricity load of the air conditioning was at a high level in most time periods, around 75 kW. The load level was lower only in short periods, such as 2:00 to 5:00, 8:00 to 10:00 and 18:00 to 20:00. Therefore, it was difficult to realize the energy storage and peak shifting within a day. Therefore, day-ahead energy storage and peak shifting should be adopted.

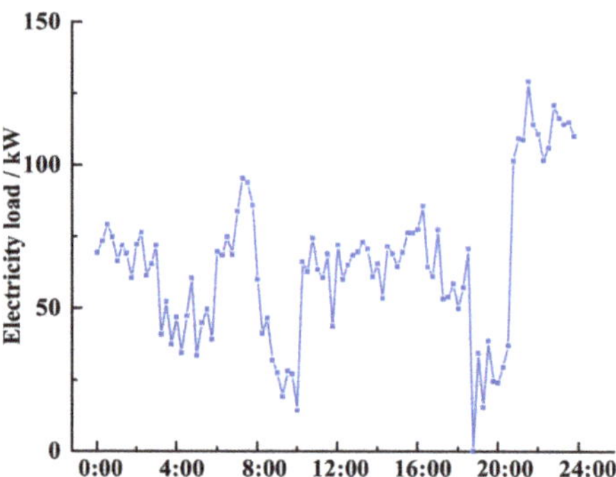

Figure 7. Load curve of the air conditioning in a heating month.

4. Peak-Load-Shifting Strategy

According to the above analysis, the number of visitors, i.e., the day type, was the key factor that affected the electricity load of the home hotel village. The scale of one home hotel was small, and thus, a cluster energy storage and peak-shifting strategy was proposed based on the load characteristics analysis. Several rural home hotels were combined into a cluster, and an energy storage tank was arranged in the cluster. Each user can absorb energy from and release energy into the energy storage tank. The control strategy was uniformly scheduled by the users' real-time electricity load.

A comparative analysis between monomer and cluster energy storage is shown in Table 3. It can be found that, for the home hotel village, the cluster energy storage for peak-load shifting is more compatible with the characteristics of a large change in the number of visitors and different peak hours of each home hotel.

Table 3. Comparison of energy storage peak-shifting strategies.

Day Type	Occupancy Rate Characteristics	Electrical Characteristics	Monomer Energy Storage	Cluster Energy Storage
Working day	The overall occupancy rate was low and the difference between each hotel was high	Peak time of electricity consumption was different	Energy storage and peak shifting did not match, the amount of equipment was large and the whole system was complex	Effective balance of energy storage and peak shifting in the cluster, and the amount of energy storage equipment was small
Weekend and holiday	The overall occupancy rate was high and similar among each hotel	Peak time of electricity consumption was similar	Individual load varied greatly and the peak-shifting system was complicated	Overall load variation was small and the flexibility was high

4.1. Cluster Division Principles

If the cluster control system is used for the energy storage and peak shifting of the cooling and heating loads, it is necessary to divide the home hotels first. Through comprehensive consideration of the system installation, control and peak-shifting effect, the principles of cluster division were put forward:

(1) The distance between buildings in the same cluster should not be too large; otherwise, the system pipeline will be too long, which will increase the system operation and maintenance cost, and will also increase the thermal dissipation and decrease the system efficiency.

(2) The buildings in the same cluster should not cross the main rural roads.

(3) The number of buildings in the same cluster should not be too much. Too many buildings will lead to a large capacity of energy storage tanks, a large occupied area and high requirements for the site.

After the buildings were combined into clusters, the peak-load-shifting strategy of the cluster should follow the following main aspects:

(1) Maximum peak shifting limit—the limit condition after peak shifting is that the peak value of the power consumption is equal to the valley.

(2) Day-ahead peak shifting is the main method, supplemented by intra-day peak shifting. If there is a temporary peak electricity load of individual home hotels, the peak shaving can be completed by the surplus energy storage of other home hotels in the same cluster.

(3) The energy storage should be carried out during the period of the valley electricity price, which was 22:00 to 8:00 of the next day in this home hotel village [41]. Villagers can get the benefits of the difference in peak and valley electricity prices.

(4) The interval between the energy storage and release should not be too long and should not span weekends and holidays. Too long of an interval leads to an increase in thermal loss in the energy storage tank and a decrease in energy storage efficiency.

4.2. Cluster Control System Form

As shown in Figure 8, the control system structure was mainly divided into an energy-using end, energy storage and a pipeline. The energy-using end is the user end, which generates heat in the valley of electricity consumption and stores the extra thermal in the energy storage tank. In the peak period of electricity consumption, the thermal energy in the energy storage tank is extracted and supplied to the users. The PCMs in the energy storage tank are responsible for storing the heat generated by the users. The energy storage tank adopts a horizontal non-pressure shell tube type, which is convenient for switching PCMs in winter and summer. Thermal insulation materials are installed on the outer wall to reduce the thermal loss of the energy storage tank.

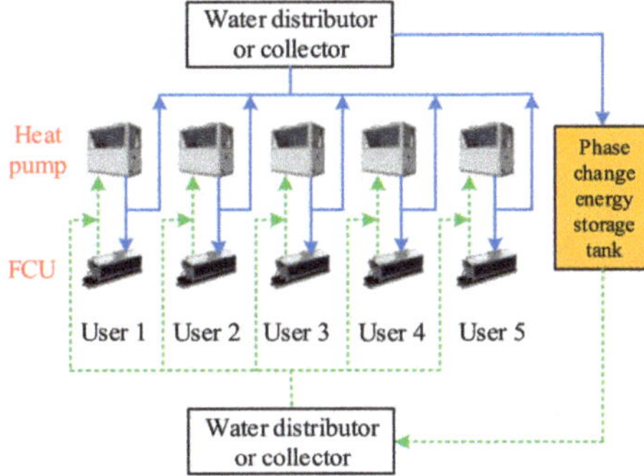

Figure 8. Theoretical framework of the control system.

The pipeline connects the energy storage end and the energy-using end, including the electrical circuit, the water inlet circuit, the water outlet circuit and other pipeline accessories. The electric circuit can turn on the users' heat pump to store thermal energy in the energy storage tank during the valley electricity consumption period. In the peak period, the water pump is turned on and the thermal energy is extracted from the energy storage tank and sent to the end users. The water inlet and outlet circuit connect users with the phase change thermal storage tank to transfer the thermal energy.

4.3. Cluster Control Logic

The whole control logic was divided into two processes, namely, thermal storage and thermal release, which constituted a cycle period. The flow chart is shown in Figure 9.

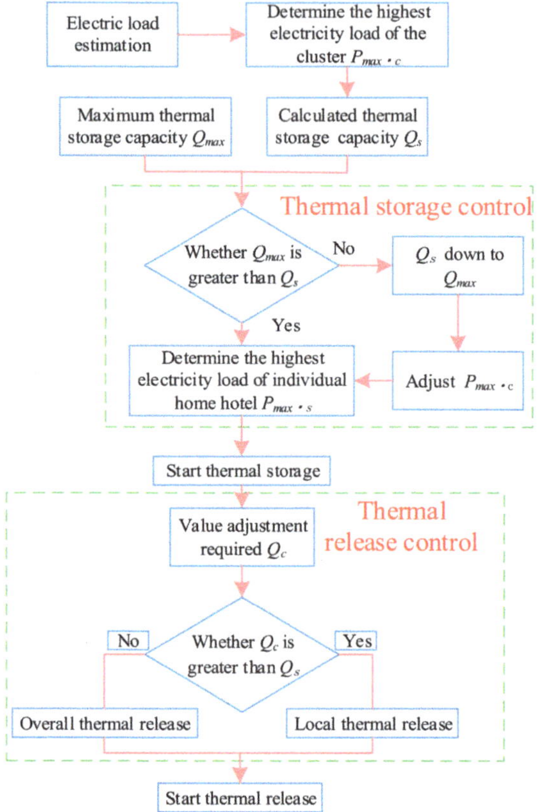

Figure 9. Control logic of the cluster control system.

4.3.1. Thermal Storage Process

(1) The maximum load after cluster peak shifting $P_{max \cdot c}$ is determined. When the electricity load after peak shifting in the valley is equal to the electricity load in the peak, the electricity load is $P_{max \cdot c}$, as shown in Equation (3).

$$\sum (P_{max \cdot c} - p_{i \cdot s}) \times \alpha = \sum (p_{i \cdot r} - P_{max \cdot c}),\tag{3}$$

where $P_{max \cdot c}$ is the highest electricity load after peak shifting under ideal conditions, kW; $p_i \cdot s$ is the load below the load of an individual home hotel during the predefined thermal storage period, kW; $p_{i \cdot r}$ is the load above the highest load of an individual home hotel during the predefined thermal release period, kW; and α is the thermal dissipation coefficient during thermal storage.

(2) The highest load of an individual home hotel Pmax·s is determined. The highest load in the cluster is divided by the number of home hotels to obtain the highest load for an individual home hotel, as shown in Equation (4).

$$P_{max\ s} = P_{max\ c}/m,\tag{4}$$

where $P_{max \cdot s}$ is the highest electricity load of an individual home hotel, which is the electricity load achieved by each home hotel in the cluster after peak shifting under ideal conditions, kW; m is the number of home hotels in the cluster.

(3) Thermal storage calibrations. The theoretically calculated thermal storage capacity (i.e., calculated thermal storage capacity Q_s, Equation (5)) is calibrated against the thermal storage capacity that can be produced by the maximum working intensity of the thermal storage tank (i.e., maximum thermal storage capacity Q_{max}, Equation (6)), and it is ensured that Q_s does not exceed Q_{max}. If Q_s does exceed Q_{max}, Q_s is reduced to Q_{max}.

$$Q_s = \sum (P_{\max \cdot s} - p_v)t, \tag{5}$$

where Q_s is the calculated thermal storage capacity, kJ; p_v is the valley electricity load of an individual home hotel in the cluster, kW; and t is the thermal storage time, s.

$$Q_{max} = \rho V \cdot (r + C_p \Delta T) \cdot \eta, \tag{6}$$

where Q_{max} is the maximum thermal storage capacity, kJ; ρ is the density of the PCM, kg/m^3; V is the limited capacity of the thermal storage, m^3; r is the phase change latent thermal energy of the PCM, kJ/kg; C_p is the specific thermal capacity at a constant pressure of the PCM, kJ/(kg·K); ΔT is the sensible temperature rise value of the PCM during the thermal storage, K; and η is the thermal storage coefficient, which represents the thermal storage capacity of the PCM under the actual working conditions (generally, it is 60–80%).

(4) Start thermal storage. Increase the valley electricity load of an individual home hotel to the peak electricity load for thermal storage.

4.3.2. Thermal Release Process

(1) Thermal release control. The measured electricity load of an individual home hotel is compared with the $P_{max \cdot s}$, and the total peak load shifting value is compared with the total thermal storage (Equation (7)). If the total load shifting value does not exceed the total thermal storage, the overall thermal release will be carried out so that the electricity load of an individual home hotel in the cluster at each moment is controlled below the $P_{max \cdot s}$. If the total load-shifting value exceeds the total thermal storage, the local thermal release will be carried out. First, the highest electricity load users will be satisfied, and at the same time, the thermal release system structure will be relatively simple.

$$Q_c = \sum (p_p - P_{max \cdot s})t, \tag{7}$$

where Q_c is the total peak-load-shifting value in the cluster based on the measured electricity load of an individual home hotel, kJ; p_p is the peak load of an individual home hotel in the cluster, kW.

(2) Start the thermal release. The thermal release starts when the electricity load of an individual home hotel exceeds p_p.

5. Operation Simulation and Analysis of Cluster Control Strategy

5.1. Analysis of the Actual Electricity Load in a Case Study

In the village, the home hotels showed a characteristic of extending to both sides with the road as the center, as shown in Figure 10. Due to the natural landforms between the mountains, the buildings were naturally divided into blocks. Five home hotels on one side of the road were selected as the study object, with the distances between every two buildings not exceeding 10 m and not crossing the main roads in the village. These five home hotels were integrated into a cluster control system.

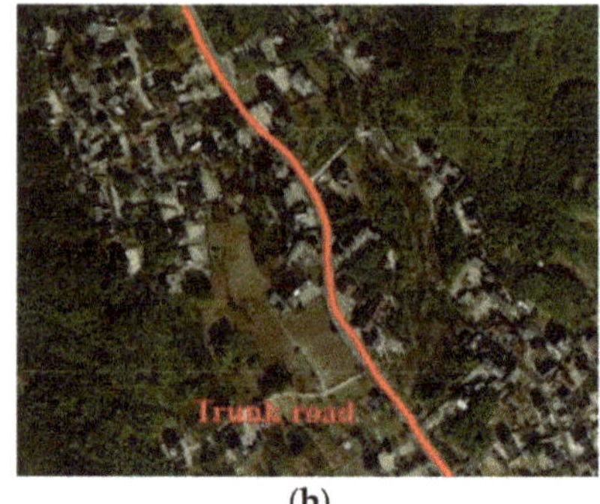

(**a**)   (**b**)

Figure 10. Topographic map of the home hotel village: (**a**) overall topographic map of the home hotel village; (**b**) local topographic map of the home hotel village.

The Spring Festival holiday and the working day one week before the holiday were selected as the research period. During the Spring Festival, the home hotel village had the highest electricity load and its peak shifting was the most difficult, and thus it was more representative.

The overall electricity load curve of each home hotel is shown in Figure 11. The daily electricity load on working days was much lower than that on the holidays. The daily electricity load on working days was about 30 kW, but the average electricity load on holidays was about 80 kW, up to 120 kW. During the holidays, the electricity load fluctuated greatly, with the lowest value of 25 kW and the highest value of 120 kW, and the peak–valley difference phenomenon was serious. Furthermore, the electricity load curves of users were relatively consistent, which showed that the peak periods of electricity consumption were similar. This will easily lead to a sharp rise in the electricity load of the same station area within a certain period, forming a peak load at the station level.

The single-user electricity load characteristics matched the overall characteristics of the heating season holidays. Therefore, the regional cluster regulation with thermal storage and peak shifting as the technical core was also applicable to this case. Since air conditioning was the main cause of the load peak, when the phase change energy storage tank was used to transfer the air conditioning load, the total load of the cluster also changed obviously.

5.2. Case Parameter Selection and Result Analysis

(1)   Treatment of the electricity load. First, the electricity load of each user in the cluster during one peak-shifting period of energy storage was obtained. The highest electricity load of the cluster was calculated to be 120.71 kW by putting the electricity load of each user together.

(2)   Determination of calculated thermal storage capacity. According to Equation (6), the calculated thermal storage capacity was 4507.5 kWh.

(3)   Determination of the maximum thermal storage capacity of a thermal storage tank. In this case, paraffin was selected as the energy storage material, which had the characteristics of large latent thermal phase change and good heat stability. The selected paraffin has 23 carbon atoms, and its phase change temperature is 47.5 °C, density is 900 kg/m^3, latent heat of phase change is 234 kJ/kg and specific heat capacity at constant pressure is 2.64 kJ/(kg·K). According to the cluster electricity load estimation, the effective thermal storage capacity of the thermal storage tank was 50 m^3, and the sensible temperature rise value of the PCM designed for thermal storage was 6 K. The heat storage coefficient was 70%. According to Equation (5), the maximum thermal storage capacity was 7,869,960 kJ = 2186 kWh.

(4)   The calculated thermal storage capacity was 4507.5 kWh, which was greater than the maximum thermal storage capacity of 2186 kWh. It is necessary to reduce the

calculated thermal storage capacity to the maximum thermal storage capacity. The maximum thermal storage load value was 74.47 kW and the minimum thermal release load value was 148 kW.

(5) The maximum thermal storage load of a single user was calculated to be 14.72 kW, and the minimum thermal release load of a single user was 46.17 kW.

(6) The results of the calculated thermal storage and thermal demand of every household are shown in Table 4.

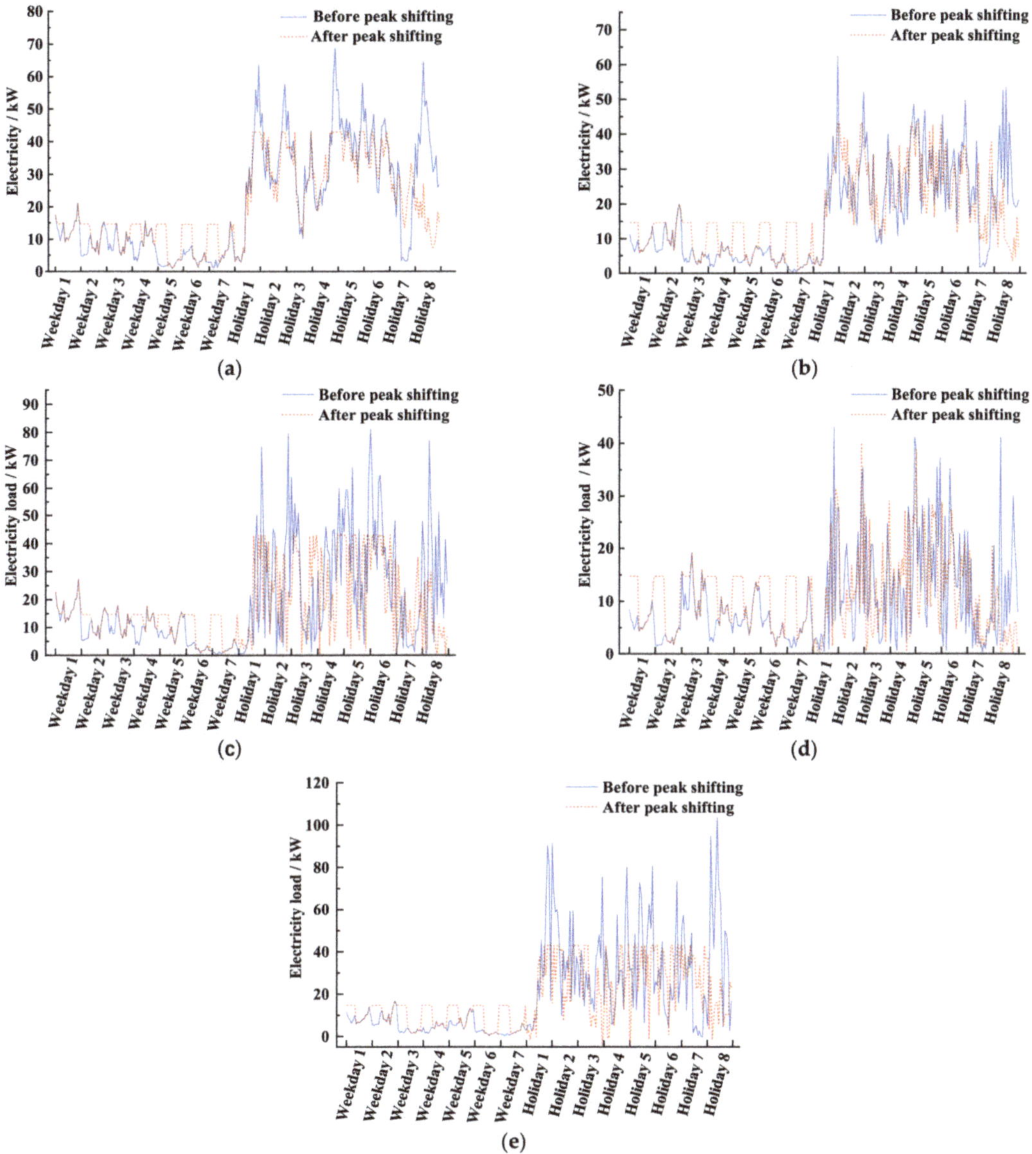

Figure 11. Peak-load changes of the home hotel cluster during the Spring Festival holidays and seven days before the holiday before and after peak shifting: (**a**) user 1; (**b**) user 2; (**c**) user 3; (**d**) user 4; (**e**) user 5.

Table 4. Thermal storage and thermal demand of the home hotels cluster.

	User 1	User 2	User 3	User 4	User 5
Thermal storage	1603.04	1906.58	1400.22	1751.69	2074.75
Thermal demand	1299.71	76.23	2096.42	16.28	5248.82

Unit: kW.

The results of the overall electricity load change after the peak shifting of the home hotels cluster are shown in Figure 11. Compared with that before the peak shifting, the peak load during holidays was significantly reduced. The highest value of the electricity load dropped from 120 kW to 46.72 kW, with a decrease of about 61.1%. The electricity load was transferred to the valley period of the working days before the holiday when the electricity load was low, and the difference in electricity load during the day was largely alleviated.

6. Discussion

According to the above analysis, the PCM energy storage tank required for the load peak regulation was about 50 m^3 for a cluster of five users. The reconstruction cost of the 50 m^3 energy storage tank system was about USD 100,000, and the PCM cost was about USD 200,000. After the completion of the transformation, the operation cost can be reduced by USD 18,000 per year, and a government subsidy of USD 11,000 per year could be obtained. The investment return period was 10.3 years, which is acceptable. Furthermore, the period can be further shortened after quantity production, which has great investment potential.

According to the field survey results, the village had a total of 470 home hotels. The home hotels were transformed from ordinary rural houses, and thus, they had the characteristics of similar scale (between 300–600 m^2), the same equipment type (the main electrical equipment is air conditioning) and similar occupancy rates (affected by holidays and weekends). Therefore, the proposed peak-shifting system is theoretically applicable to all home hotels in the village. If the system was used for peak load regulation for all the home hotels, according to the case analysis results, the maximum load was reduced from the original 186.74 kW to 72.64 kW, and the peaking shifting effect was very significant. This strategy could also effectively absorb renewable energy and avoid the waste of resources caused by expanding the power grid.

The proposed strategy is only applicable to districts with significant air conditioning loads. Furthermore, the concentrated rural home hotels are more applicable. For districts with short heating and cooling periods, its effect and economy will be reduced. The strategy is not applicable to small or scattered rural home hotel buildings because they cannot be divided into a cluster well, which is not conducive to mutual energy storage.

7. Conclusions

The prominent contribution of this study was the proposal of a novel peak-shaving strategy for rural home hotels cluster, which can relieve the difference between the rapid developments of rural tourism and the difficulty of power grid upgrading for a rural area. The conclusions were as follows:

(1) The electricity loads of the home hotel village were greatly affected by the number of visitors, especially on weekends and holidays in the heating and cooling months. The electricity loads in the transition months were less affected by the number of visitors, and the differences in electricity loads on working days, weekends and holidays were small. The main peak electricity loads of the village were generated by the heating and cooling air conditioning equipment of visitors.

(2) The cluster regulation peak-load-shifting system is applicable to the building clusters with similar load characteristics, which are difficult to regulate for single buildings. The building clusters in rural areas show obvious aggregation phenomena according to the natural geographical conditions. This system will reduce the peak electricity

load of the cluster and achieve the purpose of peak load shifting on the demand side of the rural home hotel industry.

(3) The simulation operation of the cluster peak-load-shifting system was carried out through an actual case study. After peak load shifting, the highest peak electricity load was reduced from 120 kW to 46.72 kW, and the reduction rate was 61.1%. The valley and peak values of the electricity load were 14.72 kW and 46.17 kW, respectively, and its peak-to-valley difference was 31.45 kW, which was only 28.6% of that before peak shifting.

The proposed peak-load-shifting system and energy storage can effectively reduce the difference between peak and valley electricity loads and provide technical support for a rural home hotel electricity peak-shifting system.

Author Contributions: Conceptualization, W.L. and J.W.; methodology, Z.L.; software, W.L.; validation, Y.L., R.L. and Y.T.; formal analysis, Y.L.; investigation, Y.L.; resources, W.L.; data curation, Y.L.; writing—original draft preparation, Y.L.; writing—review and editing, W.L.; visualization, Y.T.; supervision, J.W.; project administration, W.L.; funding acquisition, W.L. All authors have read and agreed to the published version of the manuscript.

Funding: This research was funded by the Science and Technology Project (CHECKJ22-01-03) of China Huadian Technology and Industry Co., Ltd., and the Key R&D and Promotion Project (no. 222102320113) of the Department of Science and Technology of Henan Province, China.

Data Availability Statement: Data are available on request due to restrictions. The data presented in this paper are available on request from the corresponding author.

Acknowledgments: This article was supported by Huadian Zhengzhou Mechanical Design Institute Co., Ltd. and Zhengzhou University.

Conflicts of Interest: The authors declare no conflict of interest.

Nomenclature

C_p	Specific thermal capacity at constant pressure	kJ/(kg·K)
m	Number of home hotels in the cluster	
n	Number coefficient of visitors	
P	Total electricity load	kW
P_{AC}	Electricity load of the air conditioning	kW
$p_{i·r}$	Load above the highest load of individual home hotel during the predefined thermal release	kW
$p_{i·s}$	Load below the highest load of an individual home hotel during the predefined thermal storage	kW
$P_{max·c}$	Highest electricity load of the cluster	kW
$P_{max·s}$	Highest electricity load of individual home hotel	kW
p_p	Peak load of an individual home hotel in the cluster	kW
P_{SH}	Total daily-life load of visitors and residents	kW
$P_{SH·J}$	Load of a no-visitor day	kW
p_v	Valley electricity load of an individual home hotel in the cluster	kW
Q_c	Total peak-load-shifting value in the cluster based on the measured electricity load of individual home hotel	kJ
Q_{max}	Maximum thermal storage capacity	kJ
Q_s	Calculated thermal storage capacity	kJ
r	Phase change latent heat	kJ/kg
t	Thermal storage time	s
V	Limited capacity of thermal storage	m^3
ΔT	Sensible temperature rise value of PCM during thermal storage	K
α	Thermal dissipation coefficient during thermal storage	
η	Thermal storage coefficient	
ρ	Density	kg/m^3
PCM	Phase change material	

References

1. Zhou, Y.; Nianping, L.; Yongga, A. Investigation in winter environment of rural residents in targeted poverty alleviation areas. *Build. Sci.* **2022**, *38*, 44–50.
2. Niu, H.; Minghao, Y.; Tianjun, J.; Hancheng, L.; Jiyuan, L. Linear optimal operation model and algorithm for active distribution network in rural areas. *Trans. Chin. Soc. Agric. Eng.* **2013**, *29*, 190–197.
3. Chang, D.; Juan, S.; Lili, Q.; Peng, Z. Smart emergency management system in wind farms and its application: 5G technology empowerment. *China Saf. Sci. J.* **2022**, *32*, 57–67.
4. Cao, Z.; Kaiyun, Z.; Jiali, Z.; Gaoming, L.; Min, Y.; Shun, T.; Yuancheng, C.; Shijie, C.; Weixin, Z. Patent analysis of fire-protection technology of lithium-ion energy storage system. *Energy Storage Sci. Technol.* **2022**, *11*, 2664–2670.
5. Arnaudo, M.; Monika, T.; Pablo, P.; Edmund, W.; Björn, L. Heat Demand Peak Shaving in Urban Integrated Energy Systems by Demand Side Management—A Techno-economic and Environmental Approach. *Energy* **2019**, *186*, 115887. [CrossRef]
6. Ren, H.; Yongjun, S.; Ahmed, K.; Albdoor, V.V.; Tyagi, A.K. Pandey, and Zhenjun Ma. Improving Energy Flexibility of a Net-zero Energy House Using a Solar-assisted Air Conditioning System with Thermal Energy Storage and Demand-side Management. *Appl. Energy* **2021**, *285*, 116433. [CrossRef]
7. Yang, L.; Shiming, D.; Guanyu, F.; Weilin, L. Improved indoor air temperature and humidity control using a novel direct-expansion-based air conditioning system. *J. Build. Eng.* **2021**, *43*, 102920. [CrossRef]
8. Tai, Y.; Jinqi, P.; Rongxin, Y.; Nianping, L.; Xiufeng, P. Research on the adaptablity of demand response for fan-coil air-conditioning system. *Build. Sci.* **2022**, *38*, 195–201+208.
9. Mousavi, S.B.; Adib, M.; Soltani, M.; Razmi, A.R.; Nathwani, J. Transient Thermodynamic Modeling and Economic Analysis of an Adiabatic Compressed Air Energy Storage (A-CAES) Based on Cascade Packed Bed Thermal Energy Storage with Encapsulated Phase Change Materials. *Energy Convers. Manag.* **2021**, *243*, 114379. [CrossRef]
10. Ben, K.; Nidhal, R.A.; Bantan, L.K.; Mohamed, O. Performance Investigation of a Vertically Configured LHTES via the Combination of Nano-enhanced PCM and Fins: Experimental and Numerical Approaches. *Int. Commun. Heat Mass Transf.* **2022**, *137*, 106246.
11. Al-Najjar, H.M.T.; Mahdi, J.M.; Bokov, D.O.; Khedher, N.B.; Alshammari, N.K.; Catalan Opulencia, M.J.; Fagiry, M.A.; Yaïci, W.; Talebizadehsardari, P. Improving the Melting Duration of a PV/PCM System Integrated with Different Metal Foam Configurations for Thermal Energy Management. *Nanomaterials* **2022**, *12*, 423. [CrossRef]
12. Zhao, Y.; Zhao, C.Y.; Markides, C.N.; Wang, H.; Li, W. Medium- and High-Temperature Latent and Thermochemical Heat Storage Using Metals and Metallic Compounds as Heat Storage Media: A Technical Review. *Appl. Energy* **2020**, *280*, 115950. [CrossRef]
13. Kwak, Y.; Hyunah, S.; Seongeun, M.; Kwanhee, L.; Jaewon, K.; Yongha, P.; Hyangsoo, J.; Hyuntae, S.; Jong, H.H.; Suk, W.N.; et al. Investigation of a Hydrogen Generator with the Heat Management Module Utilizing Liquid-gas Organic Phase Change Material. *Int. J. Energy Res.* **2021**, *45*, 10378–10392. [CrossRef]
14. Su, W.; Jo, D.; Georgios, K. Review of Solid–liquid Phase Change Materials and Their Encapsulation Technologies. *Renew. Sustain. Energy Rev.* **2015**, *48*, 373–391. [CrossRef]
15. Huang, Z.; Guang, L.; Arezoo, M. Ardekani. A Consistent and Conservative Phase-Field Model for Thermo-gas-liquid-solid Flows including Liquid-solid Phase Change. *J. Comput. Phys.* **2022**, *449*, 110795. [CrossRef]
16. BABA, M.; Kosei, N.; Daiki, O.; Takuto, S.; Masatoshi, T.; Noboru, Y. Temperature Leveling of Electronic Chips by Solid-solid Phase Change Materials Compared to Solid-liquid Phase Change Materials. *Int. J. Heat Mass Transf.* **2021**, *179*, 121731. [CrossRef]
17. Chen, Y.; Qing-hui, J.; Ji-wu, X.; Xin, L.; Bing-yang, S.; Jun-you, Y. Research Status and Application of Phase Change Materials. *Cailiao Gongcheng J. Mater. Eng.* **2019**, *47*, 1.
18. Zhao, M.; Yuang, Z.; Bingtao, T. Research Process in Polyurethane Form-stable Composite Phase Change Materials. *Jing Xi Hua Gong* **2020**, *37*, 2182.
19. Zhang, Z.; Changlin, Z. Study on the Thermal Storage Performance for Carbon Nanotubes-boron Nitride/Myristic Acid Composite Phase Change Material. *Zhongguo Dianji Gongcheng Xuebao* **2021**, *13*, 4585.
20. Karaipekli, A.; Ahmet, S.; Alper, B. Thermal Regulating Performance of Gypsum/(C18–C24) Composite Phase Change Material (CPCM) for Building Energy Storage Applications. *Appl. Therm. Eng.* **2016**, *107*, 55–62. [CrossRef]
21. Aljehani, A.; Siddique, A.K.; Razack, L.N.; Said, A.-H. Design and Optimization of a Hybrid Air Conditioning System with Thermal Energy Storage Using Phase Change Composite. *Energy Convers. Manag.* **2018**, *169*, 404–418. [CrossRef]
22. Sun, T.; Zhipeng, L.; Shan, Z.; Pan, E.; Li, H.; Wang, Z.; Peng, D.; Zhao, L.; Fan, L.; Wang, Y.; et al. Optimization operation of integrated energy system based on load shift function of a new type of phase change heat storage electric heater. *Therm. Power Gener.* **2021**, *50*, 141–147.
23. Wang, C.; Shuo, L.; Xuefeng, D. The study on application of phase change energy storage technology in clean heating. *Huadian Technol.* **2020**, *42*, 91–96.
24. Liu, D. *Study on Technology of Solar Phase Change Regenerative Heating System*; Inner Mongolia University of Science & Technology: Baotou, China, 2020.
25. Riahi, A.; Hassan, J.M.; Soheil, K.; Mohammad, B.S. Performance Analysis and Transient Simulation of a Vapor Compression Cooling System Integrated with Phase Change Material as Thermal Energy Storage for Electric Peak Load Shaving. *J. Energy Storage* **2021**, *35*, 102316. [CrossRef]

26. De Falco, M.; Marco, S.; Alessandro, Z. Experimental Investigation of a Multi-kWh Cold Storage Device Based on Phase Change Materials. *J. Energy Storage* **2021**, *41*, 102883. [CrossRef]
27. Hu, Y.; Per Kvols, H.; Christian, D.; Asger, S.S.; Pierre, J.C.; Vogler, F.; Kim, K. Experimental and Numerical Study of PCM Storage Integrated with HVAC System for Energy Flexibility. *Energy Build.* **2022**, *255*, 111651. [CrossRef]
28. Koželj, R.; Urška, M.; Eva, Z.; Uroš, S.; Rok, S. An Experimental and Numerical Analysis of an Improved Thermal Storage Tank with Encapsulated PCM for Use in Retrofitted Buildings for Heating. *Energy Build.* **2021**, *248*, 111196. [CrossRef]
29. Nazemi, S.D.; Mohsen, A.J.; Esmat, Z. An Incentive-Based Optimization Approach for Load Scheduling Problem in Smart Building Communities. *Buildings* **2021**, *11*, 237. [CrossRef]
30. Lai, J.; Hong, Z.; Wenshan, H.; Dongguo, Z.; Liang, Z. Smart Demand Response Based on Smart Homes. *Math. Probl. Eng.* **2015**, *2015*, 912535. [CrossRef]
31. Guelpa, E.; Giulia, B.; Adriano, S.; Vittorio, V. Peak-shaving in District Heating Systems through Optimal Management of the Thermal Request of Buildings. *Energy* **2017**, *137*, 706–714. [CrossRef]
32. Wu, L. Comprehensive Evaluation and Analysis of Low-carbon Energy-saving Renovation Projects of High-end Hotels under the Background of Double Carbon. *Energy Rep.* **2022**, *8*, 38–45. [CrossRef]
33. Wang, H.; Zhikun, D.; Rui, T.; Yongbao, C.; Cheng, F.; Jiayuan, W. A Machine Learning-based Control Strategy for Improved Performance of HVAC Systems in Providing Large Capacity of Frequency Regulation Service. *Appl. Energy* **2022**, *326*, 119962. [CrossRef]
34. Sæther, G.; Del Granado, P.C.; Salman, Z. Peer-to-peer Electricity Trading in an Industrial Site: Value of Buildings Flexibility on Peak Load Reduction. *Energy Build.* **2021**, *236*, 110737. [CrossRef]
35. Zhang, B.; Zhaoying, W.; Minjie, Z. Characteristics and driving forces of the mixed use of rural settlement land. *Trans. Chin. Soc. Agric. Eng.* **2022**, *38*, 267–275.
36. Xu, H.; Xingfa, Z. New endogenous development of rural tourism communities: Internal logic, multiple dilemmas and practical exploration. *Mod. Econ. Res.* **2022**, *481*, 114–123.
37. Wei, G.; Yu, F.; Yuqian, L.; Juanjuan, C. Energy-Saving Retrofit Strategy of Beijing Rural House Based on Software Simulation. *Build. Energy Effic.* **2020**, *48*, 84–89.
38. Zhu, L.; Binghua, W.; Yong, S. Multi-objective Optimization for Energy Consumption, Daylighting and Thermal Comfort Performance of Rural Tourism Buildings in North China. *Build. Environ.* **2020**, *176*, 106841. [CrossRef]
39. Gutierrez, R.; Alejandro, J.; Nini, J.B.; Jose, M.G.M. Validity of Dynamic Capabilities in the Operation Based on New Sustainability Narratives on Nature Tourism SMEs and Clusters. *Sustainability* **2020**, *12*, 1004.
40. D'Agostino, D.; Ilaria, Z.; Cristina, B.; Paolo, C. Economic and Thermal Evaluation of Different Uses of an Existing Structure in a Warm Climate. *Energies* **2017**, *10*, 658. [CrossRef]
41. Zhejiang Provincial Development and Reform Commission. Notice of the Provincial Development and Reform Commission on Further Improving the Province's Time of Use Tariff Policy. Available online: https://fzggw.zj.gov.cn/art/2021/9/10/art_122912 3366_2354986.html (accessed on 9 September 2021).

Article

Cooling and Water Production in a Hybrid Desiccant M-Cycle Evaporative Cooling System with HDH Desalination: A Comparison of Operational Modes

Lanbo Lai [1], Xiaolin Wang [1,*], Gholamreza Kefayati [1] and Eric Hu [2]

1 School of Engineering, University of Tasmania, Hobart, TAS 7001, Australia
2 School of Mechanical Engineering, the University of Adelaide, Adelaide, SA 5005, Australia
* Correspondence: xiaolin.wang@utas.edu.au

Highlights:

What are the main findings?

- A hybrid desiccant M-cycle cooling system with an HDH unit is proposed for simultaneous cooling and water production.
- Three typical operational modes are analysed and compared in terms of cooling and water production.

What is the implication of the main finding?

- The recirculation mode exhibited a superior cooling performance than the other two modes.
- The water production rates and system COP were similar among the three modes.

Citation: Lai, L.; Wang, X.; Kefayati, G.; Hu, E. Cooling and Water Production in a Hybrid Desiccant M-Cycle Evaporative Cooling System with HDH Desalination: A Comparison of Operational Modes. *Processes* **2023**, *11*, 611. https://doi.org/10.3390/pr11020611

Academic Editors: Yabin Guo, Zhanwei Wang, Yunpeng Hu and João M. M. Gomes

Received: 26 January 2023
Revised: 14 February 2023
Accepted: 15 February 2023
Published: 16 February 2023

Abstract: In this paper, the cooling and freshwater generation performance of a novel hybrid configuration of a solid desiccant-based M-cycle cooling system (SDM) combined with a humidification–dehumidification (HDH) desalination unit is analysed and compared in three operational modes: ventilation, recirculation, and half recirculation. The HDH unit in this system recycles the moist waste air sourced from the M-cycle cooler and rotary desiccant wheel of the SDM system to enhance water production. A mathematical model was established and solved using TRNSYS and EES software. The results of this study indicate that the recirculation mode exhibited superior cooling performance compared to the other two modes, producing up to 7.91 kW of cooling load and maintaining a supply air temperature below 20.85 °C and humidity of 12.72 g/kg under various ambient conditions. All the operational modes showed similar water production rates of around 52.74 kg/h, 52.43 kg/h, and 52.14 kg/h for the recirculation, half-recirculation and ventilation modes, respectively, across a range of operating temperatures. The recirculation mode also exhibited a higher COP compared to the other modes, as the environmental temperature and relative humidity were above 35 °C and 50%. However, it should be noted that the implementation of the recirculation mode resulted in a higher water consumption rate, with a maximum value of 5.52 kg/h when the inlet air reached 45 °C, which partially offset the benefits of this mode.

Keywords: M-cycle; evaporative cooling; solid desiccant; humidification–dehumidification desalination; water production

1. Introduction

The increasing demand for thermal comfort and water, driven by economic development and population expansion, has created significant challenges related to energy and water efficiency for sustainable development [1]. The International Energy Agency (IEA) has reported that 10% of global electricity is used for cooling, which contributes to 10% of global greenhouse gas emissions. This value is expected to be tripled by 2050 [2].

In addition, freshwater is becoming a scarce resource as more countries are facing water stress [3]. Currently, around 20% of the world's population lives in areas with water scarcity, and the situation is expected to worsen, with up to 40% of the global population facing water shortage by 2030 [4].

Indirect evaporative cooling (IEC) has gained increasing research interest due to its energy efficiency and environmentally friendly features, making it a potential replacement for mechanical vapour compression systems [5]. The IEC can significantly reduce the supplied air temperature to approximate the ambient wet-bulb temperature (WBT) without augmenting the moisture content [6]. This technology eliminates the need for chemical refrigerants and mechanical compressors, as the cooling process is completed by water evaporation [7]. In 2002, a novel thermodynamic cycle of IEC was developed by Maisotsenko, also known as the M-cycle [8]. In this cycle, a fraction of the pre-cooled supply air is redirected to wet channels as the working air, allowing the product air to be cooled below the WBT and towards the dew point temperature (DPT) of the incoming air at a constant humidity level [9,10]. However, the M-cycle's performance is greatly influenced by the ambient relative humidity. The literature suggests that the M-cycle is most effective when the ambient relative humidity is below 70% [11,12]. Hybridising the M-cycle cooler and a solid desiccant wheel is often regarded as an effective approach to improve the M-cycle cooler's performance in humid environments [12,13].

Goldsworthy and White [14] optimised a solid desiccant evaporative cooling system by changing the secondary/primary ratio of the IEC and the process-to-regeneration ratio of the desiccant wheel. The results showed that the highest COP could reach more than 20, while the secondary/primary ratio of the IEC was 0.67, and the process-to-regeneration ratio of the desiccant wheel was 0.3. Gao et al. [15] numerically evaluated a solid desiccant-based IEC cooling system. It was found that the optimum inlet air humidity ratio and temperature should be no more than 18 g/kg and 35 °C, respectively. Three novel two-stage desiccant cooling systems integrated with the Maisotsenko cooling cycle were developed and analysed by Gadalla and Saghafifar [16]. They concluded that the two-stage Maisotsenko desiccant system with inlet air pre-cooling was effective in minimising the energy usage of air conditioning systems in humid environments. This system was recommended for use in buildings during peak times. Lin et al. [17] examined the impact of removing humidity from the inlet air on the M-cycle cooler's cooling performance. Their results indicated a 70–135% cooling performance and energy efficiency increment after introducing a dehumidification process for the M-cycle. Kashif Shahzad et al. [18] conducted the first experimental work on integrating a solid dehumidification system with an M-cycle cooler (cross-flow type). The proposed system was tested and assessed with a traditional solid desiccant-based cooling system under various inlet air parameters. It was observed that the investigated system was 60% to 65% more efficient in thermal COP than the conventional system. Pandelidis et al. [19] presented four novel desiccant cooling systems based on pre-cooling the inlet air with different M-cycle IECs. It was found the most efficient configuration consisted of a cross-flow M-cycle as a pre-cooler and a regenerative M-cycle as a post-cooler. Delfani and Karami [20] performed research on solar-powered desiccant M-cycle IEC systems using TRNSYS software. They proposed three novel configurations and evaluated them under three different weather conditions. Their findings proved that integrating the M-cycle with solar desiccant cooling systems showed better cooling performance than that hybrid with conventional evaporative coolers. The results also showed that the configuration with two M-cycle coolers in moderate and humid conditions reached the highest COP, 0.728 on average. Harrouz et al. [21] explored two desiccant-assisted evaporative cooling systems for the poultry industry in hot and humid climates. It was reported that the system using the M-cycle could maintain indoor thermal comfort and air quality with an operation cost that was 35% less than the system with a direct evaporative cooler (DEC). Lai et al. [22] numerically studied the influence of the recirculation air ratio on the performance of a solar-powered solid desiccant M-cycle system. The simulation results highlighted that the product air humidity ratio and temperature could be lowered by 41.9% and 23.1%, respectively, as the return air ratio varied from 0% to 60%.

Previous studies have indicated that the M-cycle cooler is a vital option for sustainable cooling. The integration of a solid dehumidification process, which removes the moisture from the inlet air, can further extend the applicability of this technology in humid regions. However, these systems consume a large amount of water, which might add water scarcity for remote communities. On the other hand, desalinating seawater or brine water offers a possible way to overcome water scarcity. Among the existing desalination technologies, the humidification–dehumidification (HDH) desalination method showed some advantages, including a simpler structure, lower capital and maintenance costs, and compatibility with low-grade heat energy [23]. These features make it suitable for providing freshwater to residents in remote areas [24]. Therefore, hybridising the M-cycle cooling with HDH desalination for co-producing a cooling load and water has received significant research interest in these years.

Efforts have been made to harness the humid and warm waste air stream from the wet channel of the M-cycle cooler, which is typically released into the environment, to boost the efficiency of the HDH unit. Kabeel et al. [25] performed an experimental analysis on a new hybrid system consisting of an HDH desalination unit and an IEC cooler, powered by solar thermal energy. The results suggested that the developed system performed well in remote areas of Egypt and could produce more than 400 W of cooling load and 38 L of freshwater daily. Similarly, Chen et al. [26] combined an IEC with an HDH desalination unit in which the purged air from the IEC is used as the inlet air of the HDH to enhance freshwater productivity, and they compared the proposed system with other HDH cycles. Their findings revealed that the proposed system had better performance than the other configurations, with the water production rate and gain–output ratio (GOR) ranging from 25–125 L/h and 1.6–2.5, respectively. Abdelgaied et al. [27] conducted an experimental test that utilised a latent heat energy storage system to improve the performance of a hybrid IEC cooler and HDH system. The results showed that the studied hybrid system produced 14.6% more water than the original system, up to 241.7 L/day. On the other hand, the operation of a desiccant unit also produces hot and humid exhaust air during the regeneration process. Therefore, attempts have been made to reuse the desiccant exhaust air. Kabeel et al. [28] numerically examined a hybrid desiccant cooling system and an HDH desalination system. They concluded that the proposed system was feasible in hot and humid areas. Wang et al. [24] performed a numerical assessment of an innovative solid desiccant-assisted cooling system combined with an HDH desalination unit. In their developed system, the waste air from the regeneration process of the desiccant wheel was recycled as the inlet air of HDH. The authors reported that the developed system achieved the best performance when the air mass flow rate was 0.78 kg/s, resulting in the highest water production rate of 4.9 L/h.

In sum, it can be seen that the inclusion of a solid dehumidification unit significantly enhances the performance of the M-cycle, making evaporative cooling technology more practical in humid regions. Additionally, implementing HDH desalination technology is a practical approach to satisfying potable water needs. However, research on the combination of solid desiccant evaporative cooling systems and HDH desalination units is still insufficient. Recently, Lai et al. [29] investigated a new hybrid arrangement of a solid desiccant-assisted M-cycle cooling system combined with an HDH desalination unit that simultaneously utilised the two humid waste air streams from the M-cycle's wet channels and the desiccant wheel for water production. However, the influence of the operation strategy, comprising ventilation, recirculation, and half-recirculation modes, on the system's cooling and water production performance was not reported. To address this research gap, a numerical model was developed using TRNSYS software to explore the performance of a hybrid system that contains a solid desiccant-assisted M-cycle system and an HDH desalination unit under different operating modes. A comparative analysis was undertaken to assess the performance differentials of the developed system among its ventilation, recirculation, and half-recirculation modes, in the context of concurrent cooling and freshwater production. In this step, we also evaluated the impact of the operational parameters on the product air conditions (humidity and temperature), water consumption and production rates, cooling capacity, and COP for all three modes.

2. System Description

In this study, a co-producing cooling and water system was numerically investigated under ventilation, recirculation, and half-recirculation modes. The studied system is based on solid desiccant/evaporative cooling and HDH desalination technologies, as reported in our previous study [29]. A detailed explanation of the system's working principle for each operation mode is presented in the following sections.

2.1. Ventilation Mode

Figure 1 presents the schematic diagram of the proposed system (SDM-HDH) in ventilation mode and its working process on a psychrometric chart. There are two primary cycles in the SDM-HDH—the SDM cooling cycle and the HDH desalination cycle. The process of dehumidifying and cooling ambient air in the described system involves several steps. First, a desiccant wheel (1–2) is utilised to dehumidify the ambient air. The resulting dry, hot air is then cooled by flowing over a sensible heat wheel (2–3). The supply air is further cooled by the M-cycle cooler (3–4) before being distributed to the room for thermal comfort. To maintain the efficiency of the dehumidification process, a regeneration process is implemented. This involves mixing indoor and ambient air to provide a sufficient flow rate for the regeneration air, which flows over a heat wheel and exchanges heat with the hot supply air (5–6). The solar heater and auxiliary heater (6–7) further increase the temperature of the air stream to the required level for regeneration. The heated air stream then passes through the desiccant wheel (7–8) and removes the absorbed water. In the HDH desalination cycle, the inlet air (10) consists of the moist waste air from the rotary desiccant wheel (8) and the M-cycle cooler (9). This mixed air is sent to the humidifier, where it is further humidified and heated by the heated brine (10–11). The humid and hot airflow is then chilled and condensed in the dehumidifier by coming into contact with cold feed brine. The feed brine is pre-heated and then heated up to the required temperature level through solar and auxiliary heaters before being sprayed into the humidifier to complete the cycle.

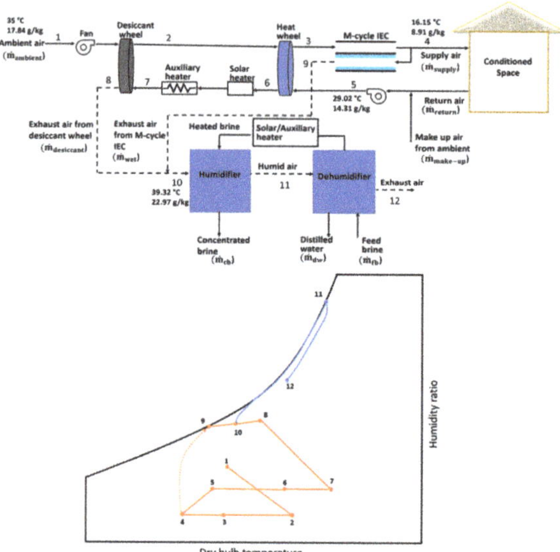

Figure 1. Schematic and psychrometric diagram of SDM-HDH in ventilation mode.

2.2. Recirculation and Half-Recirculation Mode

Figure 2 presents the operation of the SDM-HDH in recirculation mode and its process on a psychrometric diagram. In this mode, the return air from the conditioned space (6) is mixed with the ambient air (1) prior to undergoing dehumidification. The mixed air stream is then dried by a desiccant wheel (2–3) and pre-cooled by a rotary heat wheel (3–4). Subsequently, the air stream is further cooled by an M-cycle cooler before sending to the indoor space. In the regeneration part, the regeneration air is sourced from the environment, while the air from the conditioned space is used as recirculation air. The regeneration process is similar to that in the ventilation mode, with the exception of the air source. In the HDH water generation cycle, the exhaust air from the M-cycle (11) is combined with the desiccant wheel exhaust (10) and sent to the HDH to complete the process (12–14). In half-recirculation mode, the system combines 50% of the return air from the conditioned space with ambient air to form the process air stream. The remaining indoor air is used for the desiccant wheel regeneration. Aside from these distinctions, the other processes are the same as in recirculation mode.

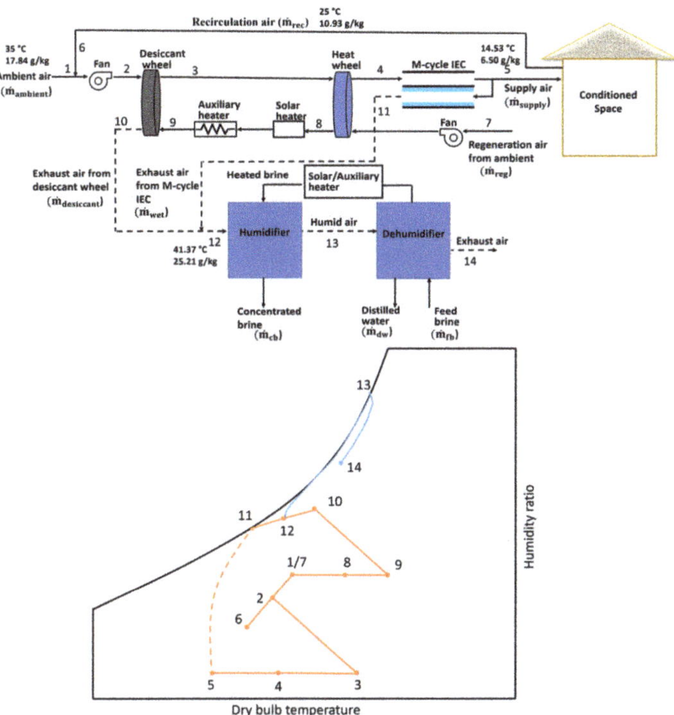

Figure 2. Schematic and psychrometric diagram of SDM-HDH in recirculation mode.

3. Mathematical Modelling

To compare the cooling performance and water productivity of the developed system in the ventilation, recirculation, and half-recirculation modes, commercial software TRNSYS 18 was utilised. TRNSYS is a widely used transient system modelling program that is particularly well-suited for air conditioning and renewable energy systems [30]. It features a flexible graphical interface that allows users to add and link system components easily, and its built-in library and supplementary library, TESS, includes more than 650 validated modules that can cover most of the commonly used components in this study, such as desiccant wheels, heat wheels, pumps, and fans [30,31]. Additionally, external software can be integrated with TRNSYS through the Calling External Programs function, allowing for the creation of new components. In this study, the M-cycle cooler and the HDH desalination unit were built using EES and connected to TRNSYS through the Calling External Programs function. The core components are listed in Table 1, and the detailed simulation of these components is explained and shown in the following sections.

Table 1. Descriptions of the main TRNSYS elements used in the simulation.

TRNSYS Component	Key Inputs	Description
Type 1716b–Desiccant wheel	Humidity mode: 1 ε_{F_1}: 0.05 ε_{F_2}: 0.95	High-performance rotary desiccant dehumidifier
Type 760–Heat wheel	Sensible effectiveness: 0.85 Humidity mode: 1	Air-to-air sensible rotary heat exchanger
Type 6–Electrical heater	Target temperature: 70 °C	Electrical heater, which heats the regeneration airflow to the target temperature
Type 112a–Fan blower	Airflow rate: 660 kg/h Efficiency: 0.9	Constant flow rate fan blower
Type 66–M-cycle IEC	Working/intake air ratio: 0.4 Height (Channel): 5 mm Height: 0.48 m Width: 0.28 m Length: 0.9 m	M-cycle-based indirect evaporative cooler
Type 66–HDH unit	ε_{hum}: 0.8 ε_{deh}: 0.8	HDH desalination unit
Type 65–Online plotter	–	Present the user-selected outputs

3.1. Rotary Desiccant Wheel

The rotary desiccant wheel plays a crucial role in the dehumidification process of the entire system. In this study, silica gel was chosen as the adsorbent material due to its low cost and good adsorption performance [32,33]. To predict the performance of the desiccant wheel, TRNSYS 18 used component Type 1716b, which is based on modelling equations developed by Jurinak [33]. These equations are derived from the potential functions F_1 and F_2, as shown in Equations (1) and (2).

$$F_1 = \frac{-2865}{T^{1.49}} + 4.34\omega^{0.8644} \tag{1}$$

$$F_2 = \frac{T^{1.49}}{6360} - 1.127\omega^{0.07969} \tag{2}$$

where T represents the airflow temperature, and ω represents the airflow humidity.

3.2. Rotary Heat Wheel

The heat wheel functions as a rotary heat exchanger, enabling sensible heat exchange between two air flows. It is available in TRNSYS as Type 760, and a constant effectiveness minimum capacitance model was used to set up the numerical models. The minimum capacitance is shown in Equation (3).

$$C_{min} = MIN\left(\dot{m}_1 c_{p1}, \dot{m}_2 c_{p2}\right) \tag{3}$$

where $\dot{m}_1$ and $\dot{m}_2$ are the mass flow rates of each airflow, c_{p1} and c_{p2} are the values of the specific heat of each working airflow. Then, the maximum sensible heat that can be transferred between air streams can be determined based on the minimum capacitance:

$$\dot{Q}_{sens} = \varepsilon_{sens} C_{min} |T_{1,\,in} - T_{2,\,in}| \tag{4}$$

where $\dot{Q}_{sens}$ is the maximum possible heat transfer rate between the two air flows inside the heat wheel, and ε_{sens} denotes the effectiveness of the rotary heat wheel.

3.3. M-Cycle Cooler

The M-cycle cooler is another crucial system component that produces a sensible cooling load. Since this component could not be found in the TRNSYS library, a mathematical model was created by considering the mass and heat balance equations inside the M-cycle cooler through the use of external software, EES. To integrate this model with TRNSYS, Type 66 was utilised in the TRNSYS interface to call the external program. The following assumptions were made in the derivation of the model:

- There is no heat loss to the surroundings;
- The wet channel's surface is completely wetted;
- The fluid properties within each control volume are uniform;
- The fluid flow is assumed to be incompressible;
- There is no temperature difference between the wet and dry channel surfaces.

In the dry channel of the M-cycle, the product air transfers sensible heat to the adjacent wet channel through the channel wall and a thin water film. The energy balance of a control volume in the dry channel can be described as follows:

$$\dot{m}_{dry} c_{p,\,dry} dT_{dry} = h_{c,dry}\left(T_{dry} - T_{wf}\right) dA \tag{5}$$

where $c_{p,\,dry}$ denotes the dry channel air's specific heat, $h_{c,dry}$ represents the heat transfer coefficient of the air inside the dry channel, T_{dry} denotes the air temperature inside the dry channel, and T_{wf} is the water film temperature in the wet channel.

Heat transfer occurs in the dry channels as the sensible heat is transferred to the adjacent wet channel, causing the water to evaporate and carry away the heat. Therefore, both mass and heat transfer must be considered inside the wet channel.

$$\dot{m}_{wet} c_{p,\,wet} dT_{wet} = h_{c,wet}(T_{wf} - T_{wet}) dA + Lh_m(\omega_{wf} - \omega_{wet}) dA \tag{6}$$

$$\dot{m}_{wet} d\omega_{wet} = h_m(\omega_{wf} - \omega_{wet}) dA \tag{7}$$

where $c_{p,\,wet}$ means the wet channel air's specific heat, $h_{c,wet}$ represents the coefficient of the heat transfer of the moist air along the wet channel, L means the latent enthalpy of the vaporisation of water, h_m denotes the convective mass transfer coefficient, and ω_{wet} is the moist air humidity along the wet channel.

3.4. HDH Desalination Unit

The HDH desalination unit generates freshwater for the system. It is modelled using EES and connected to TRNSYS through Type 66, similar to the M-cycle model. The performance of the HDH unit, including the humidifier and dehumidifier, is predicted using the heat and mass balance equations within the unit. The assumptions and modelling equations used in this process are presented below [26,34–36]:

- Fluid properties are uniform in each part of the unit;
- The heat and mass transfer between the HDH and the ambient air are assumed to be minimal;
- The effectiveness of the HDH unit is assumed to be a constant value;
- Upon exiting the humidifier and dehumidifier, the air is assumed to be fully saturated.

$$\dot{m}_{fb} - \dot{m}_{10}(\omega_{11} - \omega_{10}) = \dot{m}_{cb} \tag{8}$$

$$\dot{m}_{fb} h_{hb,in} - \dot{m}_{cb} h_{cb} = \dot{m}_{10}(h_{11} - h_{10}) \tag{9}$$

$$\varepsilon_{hum} = \max\left\{\frac{h_{hb,in} - h_{cb}}{h_{hb,in} - h_{cb,idea}}, \frac{h_{11} - h_{10}}{h_{11,idea} - h_{10}}\right\} \tag{10}$$

where $\dot{m}_{fb}$ is the cold feed brine flowrate, $\dot{m}_{cb}$ is the concentrated brine flowrate, ε_{hum} is the humidifier's effectiveness, and $h_{hb,in}$ and h_{cb} represent the enthalpy of the heated and concentrated brine, respectively.

The heat and mass balance equations for the humidifier are presented in Equations (8) and (9), respectively. Equation (10) indicates the effectiveness of the humidifier. Similarly, the heat and mass balance equations for the dehumidifier are expressed in Equations (11) to (13).

$$\dot{m}_{10}(\omega_{11} - \omega_{12}) = \dot{m}_{dw} \tag{11}$$

$$\dot{m}_{10}(h_{11} - h_{12}) - \dot{m}_{dw}h_{dw} = \dot{m}_{fb}\left(h_{fb,out} - h_{fb,in}\right) \tag{12}$$

$$\varepsilon_{deh} = \max\{\frac{h_{fb,out} - h_{fb,in}}{h_{fb,out,idea} - h_{fb,in}}, \frac{h_{11} - h_{10}}{h_{11} - h_{10,idea}}\} \tag{13}$$

where $\dot{m}_{dw}$ represents the flowrate of distilled water, h_{dw} is the freshwater enthalpy, and ε_{deh} is the dehumidifier's effectiveness.

4. Performance Index

The water productivity of the system plays a vital role in determining the overall performance of the system. The HDH unit produces distilled freshwater, while the M-cycle cooler consumes water to produce a cooling load. Consequently, the net freshwater production rate of the entire system can be expressed as follows:

$$\dot{m}_{sys} = \dot{m}_{dw} - \dot{m}_{M-cycle} \tag{14}$$

where $\dot{m}_{sys}$ represents the net freshwater production rate, and $\dot{m}_{M-cycle}$ is the water consumption rate of the M-cycle, which can be calculated based on the humidity variation between the outlet and inlet of the wet channel.

The cooling capacity ($\dot{Q}_{cool}$) generated by the SDM is another vital performance metric that can be determined as:

$$\dot{Q}_{cool} = \dot{m}_4(h_4 - h_1) \tag{15}$$

In a traditional HDH desalination system, the GOR value is used to assess the system's water generation performance, expressed as the ratio of latent heat of generated water to the thermal energy input to the system, as shown in Equation (16). As the cooling performance was also considered in the system performance assessment, the overall system COP was adopted as the performance indicator, which can be expressed as the ratio of the water productivity and cooling load, which are the beneficial effects, to the thermal energy consumed for the regeneration process of the desiccant wheel and the heating load used in the HDH component, which is the overall consumed thermal energy.

$$GOR = \frac{\dot{m}_{dw}L}{\dot{Q}_{heat}} \tag{16}$$

$$\dot{Q}_{reg} = \dot{m}_5(h_7 - h_6) \tag{17}$$

$$\dot{Q}_{heat} = \dot{m}_{fb}\left(h_{fb,out} - h_{hb,in}\right) \tag{18}$$

$$COP = \frac{\dot{m}_{sys}L + \dot{Q}_{cool}}{\dot{Q}_{reg} + \dot{Q}_{heat}} \tag{19}$$

5. Model Verification

5.1. SDM Model Verification

As the M-cycle model was verified in our previously published work [22], and the other components in the SDM were available in the TRNSYS library, the SDM was validated by comparing its dehumidification and cooling performance with the published experimental data from reference [18] under identical conditions, with inlet air temperatures ranging from 25 to 45 °C and the inlet air humidity levels varying from 12 to 18 g/kg. As shown in Figure 3, the maximum relative errors between the simulation and experimental results were 11.35% for the dehumidification performance and 7.56% for the supply air temperature. These results suggest that the developed model is effective in predicting the cooling and dehumidification performance of the system.

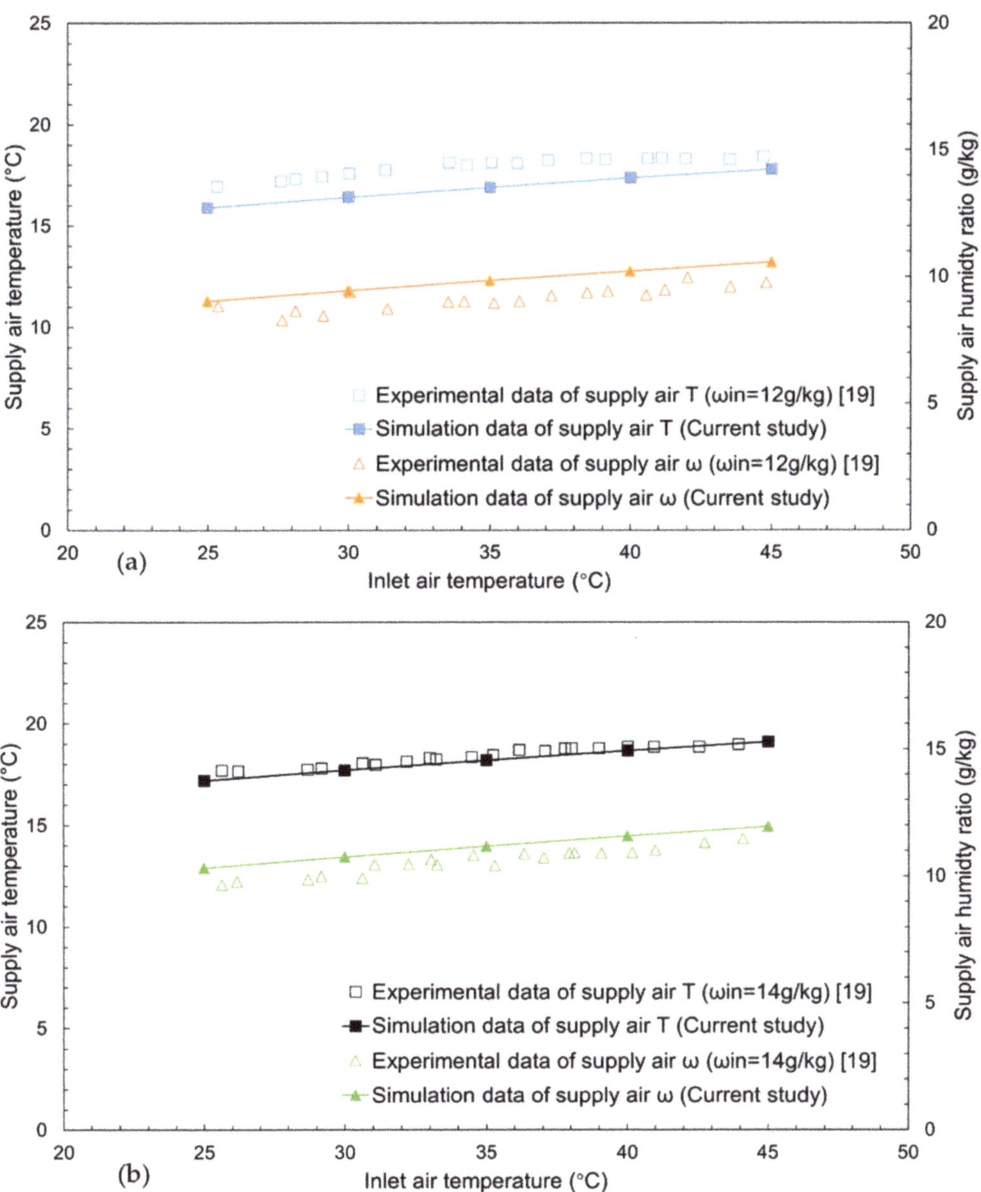

Figure 3. *Cont.*

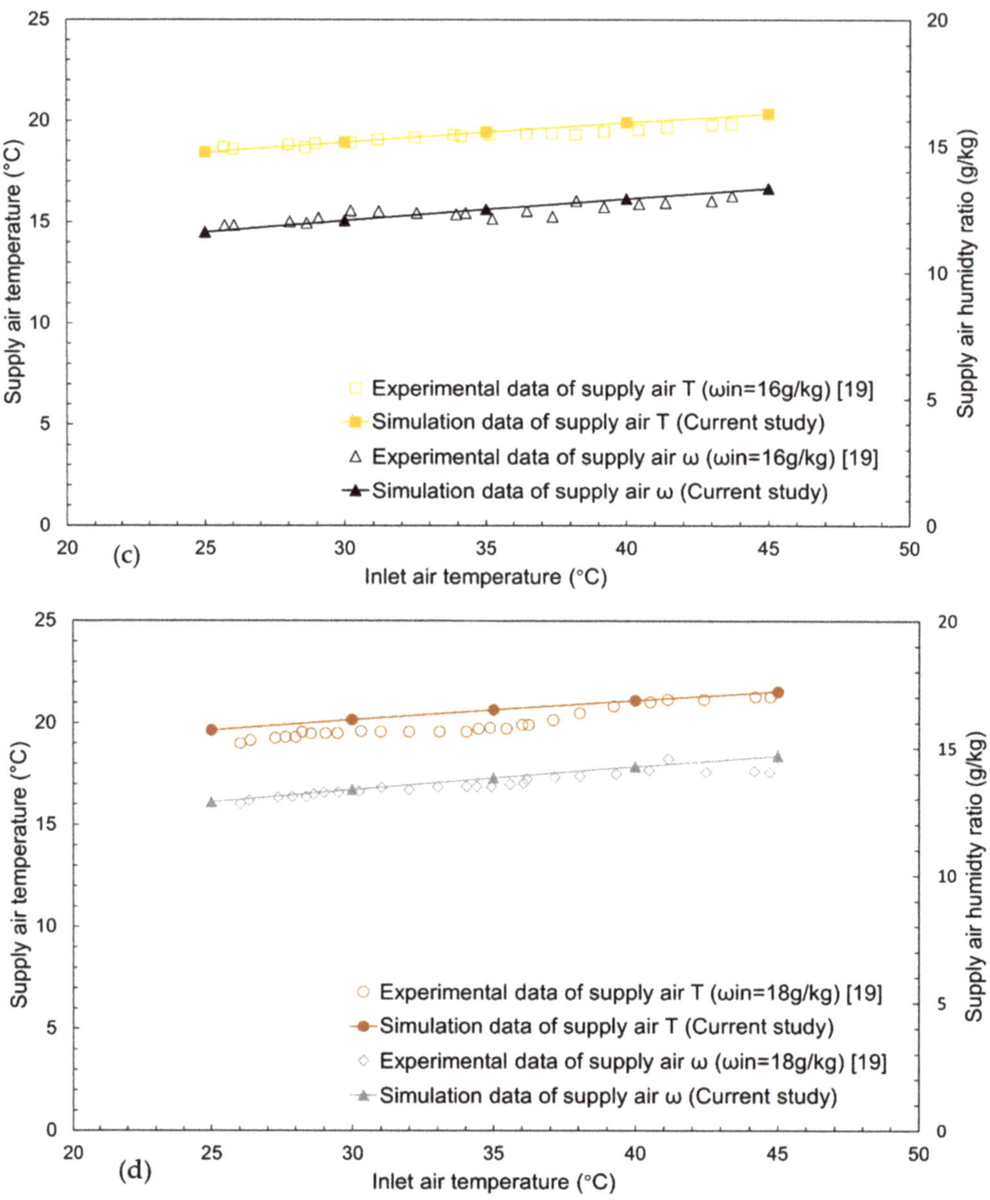

Figure 3. SDM model verification with experimental data from Ref. [18] at four different inlet air humidity levels: (**a**) ω_{in} = 12 g/kg; (**b**) ω_{in} = 14 g/kg; (**c**) ω_{in} = 16 g/kg; (**d**) ω_{in} = 18 g/kg.

5.2. HDH Model Verification

The model of HDH is verified by comparing the gain–output ratio (GOR) with the published data from reference [34]. Figure 4 shows the variation in the GOR between the current work and the published data at three different heating source levels, ranging from 60 °C to 80 °C, and various seawater-to-airflow rate ratios. It can be observed that the maximum discrepancy between these two studies is 4.7%, which demonstrates that the developed model can be used to predict HDH performance.

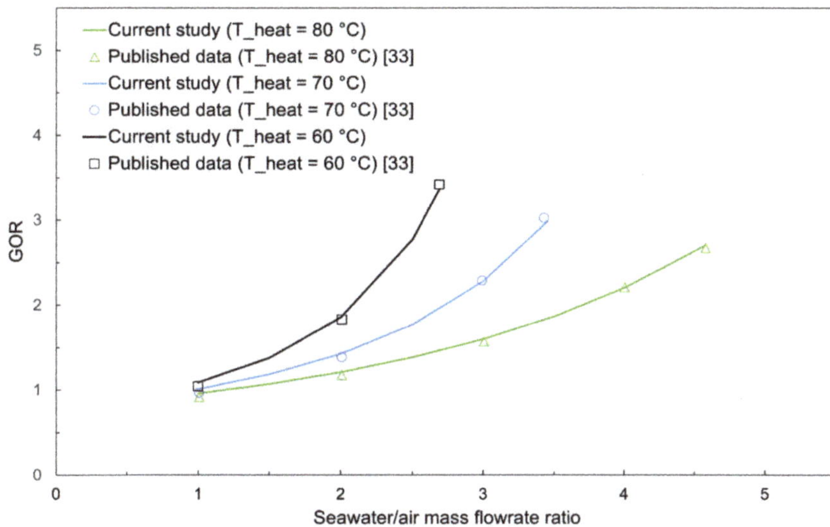

Figure 4. HDH model verification with published data [34].

6. Results and Discussions

In this study, the M-cycle cooling-based system combined with an HDH desalination unit was numerically evaluated and compared among the ventilation, recirculation and half-recirculation modes in terms of cooling performance, freshwater production rate, and overall system COP under different working conditions, using TRNSYS and EES. The proposed system integrates the SDM with the HDH unit, and the process air of the HDH component is a mixture of the exhaust of the desiccant wheel and M-cycle. The initial operating parameter settings are given in Table 2.

Table 2. Initial working conditions of the model.

Key Parameters	Value
Outdoor temperature	35 °C
Outdoor humidity ratio	50%
Process airflow rate	660 kg/h
Regeneration airflow rate	660 kg/h
Desiccant wheel regeneration temperature	70 °C
HDH unit heat source temperature	55 °C
Indoor temperature	25 °C
Indoor humidity ratio	55%

6.1. Cooling Performance Analysis

6.1.1. Comparison of Air Cooling Conditions under Various Ambient Temperatures

Figure 5 shows the comparison of product air states under different ambient air temperatures. It was observed that the SDM-HDH operating in the recirculation modes generated a supply of air that was cooler and drier than that in the ventilation mode. This was due to the return air emanating from the conditioned room, which tended to have a lower temperature and humidity than the ambient air. As presented in Figure 5a, the product air temperature of the system was comparable among the three operational modes, with a value of around 12 °C, while the temperature of the outdoor air was 30 °C. As the process air temperature increased from 30 to 45 °C, there was a corresponding increase in the product air temperature. Specifically, in the ventilation mode, the supply air temperature rose to 26.68 °C, while in the recirculation and half-recirculation modes, it increased to 20.85 °C and 24.10 °C. In Figure 5b, it can be seen that the humidity of the product air surpassed the threshold of 20 g/kg, while the ambient air temperature surpassed 45 °C in the ventilation mode. Conversely, the recirculation mode produced a supply air stream that exhibited a humidity range of 4.457 to 12.72 g/kg, representing a dehumidification efficacy of up to 41.1% relative to the ventilation mode. These results indicate that the ventilation mode might not have been able to provide sufficient cooling at the high ambient air temperature, and recirculation was required to provide the cooling based on the ASHRAE cooling requirement.

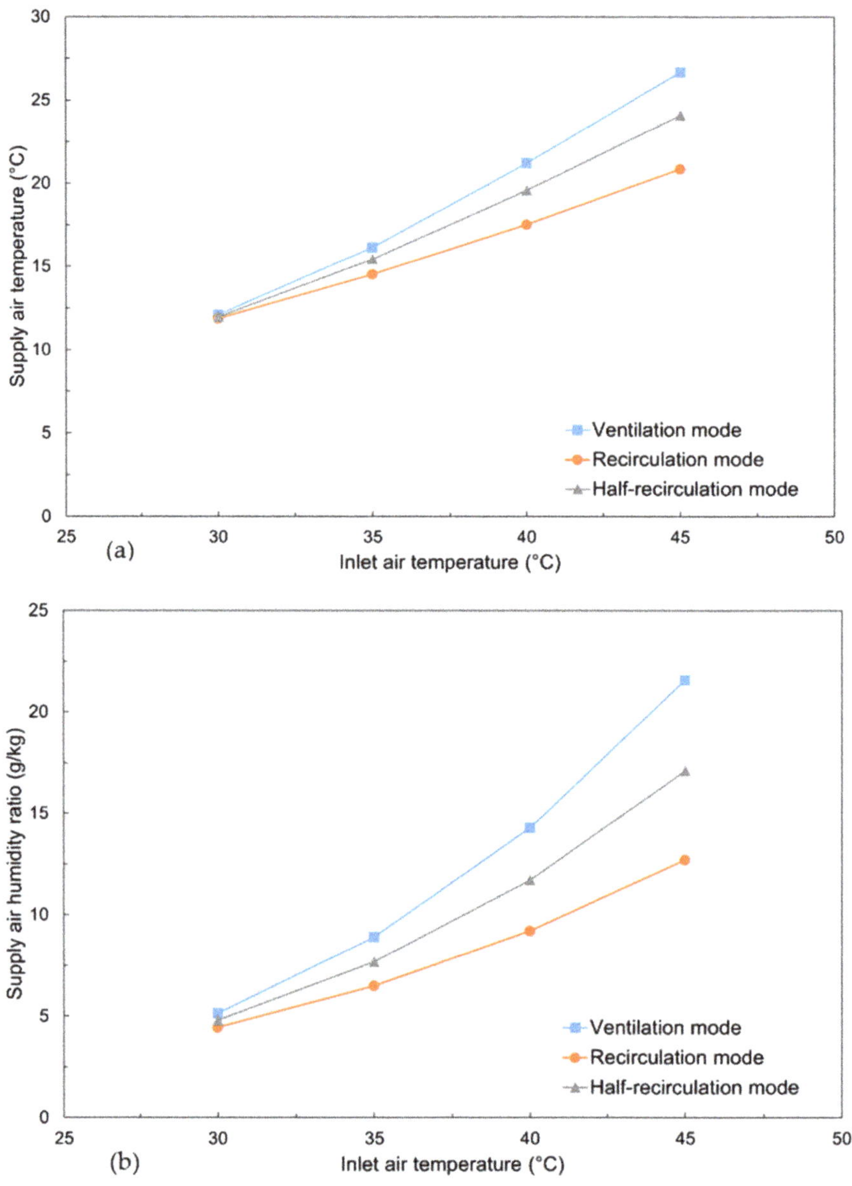

Figure 5. Effect of ambient temperature on supply air (**a**) temperature and (**b**) humidity.

6.1.2. Comparison of Product Air Conditions under Various Ambient Humidity Levels

The comparison of the product air states under different ambient air humidity levels is presented in Figure 6. It is shown that an increase in the ambient humidity ratio led to an increment in the product air temperature and humidity of all three operational modes. Figure 6a shows that, as the outdoor air humidity increased to 70%, the product air temperature of the system in the ventilation mode reached 20.99 °C. This can be attributed to the fact that the cooling efficiency of evaporative coolers is reduced in humid environments due to reduced evaporation. Similarly, the product air temperature of the system in the recirculation and half-recirculation modes also increased from 13.09 to 17.35 °C and 13.51 to 19.38 °C, respectively, as the outdoor humidity changed from 40 to 70%. As depicted in Figure 6b, the SDM system exhibited a robust latent cooling capability in all the operational modes, with the maximum humidity of the product air consistently maintained beneath 14.5 g/kg under conditions of an ambient humidity ratio of 70%. Furthermore, the supply air humidity in the system's recirculation mode was 22.3% to 30.3% lower than in the ventilation mode due to the inclusion of air from the conditioned room. These results further demonstrate that the recirculation mode is important to provide sufficient cooling for highly humid regions.

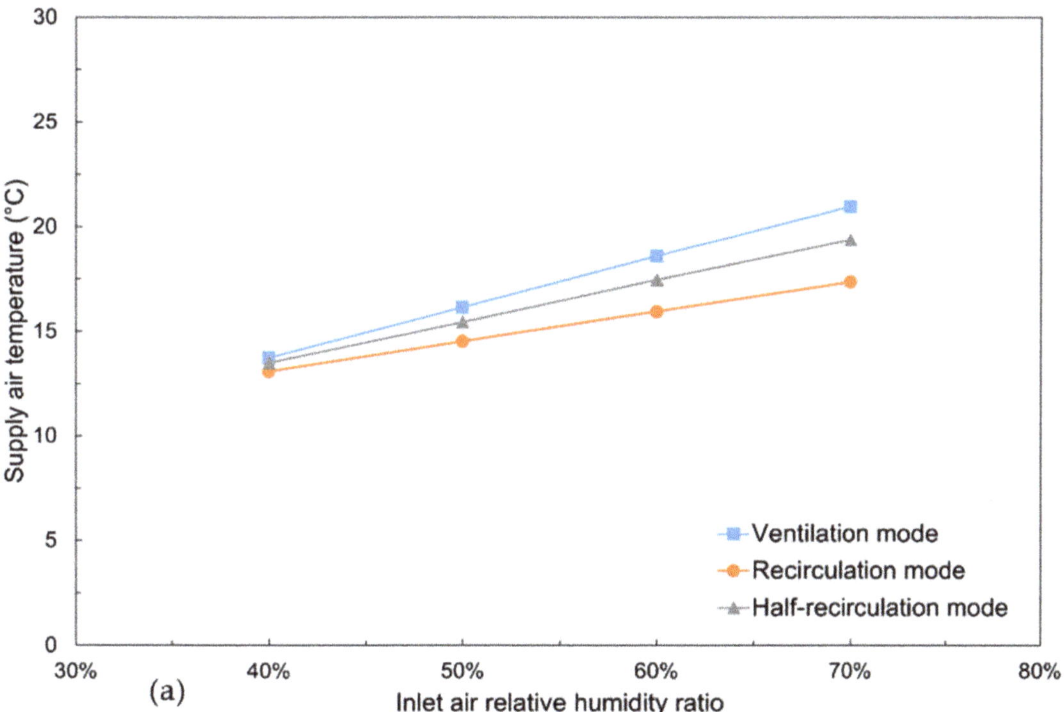

Figure 6. *Cont.*

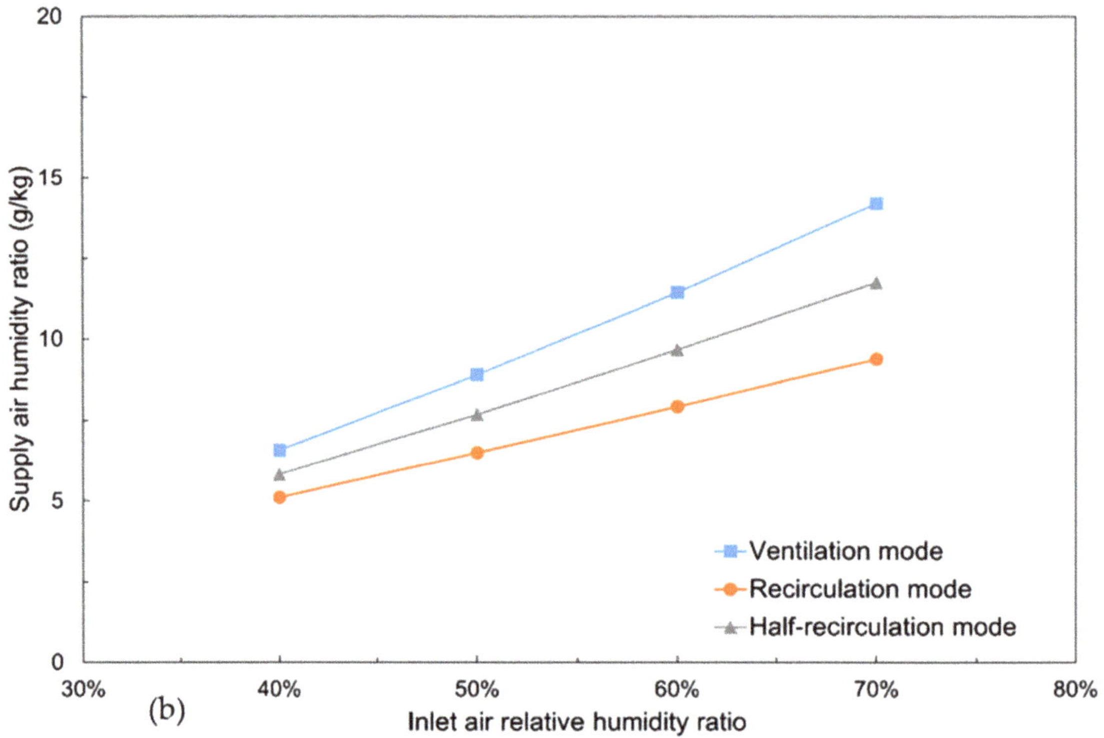

Figure 6. Effects of the ambient humidity on supply air (**a**) temperature and (**b**) humidity.

6.1.3. Comparison of Cooling Capacity under Various Ambient Conditions

In Figure 7, a comprehensive representation of the fluctuations in the cooling capacity of the proposed system, as it functions in three distinct modes, is presented in correlation to variations in the outdoor air conditions. The results reveal that the incorporation of the return air with the ambient air as the new supply air stream significantly enhanced the system's cooling performance. Figure 7a illustrates that the SDM in the ventilation mode generated 4.31 kW of cooling capacity at an ambient temperature of 30 °C, which slightly increased by 10.4% to reach 4.76 kW when the outdoor temperature reached 45 °C. However, when the system operated in the recirculation and half-recirculation modes, the cooling capacity exhibited a marked increase from 4.51 to 7.91 kW and 4.39 to 6.30 kW, respectively, within the same temperature range. The trends depicted in Figure 7b for various humidity levels demonstrate that the cooling capacity for all the modes increased as the environmental relative humidity ratio rose. As the environmental relative humidity increased from 40% to 70%, the cooling capacity for the ventilation, recirculation, and half-recirculation modes increased correspondingly, rising from 4.52 to 4.68 kW, 4.99 to 6.44 kW, and 4.38 to 5.55 kW, respectively. These results show that the system size can be reduced if the recirculation mode is used to provide cooling.

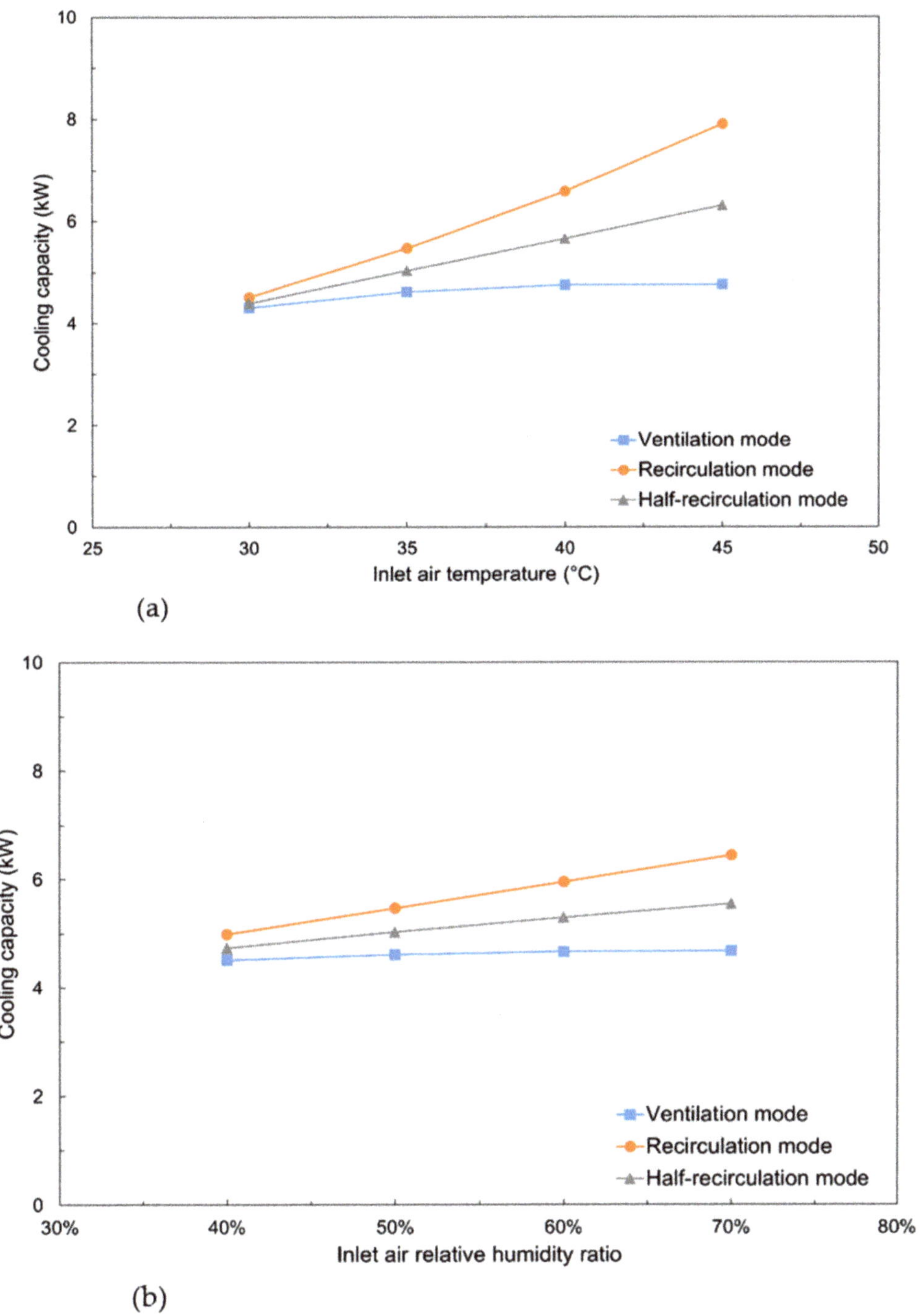

Figure 7. Effects of ambient air (**a**) temperature and (**b**) humidity on cooling capacity.

6.2. Water Production Analysis

6.2.1. Comparison of Water Usage Rate under Various Ambient Conditions

The SDM is a cooling system that relies on the water evaporation phenomenon to produce a cooling effect. In this case, the water consumption rate is also an important factor to consider in the operation and investigation of the system. Figure 8a illustrates that as the inlet air temperature increased, the water consumption rate in the ventilation mode decreased. This is because the desiccant wheel, which is a key component of the SDM, exhibited reduced dehumidification performance in hot and humid conditions, resulting in a reduced ability to evaporate water. In contrast, when the recirculation or the half-recirculation mode is implemented, the temperature and humidity ratio of the mixed air stream can be effectively reduced, which leads to a growth in the water usage rate. The highest water consumption rate was 5.52 kg/h when the system was operated in full recirculation mode at an ambient air temperature of 45 °C. The data depicted in Figure 8b demonstrate that as the humidity of the outdoor air increased, all the operational modes of the SDM exhibited a decrease in water consumption. Specifically, the freshwater usage rate dropped from 3.91 to 2.87 kg/h in the ventilation mode, from 4.99 to 4.25 kg/h in the recirculation mode, and from 4.47 to 3.58 kg/h in the half-recirculation mode. This reduction in water consumption can be attributed to the inhibitory effect of high humidity on the evaporation of water. It is clear that although the cooling capacity of the recirculation mode was higher, the water consumption rate was also much higher.

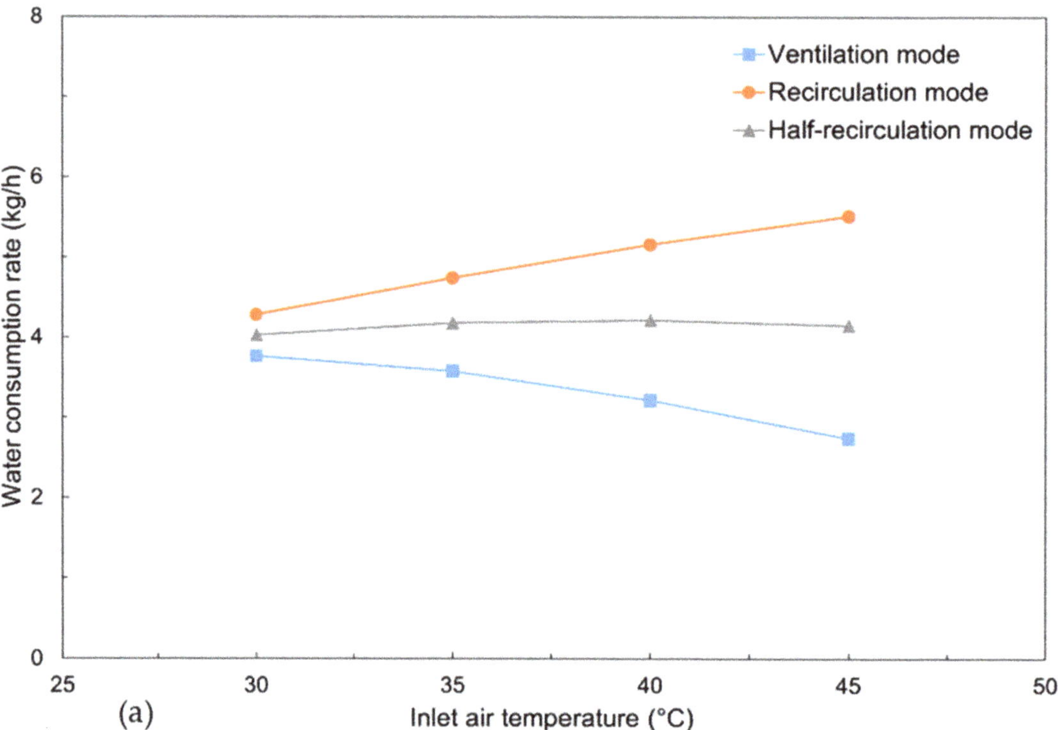

Figure 8. *Cont.*

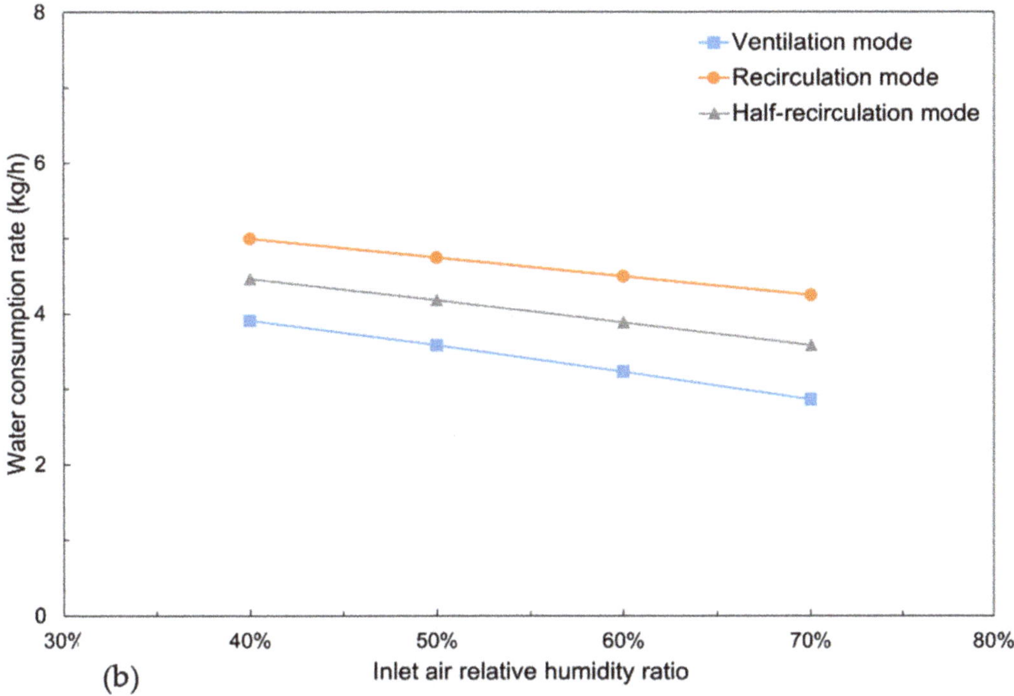

Figure 8. Effects of outdoor air (**a**) temperature and (**b**) humidity on water consumption rate.

6.2.2. Comparison of Water Production under Various Ambient Conditions

The comparative analysis depicted in Figure 9 reveals the impact of outdoor air conditions on the water productivity of the proposed system in different operational modes. The results indicate a relatively stable water productivity, with a minor increase observed as the ambient air became warmer and more humid. Figure 9a shows that the recirculation mode had the highest mean water productivity at 52.74 kg/h, followed by the half-recirculation mode with a mean of 52.43 kg/h, and the ventilation mode with the lowest mean of 52.14 kg/h. With the outdoor air temperature climbing to 45 °C, the water production rate for the recirculation mode slightly increased to 54.03 kg/h, representing a 1.12 kg/h increment compared to the ventilation mode. Figure 9b shows that the recirculation mode yielded a higher quantity of water at various humidity levels in comparison to the other two modes. For instance, when the outdoor humidity ratio reached 70%, the water production rates for the recirculation, half-recirculation, and ventilation modes were 53.06 kg/h, 52.68 kg/h, and 52.3 kg/h, respectively. These results show that the increase in water production was very small in the recirculation model. As presented in the water consumption section, the water consumption in the recirculation mode was much higher than in the ventilation mode. Therefore, it was important to evaluate the net water production in these three operation modes. As shown in Figure 9, upon considering the water consumption of each mode, it is apparent that the net water productivity of the ventilation mode surpassed that of the two recirculation modes at various environmental conditions, with a maximum difference of 1.65 kg/h at the outdoor air temperature of 45 °C. This is because the recirculation modes consumed more water compared to the ventilation mode under the same conditions, as demonstrated in Figure 8. These findings indicate that, although the two recirculation modes exhibited a slight advantage in the generation of freshwater over the ventilation mode, this advantage was offset by the higher water usage rate in the recirculation modes.

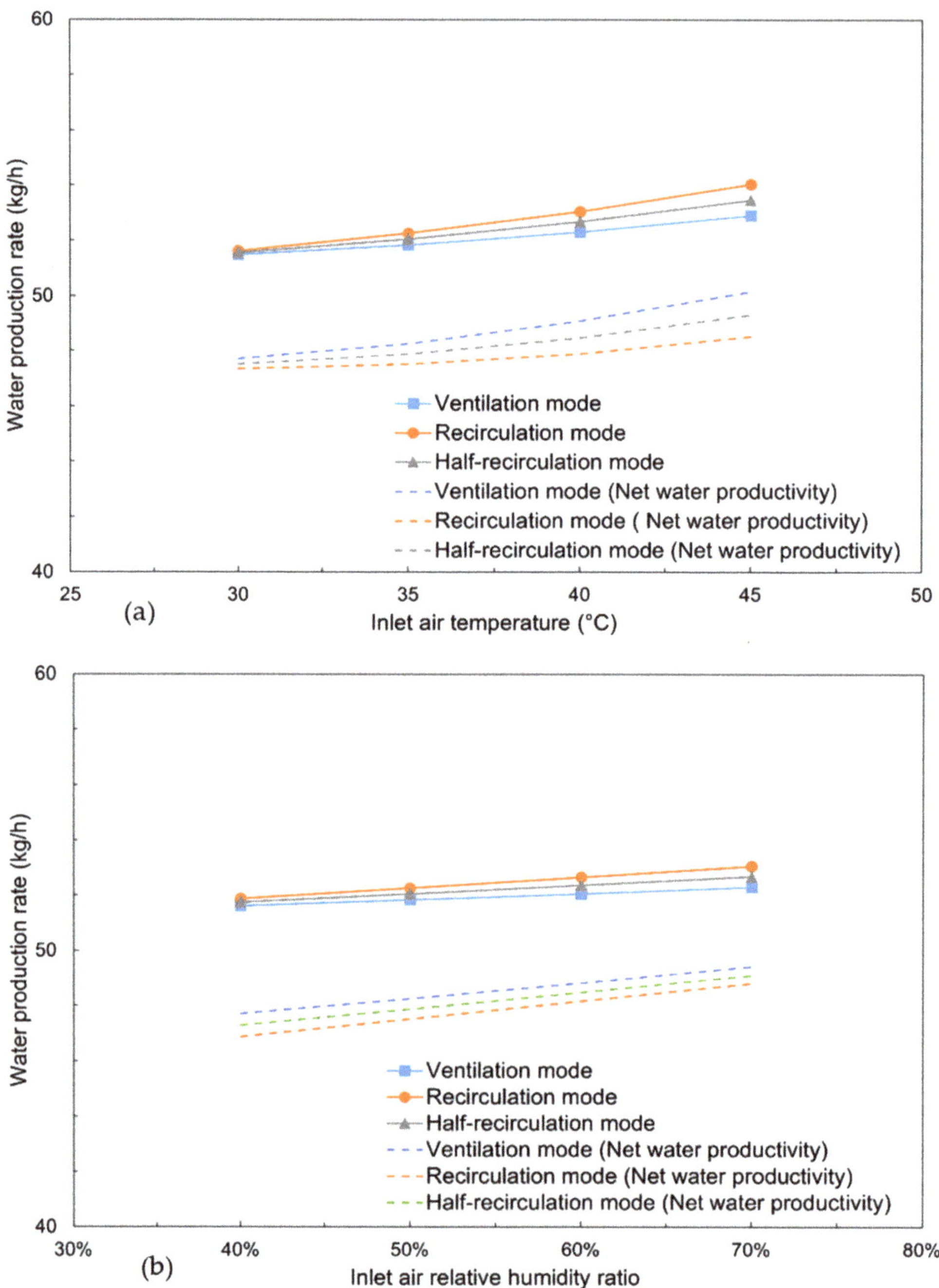

Figure 9. Effects of outdoor air (**a**) temperature and (**b**) humidity on water productivity.

6.2.3. Comparison of Water Production Rate at Various Heat Source Temperatures

The heat source temperature of the HDH unit plays a critical role in the system's water productivity, and it was therefore essential to consider its impact when assessing the overall performance of the whole system. As demonstrated in Figure 10, an increase in the heat source temperature led to an augmented yield of distilled water. The water production rate for the ventilation, recirculation, and half-recirculation modes increased from 51.6 to 148.5 kg/h, 52.3 to 149.1 kg/h, and 52.05 to 148.9 kg/h, respectively, when the temperature of the heat source increased from 55 to 70 °C. This can be attributed to the fact that a higher heat source temperature enhanced the brine water evaporation potential within the humidifier of the HDH desalination unit. It is worth noting, however, that an increase in the heat source temperature also necessitated additional energy input for the purpose of heating the brine water. Additionally, it is evident that there was no discernible difference between the water production rate of the three operational modes at different heat source temperatures given that the inlet air conditions for the HDH unit were comparable at the selected working conditions (inlet air temperature of 35 °C and inlet air humidity ratio of 50%).

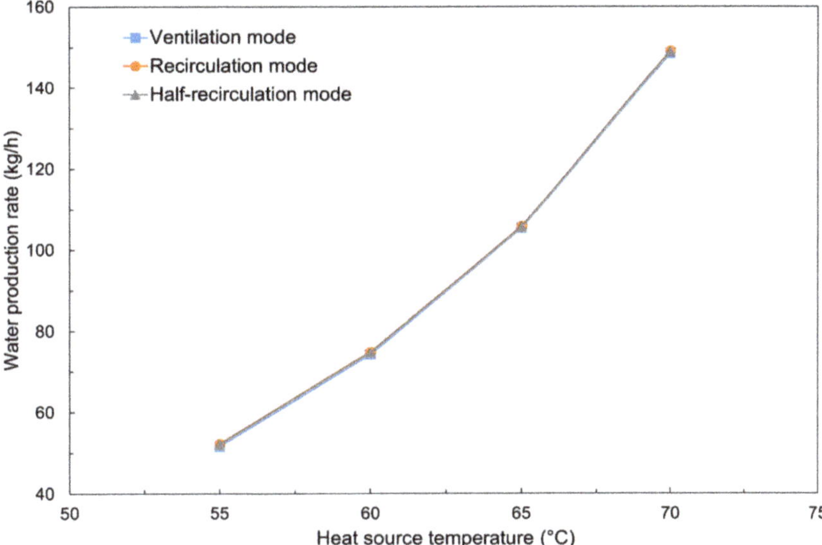

Figure 10. Effect of heat source temperature on water production rate.

6.3. Comparison of Overall System Performance Analysis (COP)

Figure 11 shows the change in the COP of the analysed system with respect to the various outdoor air conditions. It can be seen that the COP of all three modes increased slightly as the inlet air temperature and humidity ratio grew. This is because the cooling capacity and water productivity increased as the outdoor air became hot and humid for the three operational modes. It is imperative to observe that the COP for the recirculation and half-recirculation mode surpassed the ventilation mode when the inlet air temperature exceeded 35 °C and the inlet air humidity ratio was above 50%. However, the difference was very small. The maximum COP reached 0.44 at the inlet air temperature of 45 °C in the recirculation mode. This was mainly due to the more rapid growth in its cooling capacity in the recirculation modes in warmer and more humid conditions compared to the ventilation mode.

When considering the cooling capacity, water production, and system COP, the recirculation operating mode showed a slightly superior performance. This operating mode can provide the cooling requirement for users with a better system COP even at high ambient temperatures and relative humidity levels. Meanwhile, the net water production rate did not show much difference with the ventilation mode.

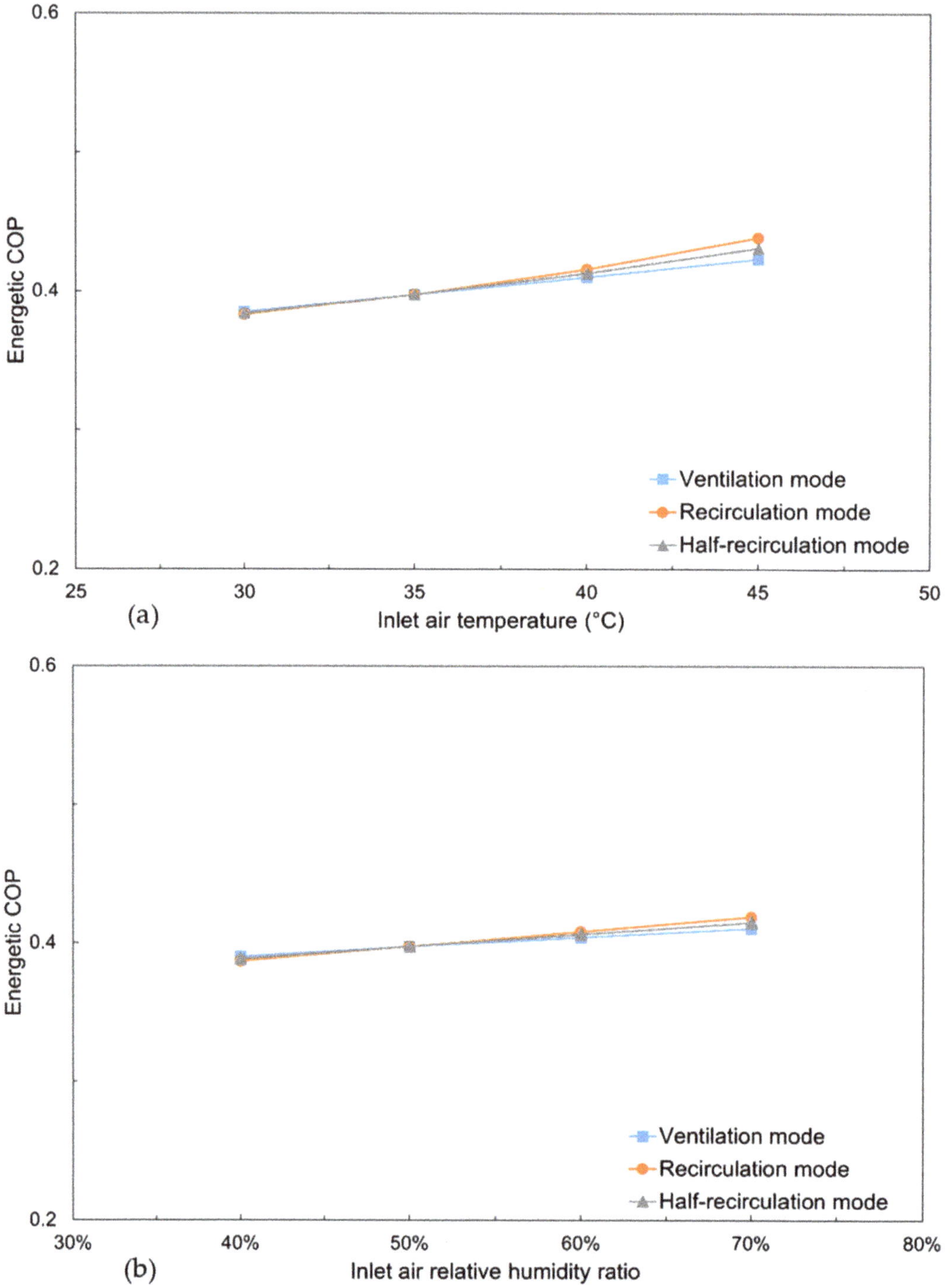

Figure 11. Effects of (**a**) ambient temperature and (**b**) humidity ratio on energetic COP.

7. Conclusions and Future Work

In the present study, a solid desiccant-based M-cycle cooling system (SDM) integrated with a humidification–dehumidification (HDH) unit was investigated under different working conditions through numerical simulation. To gain a deep understanding of the system, three operational modes were compared and assessed with respect to the cooling efficiency and water generation performance: ventilation, recirculation, and half-recirculation. The main findings of the study can be summarised as follows:

(1) The inclusion of the return air from a conditioned space in the recirculation and half-recirculation modes resulted in a superior cooling performance compared to the ventilation mode. In the recirculation mode, the system was able to generate up to 7.91 kW of cooling load and maintain the supply air temperature and humidity beneath 20.85 °C and 12.72 g/kg under various ambient conditions;

(2) The SDM-HDH system could effectively utilise the exhaust air from the solid desiccant M-cycle cooling system to generate water. All three operational modes exhibited an excellent water production rate, with average values of 52.74 kg/h, 52.43 kg/h, and 52.14 kg/h for the recirculation, half-recirculation, and ventilation modes, respectively, across a range of operating temperatures. In addition, adjusting the heat source temperature of the HDH component from 55 to 70 °C resulted in an increase in the water productivity of all the operational modes, with values ranging from 51.56 to 148.50 kg/h, 52.05 to 148.9 kg/h, and 52.26 to 149.1 kg/h for the ventilation, half-recirculation, and recirculation modes, respectively;

(3) It should be noted that the implementation of the recirculation mode resulted in the highest water consumption rate, with a value of 5.52 kg/h when the inlet air reached 45 °C, which partially offset the benefits of this mode. However, all three operating modes could provide net water production of up to 48 kg/h after considering the M-cycle water consumption;

(4) The overall energetic COP of all the operational modes increased slightly as the inlet air temperature and humidity ratio increased. The recirculation mode demonstrated a slightly better COP compared to the other two modes at ambient temperatures above 35 °C and the ambient humidity ratio above 50%. The maximum value of COP in the recirculation mode was 0.44 at the ambient temperature of 45 °C;

(5) In summary, this study has shown that the proposed system has good potential for producing cooling and water simultaneously. However, this study was based on the numerical method, and further experimental research and related economic analysis of the system should be conducted to understand the system better and evaluate the payback period. Moreover, future research could also focus on model improvements based on second-law analysis.

Author Contributions: Conceptualization, L.L. and X.W.; supervision, X.W., G.K. and E.H.; writing—L.L.; writing—review and editing, X.W., G.K. and E.H. All authors have read and agreed to the published version of the manuscript.

Funding: This research received no external funding.

Data Availability Statement: Data will be available on request.

Conflicts of Interest: The authors declare no conflict of interest.

Nomenclature

heat exchange area, m^2	A
coefficient of performance	COP
specific heat at constant pressure, kJ/kg·K	c_p
dew point temperature	DPT
potential functions 1 and 2 for desiccant wheel	F_1, F_2
enthalpy of moist air, J/kg	h
coefficient of heat transfer, W/m^2·K	h_c
humidification–dehumidification	HDH
mass transfer coefficient, m/s	h_m
latent enthalpy of vaporisation of water, J/kg	L
mass flow rate, kg/s	$\dot{m}$
heat transfer rate, kW	$\dot{Q}$
solid desiccant-based M-cycle cooling	SDM
temperature, °C	T
wet-bulb temperature	WBT
Greek Letters	
effectiveness	ε
humidity ratio of moist air, g/kg	ω
density, kg/m^3	ρ
Subscripts	
concentrated brine	cb
M-cycle's dry channels	dry
distilled water	dw
dehumidifier	deh
feed brine	fb
humidifier	hum
regeneration process	reg
sensible heat	sens
M-cycle's wet channels	wet
water film on the wet channel surface	wf

References

1. Mohammed, R.H.; Rezk, A.; Askalany, A.; Ali, E.; Zohir, A.; Sultan, M.; Ghazy, M.; Abdelkareem, M.; Olabi, A. Metal-organic frameworks in cooling and water desalination: Synthesis and application. *Renew. Sustain. Energy Rev.* **2021**, *149*, 111362. [CrossRef]
2. IEA. "The Future of Cooling," Paris. 2018. Available online: https://www.iea.org/reports/the-future-of-cooling (accessed on 1 September 2022).
3. Rocchetti, A.; Socci, L. Theoretical and numerical assessment of an enhanced Humidification-Dehumidification desalination system based on Indirect Evaporative cooling and Vapour Compression Refrigeration. *Appl. Therm. Eng.* **2022**, *208*, 118194. [CrossRef]
4. Abdelkareem, M.A.; El Haj Assad, M.; Sayed, E.; Soudan, B. Recent progress in the use of renewable energy sources to power water desalination plants. *Desalination* **2018**, *435*, 97–113. [CrossRef]
5. Sajjad, U.; Abbas, N.; Hamid, K.; Abbas, S.; Hussain, I.; Ammar, S.M.; Sultan, M.; Ali, H.M.; Hussain, M.; Rehman, T.U.; et al. A review of recent advances in indirect evaporative cooling technology. *Int. Commun. Heat Mass Transf.* **2021**, *122*, 105140. [CrossRef]
6. Shahzad, M.W.; Lin, J.; Xu, B.; Dala, L.; Chen, Q.; Burhan, M.; Sultan, M.; Worek, W.; Ng, K.C. A spatiotemporal indirect evaporative cooler enabled by transiently interceding water mist. *Energy* **2021**, *217*, 119352. [CrossRef]
7. Yang, H.; Shi, W.; Chen, Y.; Min, Y. Research development of indirect evaporative cooling technology: An updated review. *Renew. Sustain. Energy Rev.* **2021**, *145*, 111082. [CrossRef]
8. Chu, J.; Xu, W.; Fu, Y.; Huo, H. Experimental research on the cooling performance of a new regenerative dew point indirect evaporative cooler. *J. Build. Eng.* **2021**, *43*, 102921. [CrossRef]
9. Shahzad, M.W.; Burhan, M.; Ybyraiymkul, D.; Oh, S.; Ng, K.C. An improved indirect evaporative cooler experimental investigation. *Appl. Energy* **2019**, *256*, 113934. [CrossRef]

10. Mahmood, M.H.; Sultan, M.; Miyazaki, T.; Koyama, S.; Maisotsenko, V.S. Overview of the Maisotsenko cycle—A way towards dew point evaporative cooling. *Renew. Sustain. Energy Rev.* **2016**, *66*, 537–555. [CrossRef]

11. Baakeem, S.S.; Orfi, J.; Mohamad, A.; Bawazeer, S. The possibility of using a novel dew point air cooling system (M-Cycle) for A/C application in Arab Gulf Countries. *Build. Environ.* **2019**, *148*, 185–197. [CrossRef]

12. Lai, L.; Wang, X.; Kefayati, G.; Hu, E. Evaporative Cooling Integrated with Solid Desiccant Systems: A Review. *Energies* **2021**, *14*, 5982. [CrossRef]

13. Chaudhary, G.Q.; Ali, M.; Sheikh, N.A.; Gilani, S.I.U.H.; Khushnood, S. Integration of solar assisted solid desiccant cooling system with an efficient evaporative cooling technique for separate load handling. *Appl. Therm. Eng.* **2018**, *140*, 696–706. [CrossRef]

14. Goldsworthy, M.; White, S. Optimisation of a desiccant cooling system design with an indirect evaporative cooler. *Int. J. Refrig.* **2011**, *34*, 148–158. [CrossRef]

15. Gao, W.; Worek, W.; Konduru, V.; Adensin, K. Numerical study on performance of a desiccant cooling system with an indirect evaporative cooler. *Energy Build.* **2015**, *86*, 16–24. [CrossRef]

16. Gadalla, M.; Saghafifar, M. Performance assessment and transient optimisation of air precooling in multi-stage solid desiccant air conditioning systems. *Energy Convers. Manag.* **2016**, *119*, 187–202. [CrossRef]

17. Lin, J.; Wang, R.; Kumja, M.; Bui, T.; Chua, K.J. Modelling and experimental investigation of the cross-flow dew point evaporative cooler with and without dehumidification. *Appl. Therm. Eng.* **2017**, *121*, 1–13. [CrossRef]

18. Shahzad, M.K.; Chaudhary, G.Q.; Ali, M.; Sheikh, N.A.; Khalil, M.S.; Rashid, T.U. Experimental evaluation of a solid desiccant system integrated with cross flow Maisotsenko cycle evaporative cooler. *Appl. Therm. Eng.* **2018**, *128*, 1476–1487. [CrossRef]

19. Pandelidis, D.; Pacak, A.; Cichoń, A.; Drąg, P.; Worek, W.; Cetin, S. Numerical and experimental analysis of precooled desiccant system. *Appl. Therm. Eng.* **2020**, *181*, 115929. [CrossRef]

20. Delfani, S.; Karami, M. Transient simulation of solar desiccant/M-Cycle cooling systems in three different climatic conditions. *J. Build. Eng.* **2020**, *29*, 101152. [CrossRef]

21. Harrouz, J.P.; Katramiz, E.; Ghali, K.; Ouahrani, D.; Ghaddar, N. Comparative analysis of sustainable desiccant—Evaporative based ventilation systems for a typical Qatari poultry house. *Energy Convers. Manag.* **2021**, *245*, 114556. [CrossRef]

22. Lai, L.; Wang, X.; Kefayati, G.; Hu, E. Performance evaluation of a solar powered solid desiccant evaporative cooling system with different recirculation air ratios. *Energy Build.* **2022**, *270*, 112273. [CrossRef]

23. Xu, H.; Jiang, S.; Xie, M.; Jia, T.; Dai, Y.J. Technical improvements and perspectives on humidification-dehumidification desalination—A review. *Desalination* **2022**, *541*, 116029. [CrossRef]

24. Wang, N.; Wang, D.; Dong, J.; Wang, H.; Wang, R.; Shao, L.; Zhu, Y. Performance assessment of PCM-based solar energy assisted desiccant air conditioning system combined with a humidification-dehumidification desalination unit. *Desalination* **2020**, *496*, 114705. [CrossRef]

25. Kabeel, A.E.; Abdelgaied, M.; Feddaoui, M.B. Hybrid system of an indirect evaporative air cooler and HDH desalination system assisted by solar energy for remote areas. *Desalination* **2018**, *439*, 162–167. [CrossRef]

26. Chen, Q.; Burhan, M.; Shahzad, M.; Ybyraiymkul, D.; Akhtar, F.; Ng, K.C. Simultaneous production of cooling and freshwater by an integrated indirect evaporative cooling and humidification-dehumidification desalination cycle. *Energy Convers. Manag.* **2020**, *221*, 113169. [CrossRef]

27. Abdelgaied, M.; Kabeel, A.; Ezat, A.; Dawood, M.K.; Nabil, T. Performance improvement of the hybrid indirect evaporative type air cooler and HDH desalination system using shell and tube latent heat energy storage tank. *Process. Saf. Environ. Prot.* **2022**, *168*, 800–809. [CrossRef]

28. Kabeel, A.E.; Abdelgaied, M.; Zakaria, Y. Performance evaluation of a solar energy assisted hybrid desiccant air conditioner integrated with HDH desalination system. *Energy Convers. Manag.* **2017**, *150*, 382–391. [CrossRef]

29. Lai, L.; Wang, X.; Kefayati, G.; Hu, E. Analysis of a novel solid desiccant evaporative cooling system integrated with a humidification-dehumidification desalination unit. *Desalination* **2023**, *550*, 116394. [CrossRef]

30. Klein, S.A.; Beckman, W.A. *TRNSYS 18: A Transient System Simulation Program*; Solar Energy Laboratory, University of Wisconsin: Madison, WI, USA, 2017; Available online: http://sel.me.wisc.edu/trnsys (accessed on 1 September 2022).

31. Fong, K.F.; Lee, C.K. Solar desiccant cooling system for hot and humid region—A new perspective and investigation. *Solar Energy* **2020**, *195*, 677–684. [CrossRef]

32. El Loubani, M.; Ghaddar, N.; Ghali, K.; Itani, M. Hybrid cooling system integrating PCM-desiccant dehumidification and personal evaporative cooling for hot and humid climates. *J. Build. Eng.* **2020**, *33*, 101580. [CrossRef]

33. Zheng, X.; Ge, T.; Wang, R.Z. Recent progress on desiccant materials for solid desiccant cooling systems. *Energy* **2014**, *74*, 280–294. [CrossRef]

34. Narayan, G.P.; Sharqawy, M.; Lienhard, V.J.; Zubair, S.M. Thermodynamic analysis of humidification dehumidification desalination cycles. *Desalination Water Treat.* **2010**, *16*, 339–353. [CrossRef]

35. Sadeghi, M.; Yari, M.; Mahmoudi, S.; Jafari, M. Thermodynamic analysis and optimisation of a novel combined power and ejector refrigeration cycle—Desalination system. *Appl. Energy* **2017**, *208*, 239–251. [CrossRef]
36. Narayan, G.P.; McGovern, R.; Zubair, S.; Lienhard, J.H. High-temperature-steam-driven, varied-pressure, humidification-dehumidification system coupled with reverse osmosis for energy-efficient seawater desalination. *Energy* **2012**, *37*, 482–493. [CrossRef]

Article

Data Analysis and Optimization of Thermal Environment in Underground Commercial Building in Zhengzhou, China

Xi Zhao [1], Cheng Li [2], Jiayin Zhu [2,*], Yu Chen [2] and Jifu Lu [2]

[1] College of Architecture, Texas A&M University, College Station, TX 77843, USA
[2] School of Water Conservancy and Civil Engineering, Zhengzhou University, Zhengzhou 450001, China
* Correspondence: zhujiayin1234@163.com

Abstract: Underground commercial buildings have received increasing attention as an emerging place of consumption. However, previous studies on underground commercial buildings have mainly focused on the impact of a specific environment on comfort or energy consumption. Few studies have been conducted from the perspective of functional use. The purpose of this paper is to investigate, in terms of functional angles, the indoor thermal environment and air quality of an underground commercial building in Zhengzhou, China, and put forward an optimal control strategy of ventilation organization. The results showed that the relative humidity of the underground shopping mall was generally above 60%, and the average temperature of 29.1 °C led to a thermal comfort problem in the catering area in summer. Meanwhile, the concentration of CO_2 exceeded the allowed figures during the peak of the customer flow rate, and $PM_{2.5}$ concentration in the catering area also exceeded the standard, by 43.3% and 33.3%, respectively. Furthermore, to solve the indoor thermal environment and air quality problems found in the field measurements, this study assessed the air distribution by adopting three different air supply schemes for the catering area. Optimization results showed that compared with the ceiling supply, the side supply scheme kept the air temperature 0.4 °C cooler in summer and 0.5 °C warmer in winter. The temperature uniformity increased by 5.4% and 3.7%, and the velocity uniformity increased by 6.5% and 8.8%, respectively. This study can provide theoretical support for thermal environment construction and ventilation organization control of underground commercial buildings.

Keywords: underground commercial buildings; thermal environment; indoor air quality; field measurement; air distribution optimization

Citation: Zhao, X.; Li, C.; Zhu, J.; Chen, Y.; Lu, J. Data Analysis and Optimization of Thermal Environment in Underground Commercial Building in Zhengzhou, China. *Processes* **2022**, *10*, 2584. https://doi.org/10.3390/pr10122584

Academic Editors: Iztok Golobič and Titan C. Paul

Received: 14 October 2022
Accepted: 22 November 2022
Published: 4 December 2022

Publisher's Note: MDPI stays neutral with regard to jurisdictional claims in published maps and institutional affiliations.

1. Introduction

The global economy continues to grow steadily, and at a high rate. Modern cities have become the most concentrated areas of human activity, growing at an unprecedented rate and scale in emerging regions of the world, of which China is a representative country [1]. China's urban population has increased by 298.5 million in the last forty years [2], and as the urban population expands, a large number of urban issues have emerged, such as land constraints [3], high energy consumption [4], ecosystem imbalance [5], urban heat island effect [6,7], environmental degradation [8], and poor air quality [9,10]. According to British Petroleum (BP) Statistical Review of World Energy, China's carbon emissions reached 9899.3 million tons in 2020, accounting for 30.7% of the world's total [11]. In this context, urban settlement environments are facing multiple challenges, forcing people to pursue the ultimate use of land resources [12,13]. The comprehensive use of urban underground space is an important measure to solve the three major crises of urban population, resources and environment [14].

Efficient and integrated use of the underground can contribute to several sustainable development goals such as saving land resources [15], reducing energy consumption [16], alleviating traffic congestion [17] and improving air quality [18], thus, the exploration and utilization of underground space is at a stage of rapid development. Data released by the Strategic Consulting Center of Chinese Academy of Engineering showed that the scale of underground space in China increased by 259 million square meters in 2020 [19], of which the commercial use in most developed cities accounted for around 25% [20]. Underground commercial facilities generally have the highest utilization efficiency among underground facilities [21]. By providing the public with commerce, social, catering, entertainment and other functions, underground commercial buildings have become an important component of expanding the capacity of commercial facilities and promoting city vitality. In addition, the combination of commercial spaces with other underground spaces including underground metro systems, underground squares and underground passages, can promote its commercial value [22].

During the past decade, many scholars have intensively investigated the environment of underground spaces from the perspectives of comfort and energy consumption. Some researchers emphasized the importance of the thermal [23], acoustic [24] and light environments [25] of underground space for human comfort and health, and revealed that the thermal sensation of customers in underground commercial buildings is different from above ground. Chun et al. [26] measured the temperatures of underground shopping malls and concluded that compared with department stores, the acceptable temperature range in underground shopping malls was broader. Ganesh analyzed the effect of the inlet diffuser orientation on the indoor thermal comfort of occupants, and found that indoor air temperature affected PMV and PPD more than air movement [27]. Li et al. [28] conducted measurements and questionnaire surveys to investigate the thermal environments of six underground shopping malls in Nanjing, China; their results indicated that customers perceived warmth to be less dramatic than the PMV predicted in underground malls. A study of Tehran metro stations showed that passengers' noses, eyes and throats became dry and uncomfortable as the relative humidity in the Tehran metro was below 30% [29].

Inadequate indoor air quality in public facility buildings has become a central issue and has caused widespread public concern [30]. Ventilation in an underground mall is even worse, where total volatile organic compounds, $PM_{2.5}$, formaldehyde, radon gas and other pollutants are more likely to concentrate in an underground space [31], putting occupants at a higher risk for poor indoor air quality and health problems [32–34]. For example, Gao and Jia [35,36] investigated the air quality of underground shopping malls and found that the concentrations of pollutants were clearly above the standard. Tao et al. [37] concluded the correlations between formaldehyde and TVOC concentrations and indoor air velocity, indoor temperature, and indoor relative humidity, through the field measurement of nine underground malls. The correlations noted were that increasing indoor ventilation rates or reducing the indoor temperature and indoor relative humidity appeared to decrease the indoor concentrations of formaldehyde and TVOC. Some studies proposed an optimal design to improve the natural ventilation [38] and light environment [39] from the perspective of saving energy. In an evaluation of ventilation performance of indoor and outdoor space, the CFD model is the most relevant method because of cost-effectiveness, informative technique, and proficiency to predict air velocity patterns and ratios in buildings [40]. In some research, CFD simulations were performed to improve occupant satisfaction [41–43] and indoor environment quality [44,45]. Chen used the CFD model to accurately reflect the changes in kitchen smoke concentration and assess kitchen air pollution, thereby improving the thermal environment [46].

Underground commercial buildings have gradually developed from utilization for a single commercial function, to modern underground complexes. The present studies have mainly focused on the impact of a specific environment on comfort or energy consumption, and few studies have been conducted from the perspective of functional use. Personnel flow and merchandise vary in different functional areas. Therefore, it is imperative to investigate the thermal environment and air quality of underground commercial buildings for different functional areas. Underground spaces mainly rely on air-conditioning and ventilation systems to create satisfying thermal environments, therefore, air distribution has a large influence on the indoor thermal environments of underground shopping malls. Analyzing the effects of air-conditioning design parameters on the building environment via simulations is, thus, an effective method for optimizing air-conditioning design schemes [47]. The problem of using air distribution to create a comfortable thermal environment, especially in underground commercial buildings, is a critical problem that needs to be solved.

This study analyses the thermal performance and air quality of a multifunctional underground commercial building in Zhengzhou, central China, to optimize air distribution. The subject of the study has been in operation for 15 years, and the higher temperatures are associated with older air conditioning equipment. Therefore, based on field measurements, this paper emphasizes the use of CFD simulations to reduce the supplied air temperature to achieve a more comfortable indoor thermal environment, and then discusses the evaluation of different airflow organization methods using average indoor air temperature, air velocity and non-uniform coefficient for the same supplied air temperature operating conditions, in order to obtain an optimal design solution strategy. Finally, this study aims to facilitate the optimization of the thermal environment of multifunctional underground commercial buildings, which will provide theoretical support for the design of airflow organization in underground commercial buildings.

2. Materials and Methods

2.1. Data Source

An underground shopping mall in Zhengzhou, in central China's Henan Province, was selected as a representative test case. The underground commercial building is located in the prime business district of Zhengzhou, close to the railway station, with high pedestrian flow. The underfround shopping mall was opened on 28 December 2008, located in the Zhengzhou Railway Station's gold business circle, with a total area of 100,000 m^2.

Field observation revealed that there were no natural vents in the mall except in the stairwells, which resulted in poor natural ventilation. The indoor and outdoor air exchange was mainly through HVAC and mechanical exhaust systems. The airflow rate of the air conditioning system was 1.8 m/s, and the air change rate was 3.1 h^{-1}. Experiments were performed on the underground floor, which mainly included three functional areas: catering, clothing, and sundry goods. In the catering area, there were sales windows on both sides, and dining areas in the middle; this area operated mainly to cook, sell and consume fried, roasted, and stir-fried based food. Shops were spread over the middle and both sides of the garment area. Business was booming and customer traffic was at a high flow. The sundry goods area was a large space, but business was slow with many closed shops and few customers. The flow of people in different area was measured during the test periods. In catering area, at the peak time of use, people sat closely together, with an upright posture. The clothing area was a popular place of business, with the peak flow of people reaching up to 180 persons/h; they were mostly walking, and in a state of slight activity. In general, the department store area was relatively deserted and the turnover of people was about 70 persons/h.

2.2. Test Procedure

To investigate the thermal environments of the different functional areas, variables such as indoor and outdoor air temperature, relative humidity, inner wall temperature and indoor wind speed were continuously monitored during summer (3 August 2020–30 August 2020) and winter (15 December 2020–19 December 2020). To evaluate the indoor air quality of the underground mall, PM$_{2.5}$, CO$_2$, formaldehyde, and TVOC concentrations were monitored in different areas during the tests.

The locations of the monitoring points in the test room were selected according to CECS402: 2015 Urban Underground Space Operation and Management Standard [48]. A typical store was selected in each region, and the measurement point was located within the store. As the interior area of each store was less than 50 m^2, only one point was situated in each store and another point was situated in the corridor, as shown in Figure 1. The height of each sampling point was consistent with the human respiratory zone, i.e., between 1.2 and 1.5 m above the floor level [49]. Therefore, the indoor air temperature and humidity measurement points were centrally located at heights of 1.5 m.

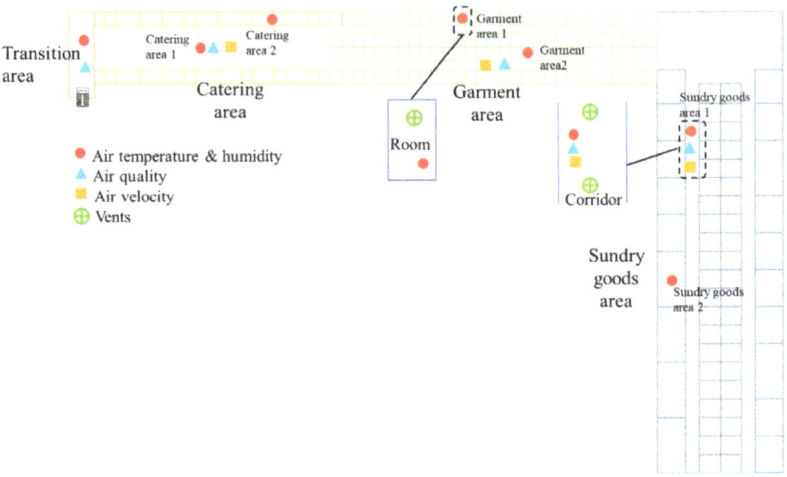

Figure 1. Arrangement of measuring points in the underground shopping mall.

The CO_2, $PM_{2.5}$, TVOC, and formaldehyde concentrations were selected as environmental pollution indicators to evaluate the air quality in underground shopping malls. Considering the instrument startup and data stability time, pollutant concentrations were collected in different areas once every hour during operation. To further analyze the indoor air, the test results were compared with the requirements stipulated in the current Chinese Standard GB/T18883-2002 Indoor Air Quality Standard [50]. The measurement instruments and parameters used are summarized in Table 1. All loggers were calibrated in the laboratory prior to testing.

Table 1. Monitored data and technical specification of measuring instruments.

Instrument Name	Monitored Parameters	Standard Value	Measurement Range	Accuracy	Photos
RC-4HC data logger	Air temperature and relative humidity	/	−30~+60 °C	±0.6 °C	
		/	0~99% RH	±5% RH	
Fluke 923 hot-wire anemometer	Air velocity	/	0.20~20.00 m/s	±5%	
SMART-126 gas detector	Formaldehyde	0.1	0~3.000 mg/m^3	±0.03 mg/m^3	
	CO_2	1000	0~5000 ppm	±45 ppm	
	PM2.5	75	0~999 μg/m^3	±20 μg/m^3	
	TVOC	0.6	0~9.999 mg/m^3	±0.03 mg/m^3	

2.3. Model Description

2.3.1. Physical Model

In considering the catering area of the underground shopping mall as the simulation object, Airpak software was used for simulation. The size of the model room was 30 m (length) × 40 m (width) × 4 m (height). The air supply mode was upper supply and upper exhaust. The size of the diffuser (air inlet) was 0.3 m × 0.3 m, with a supply speed of 1.5 m/s in a vertical downward direction. The physical model of the room is shown in Figure 2.

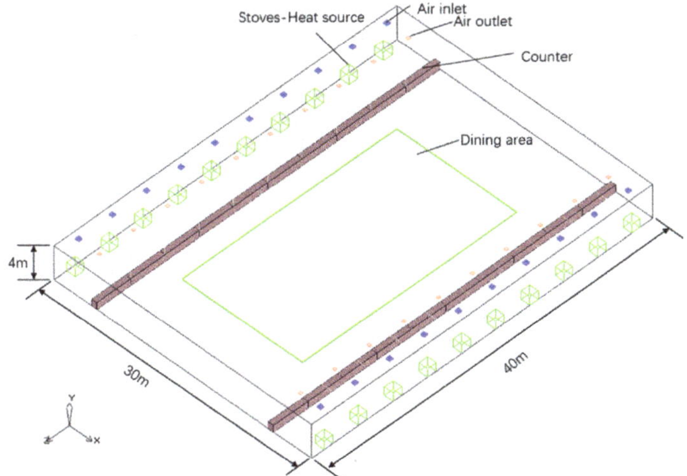

Figure 2. Physical model of the catering area.

Airpak contains five turbulence models, among which the Indoor Zero equation model is simple and fast, and can be used to quickly and accurately design and analyze indoor airflows in HVAC systems. Therefore, the Indoor Zero equation model was used to simulate the indoor air temperature in the underground shopping mall. The model used a hexahedral unstructured mesh to discretize the computational area. Local grid encryption was then carried out on the surface of the air supply and return ports, and the total number of grids was 1.02 million.

2.3.2. Mathematical Model

Airpak software includes four types of turbulence models: the mixed-length zero equation model, the Indoor Zero equation model, the RNG k-ε turbulence model, and the k-ε two-equation model. The Indoor Zero equation model can be used to quickly and accurately, in the field of HVAC, design and analyze the indoor airflow problem. Therefore, the Indoor Zero equation model was applied to simulate the air distribution of the underground shopping mall.

The governing equations are given below according to the assumptions in reference [51].

Momentum equation:

$$\frac{\partial}{\partial t}(\rho v) + \nabla \cdot (\rho v v) = -\nabla p + \nabla \cdot (\tau) + \rho g + F \tag{1}$$

Energy conservation equation:

$$\frac{\partial}{\partial t}(\rho h) + \nabla \cdot (\rho h v) = \nabla \cdot [(k + k_t)\nabla T] + S_h \tag{2}$$

where ρ is the air density, kg/m^3; v is the velocity vector, m/s; p is the pressure, Pa; F is the external volume force, N/m^3; g is the gravitational volume force, m/s^2; T is the thermodynamic temperature, K; h is the sensible enthalpy, kJ/kg; k is the thermal conductivity, W/(m·K); and S is the volumetric heat source term, W/m^3.

2.3.3. Boundary Conditions

The research object is a deeply buried underground building, and its orientation is defined according to the entrance direction. The six sides of the room were designated as ceiling, floor, front wall, back wall, left wall, and right wall. The ceiling, floor, left wall, and right wall are surrounded by soil. For deeply buried underground buildings, the stratum temperature is constant, which can be considered equal to the annual average ground temperature. Therefore, the wall temperature was set as the annual average ground temperature of Zhengzhou, i.e., 16 °C, and the ceiling exchanges heat with the outside with a coefficient of 0.7 W/m^2·°C.

There was no temperature difference between the other walls and adjacent rooms, which were in an adiabatic state. The catering area had a large space with food counters on both sides and a dining area in the middle. The kitchens, with heat sources such as stoves, were connected to the counters. The tables and counters were not the heat source, and would not be exothermic to the indoor area, therefore, did not have a remarkable influence on the indoor thermal environment. Owing to the fact that the internal structure could not be completely modeled, the stove and other heat dissipation equipment were concentrated in a 6 kW heat source in each kitchen. The dining area, personnel, and food load were evenly distributed in a 0.8 m plane heat source with a load of 72.4 W/m^2. Lighting loads were also evenly distributed along the ceiling and calculated as 14 W/m^2. Considering the influence of human body heat dissipation, human body heat dissipation was set as 92 W/person, taking into account the dining area capacity and actual site statistics, while the number of persons dining was set as 200 people.

3. Results

3.1. Description of the Underground Thermal Environment

Understanding the thermal environment of an underground shopping mall under extreme climatic conditions is the premise for studying the comfort and optimal design of an underground commercial building. According to the monitoring data from different zones, the thermal and humid environmental characteristics of an underground shopping mall were obtained.

3.1.1. Temperature

Owing to the functional differences of various areas, the indoor temperature of different area was monitored, and the influence of outdoor air temperature on indoor temperature under extreme climatic conditions was studied. Variations of indoor and outdoor air temperature are shown in Figure 3. The relationships between indoor and outdoor temperatures are shown in Figure 4. Figure 4a shows the variation of indoor temperature with outdoor temperature. The largest variation in outdoor temperature was 11.2 °C (from 23.9 to 35.1 °C) during summer, with an average value of 27.6 °C. The temperatures of the garment and sundry goods areas were stable, but the temperature of the catering area fluctuated vastly. During the monitoring period, the catering area temperatures ranged between 27.9 and 32.4 °C, with an average value of 29.1 °C. The garment area air temperature ranged between 25.6 and 28.8 °C. The temperature fluctuation of the sundry goods area was lesser still, with a variation of only 1.5 °C (26.2–27.7 °C) and an average value of 26.8 °C.

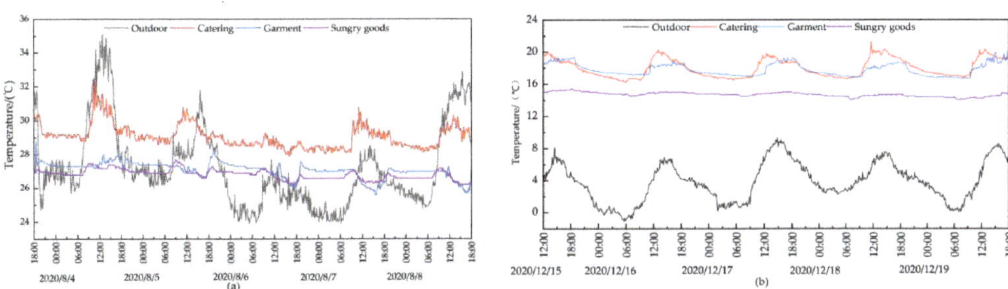

Figure 3. Variation of air temperature in summer (**a**), and winter (**b**).

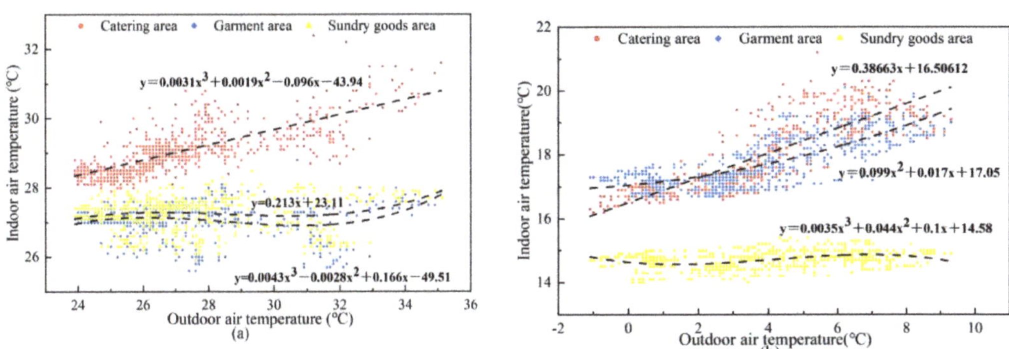

Figure 4. Correlation analysis between indoor and outdoor air temperatures in summer (**a**), and winter (**b**).

Figure 4b shows that the outdoor temperatures during winter ranged between -1.1 and 9.3 °C, with an average value of 4.2 °C. The indoor air temperature of the shopping mall was more stable in winter than in summer. The temperature of the catering area was affected by the outdoor temperature and cooking, varying from 16.2 to 21.2 °C, with an average of 18 °C. Meanwhile, the average air temperature of the garment area was 17.8 °C, ranging between 16.7 and 20.1 °C. The temperature of the sundry goods area was generally low, with small fluctuations of 1.4 °C (14–15.4 °C) and an average temperature of 14.7 °C.

When the outdoor temperature fluctuated vastly under extreme climatic conditions, the fluctuation range of the indoor temperatures was narrow, indicating that the underground shopping mall, as a semi-underground building envelope, had good heat storage capacity. Owing to the semi-closed structure and good heat insulation properties of the underground shopping mall, the soil stored and released heat effectively to limit ambient temperature fluctuations and enhance stability. Moreover, the location of the catering area was close to the transition area, which was greatly affected by external weather changes. However, the sundry goods area was located in the deep interior of the mall, which enabled the most stable temperature and was not significantly affected by outdoor temperature changes.

3.1.2. Relative Humidity

Compared with aboveground spaces, underground spaces are relatively closed off by their low terrain. Furthermore, the envelope structures of underground buildings are directly in contact with rock and soil, leading to envelope moisture transfer. These factors result in high humidity in underground spaces.

It can be seen from Figure 5 that the relative humidity in all areas was rather high during summer. The highest relative humidity in the catering area was 86.5%, due to a large amount of moisture transfer from food. At measurement point 2, the indoor dining activity had a greater impact than the outdoor relative humidity, and the peak value was observed mainly around dining times. However, the humidity trends in the garment and sundry goods areas were more influenced by the business hours of the shopping mall. The highest and lowest humidity in the garment area was 82.1% and 79.8%, respectively. The maximum and minimum humidity in the sundry goods area was 81.8% and 67.4%, respectively, among which high humidity values were observed mainly during non-business hours. During business hours (10:00–19:00), the air-conditioning devices in the shopping mall were turned on and the humidity decreased, then increased again at night.

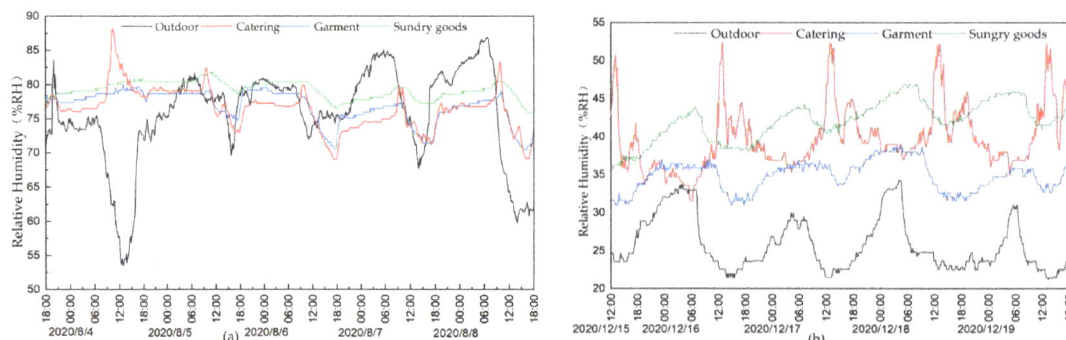

Figure 5. Variation of relative humidity in summer (**a**), and winter (**b**).

In winter, the relative humidity inside the shopping mall was generally higher than outside the mall. The humidity increased to up to 53.4% during dining times and decreased at night, and the average humidity in the catering area was 39.7%. In contrast, the humidity values in the garment and sundry goods areas decreased during business hours under the influence of air conditioning and increased at night. The fluctuation range of humidity values in the garment area was between 31% and 38.6%. Due to its location away from the entrance–exit and a lower flow of customers, the humidity in the sundry goods area was significantly higher than in the garment area, with the highest humidity being 46.9% and the lowest being 36.7%.

These results indicated that the temperature in summer was close to, or lower than, the dew point temperature of air. Owing to the semi-closed structure, the moisture was trapped, which led to high relative humidity. In winter, the humidity of the underground shopping mall was higher than outside, but still within standard requirements. Therefore, improving the natural ventilation of the underground shopping mall or designing ventilation vents may be helpful to ensure the comfort requirements of customers.

3.2. Underground Air Quality

As a public building and a place of consumption for customers, shopping malls are characterized by various shopping areas and dense streams of people. Nevertheless, air quality issues may be observed as underground commercial buildings are rather confined spaces.

3.2.1. Underground Air Pollution Level in Different Merchandise Sections

Formaldehyde and TVOC emissions are affected by many factors [52], for example, temperature and humidity, ventilation conditions, premises decoration materials, etc., and in the case of the subject of this paper, they are also influenced by the type of goods sold in the area. By analyzing the air qualities at different merchandise sections, we sought to determine the impacts of various displayed or manufactured goods on air quality.

The pollutant concentrations in different merchandise sections varied. It can be seen from Figure 6 that the CO_2 concentration in the transition and sundry goods areas was lower than the limit values, whereas it was slightly above the limit in the catering area. In this mall, the garment area had the highest CO_2 concentration, since this section often has high business activity, with customers staying for longer times. The transition and catering areas were heavily polluted with $PM_{2.5}$ because the food manufacturing process generates high levels of $PM_{2.5}$. The kitchen close to the dining area had poor ventilation, thereby leading to excess pollution in the catering area, which also contributed to the pollution in the transition area. Volatile substance concentrations, such as those of formaldehyde and TVOC in underground shopping malls, were all within the safe range, except for the catering area.

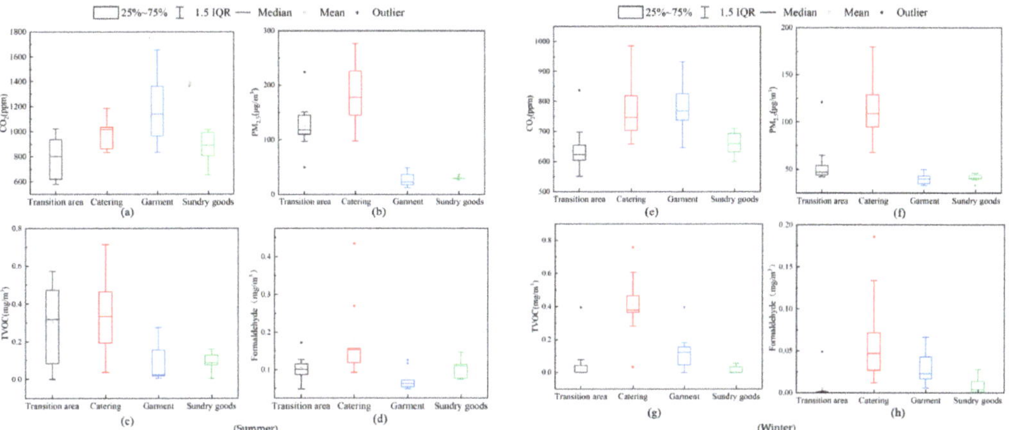

Figure 6. The indoor CO_2 (**a,e**), $PM_{2.5}$ (**b,f**), TVOC (**c,g**), and formaldehyde (**d,h**) levels of different merchandise sections.

In winter, the concentration of CO_2 in each section of the underground shopping mall met the health requirements. The concentrations of the other pollutants in the garment and sundry goods areas were also within standard limits. The $PM_{2.5}$ concentration in the catering area was severely in excess, which also increased the $PM_{2.5}$ concentration in the adjacent transition area to be much higher than the standard limit. The concentrations of TVOC and formaldehyde in the dining area were slightly above the standard values, and there was no excess in the transition area.

3.2.2. Underground Air Pollution Level at Different Time Periods

Occupant activities and occupant density affect indoor air pollutant distributions and fresh air volumes [53,54]. Trends were observed in customer flow rates throughout business hours: an increasing number of customers entered the catering area during dining times, while a customer 'rush' was observed in the garment and sundry goods areas in the afternoon. Figure 7a shows that the CO_2 concentration of the mall was lower than the limit value in all areas except the garment area. The CO_2 concentration in the garment area rapidly increased during the customer rush in the afternoon. This may be attributed to the fact that a greater number of customers shopped in the clothing area which had insufficient ventilation, thereby leading to the increase in CO_2 concentration. In the catering and transition areas, food preparation activities increased the $PM_{2.5}$ concentrations to exceed the standard. As for TVOC and formaldehyde, their increased concentrations were attributed to fuel gas burning and the production of lampblack. Such pollutants are more volatile under high-temperature conditions. Food production and heat dissipation in the catering area inevitably caused temperature rise and pollutant aggregation, which was difficult to dilute.

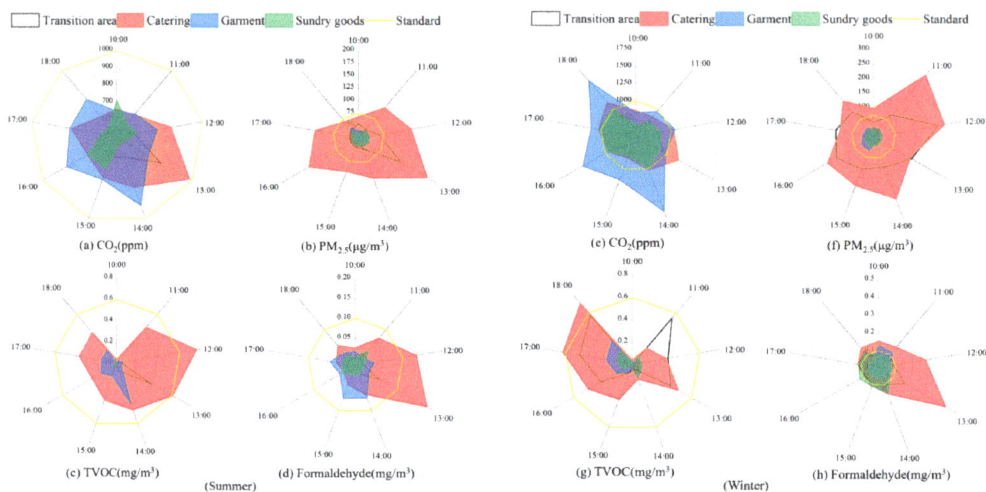

Figure 7. The indoor CO_2 (**a,e**), $PM_{2.5}$ (**b,f**), TVOC (**c,g**), and formaldehyde (**d,h**) levels at different times in summer.

As shown in Figure 7, the CO_2 concentration in the underground shopping mall in winter typically soared during business hours but did not exceed the standard; however, $PM_{2.5}$ was the prime pollutant in winter. In the catering area, the $PM_{2.5}$ concentration exceeded the standard, which also affected the air quality in the transition area. In addition, both TVOC and formaldehyde concentrations were over the standard values at mealtimes.

Underground floors have poorer indoor air quality than aboveground floors because of their enclosed structures and poor natural ventilation [55]. Pollutants in the air are, therefore, hard to dilute when the shopping mall is crowded or high levels of pollutants are generated during some activities. However, increasing indoor ventilation and decreasing indoor temperature can reduce indoor formaldehyde and TVOC concentrations [37]. Thus, by optimizing air distribution, high indoor temperature and poor air quality in the catering area may be improved.

3.3. Simulation Optimization of Ventilation and Air Distribution in Catering Area

In air-conditioned buildings, in addition to changing the set air supply temperature and speed, various air supply modes will also form different air distributions. Underground shopping malls generally adopt the air distribution form of upper supply and upper return. According to the measured analysis, under this air distribution form, the air quality of the underground shopping mall was poor, and the thermal environment of different functional areas barely arrived to a satisfactory thermal comfort. As a shopping mall area with special functions, the connecting design of the kitchen and dining area was convenient, but also caused the accumulation of heat and pollutants. In addition, without adequate ventilation, the underground building was prone to high indoor temperature, poor air quality, and unsanitary conditions during the peak period. For the catering area to have a reasonable air distribution, effective cooling and ventilation are key to improving the internal environment, so Airpak software was used simulate a change to the air distribution of the catering area to optimize the thermal environment.

3.3.1. Optimization Simulation of Air Distribution in the Catering Area

The poor thermal environment in the catering area was attributed to the high internal cooling load and supply air temperature under actual working conditions. Therefore, in the simulation, we reduced the air supply temperature in the catering area from the original supply air temperature (24 °C), to simulate three different air supply scenarios. The airflow rate of the air-conditioning system was 1.8 m/s, and the air exchange rate was 3.6 h^{-1}. Three types of air-conditioning supply schemes were designed for the catering area by varying the locations of the inlet and outlet. The three kinds of supply air paths are described in Table 2, among which Scheme 1 (upper supply upper return) was the one in practical use. The research focused on the occupied region at a height of 1.5 m above the floor in the dining area. The optimal air supply temperature was 21 °C in summer and 20 °C in winter, and the upper supply velocity was 1.5 m/s and side supply velocity was 3.5 m/s. As simulated results, the temperature, wind speed profile of Y = 1.5 m (customer activity area) and temperature profile, and velocity vector diagram of Z = 18 m (cross-section of air supply outlet in the center of the dining area) were chosen for display. The temperature and velocity uniformities of the different schemes were compared to select a more suitable air distribution mode in the catering area.

Table 2. Different scheme conditions.

Scheme 1: Actual Condition	Scheme 2: Exhaust Vents Are Arranged Horizontally (Short for Horizontal Exhaust)	Scheme 3: Side Supply Upper Return (Short for Side Supply)
The exhaust vents and air supply out-lets are arranged in the ceiling in the shape of a plum, with two columns (the direction along the depth of the room) of air supply outlets as a group, and the exhaust vents are arranged verti-cally.	The exhaust vents and air supply outlets are arranged in the ceiling in the shape of a plum, with two rows (the direction along the span of the room) of air supply outlets as a group, and the exhaust vents are arranged horizontally.	Based on the actual condition, the exhaust vents are still arranged vertically. The air supply lower outlets are arranged on the suspended ceiling of the counter on both sides as the side supply scheme.

(1)   Scheme 1: Actual condition.

In the actual condition, the diffuser delivered air vertically downwards, which greatly influenced the lower area of the air outlet. The distributions of temperature and velocity at a height of 1.5 m above the ground in the catering area are presented in Figures 8–11.

According to Figures 8a,b and 9a,b, the air velocity distributions show that in the catering area, the cold air in summer dropped and then formed a backflow near the dining tables at a height of about 0.6–2 m. In the occupied region, the wind speed was appropriate with an average value of 0.109 m/s. In winter, most airflows moved downward and mixed with cold air to form the backflow. A part of the hot air flowed along the ceiling and moved downward after colliding with the other airflows. The average indoor air velocity was about 0.119 m/s. In the actual condition, the velocity values in the occupied regions were less than 0.3 m/s, excluding the region around the air supply axis, which may cause discomfort to some customers.

Figures 10a,b and 11a,b illustrate the thermal environments in the catering area. The wall of the connecting part of the dining area and the exit was affected by the outside temperature, and the other wall was in the adiabatic state, which is connected with the soil or the adjacent room, except the wall. The heat dissipation from people and food caused high temperatures in the middle. In winter, the indoor temperature distribution was uneven, and the hot air from the air outlets was inhaled by the return air outlets on both sides before it was entirely mixed with the cold air. As seen from the longitudinal temperature distribution of the room in Figure 11a,d, the air temperature was suitable under interactions of the lower ground temperature and supply air. Therefore, there was no obvious stratification in the occupied area. The ground temperature was lower, while the roof was affected by the external environment and lighting heat dissipation, and the temperature was higher.

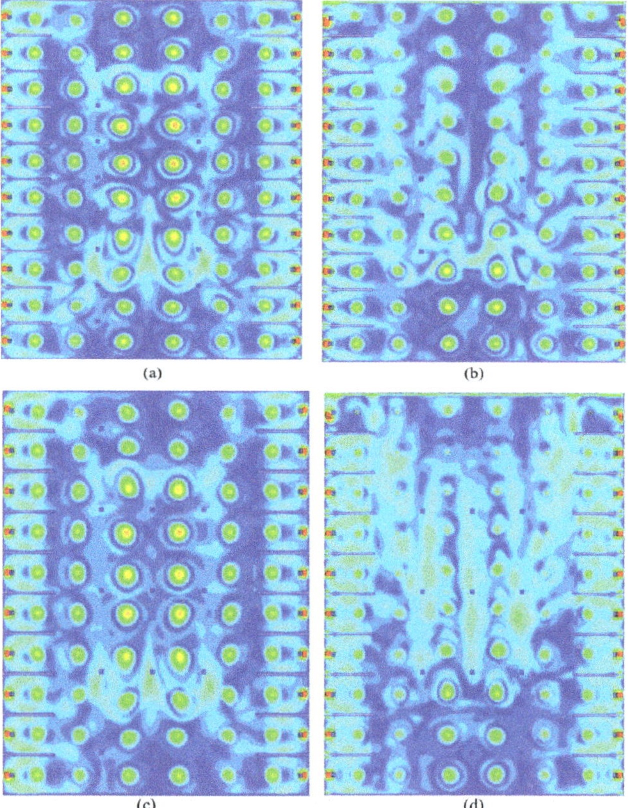

(a)   (b)

(c)   (d)

Figure 8. *Cont.*

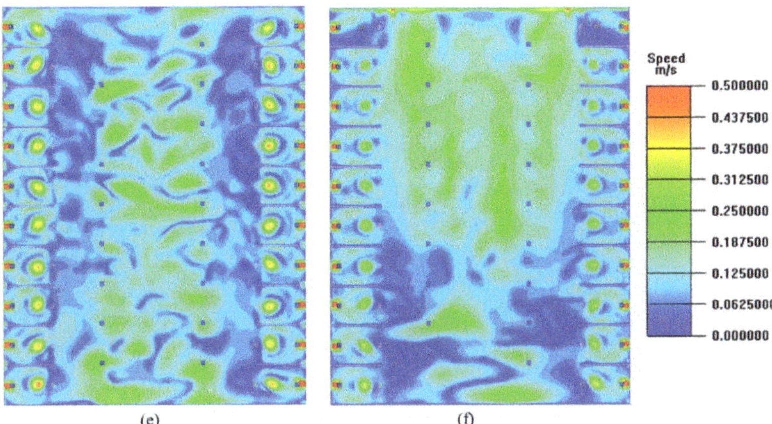

(e)　　　　(f)

Figure 8. Velocity distribution at Y = 1.5 m: Scheme 1 in summer (**a**), Scheme 1 in winter (**b**), Scheme 2 in summer (**c**), Scheme 2 in winter (**d**), Scheme 3 in summer (**e**), and Scheme 3 in winter (**f**).

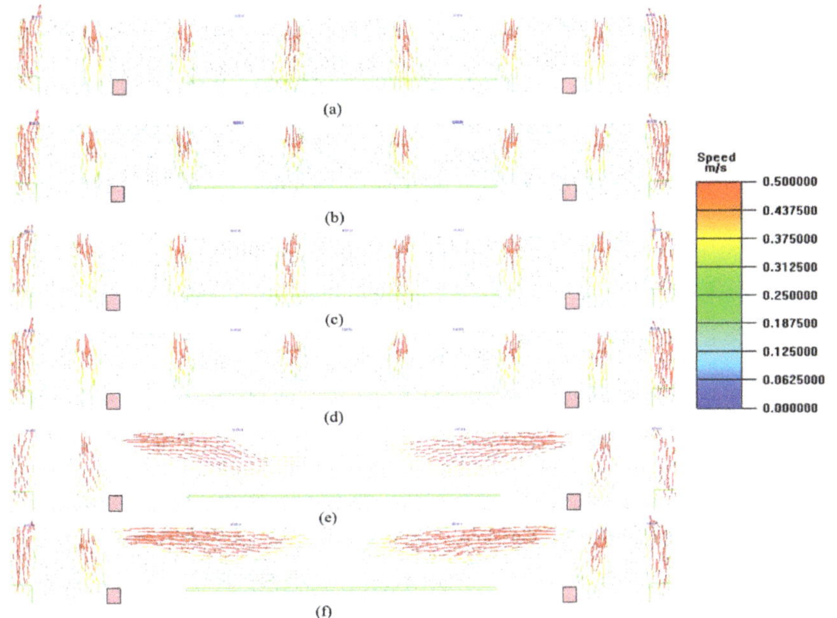

Figure 9. Vector diagram of velocity distribution at Z = 18 m: Scheme 1 in summer (**a**), Scheme 1 in winter (**b**), Scheme 2 in summer (**c**), Scheme 2 in winter (**d**), Scheme 3 in summer (**e**), and Scheme 3 in winter (**f**).

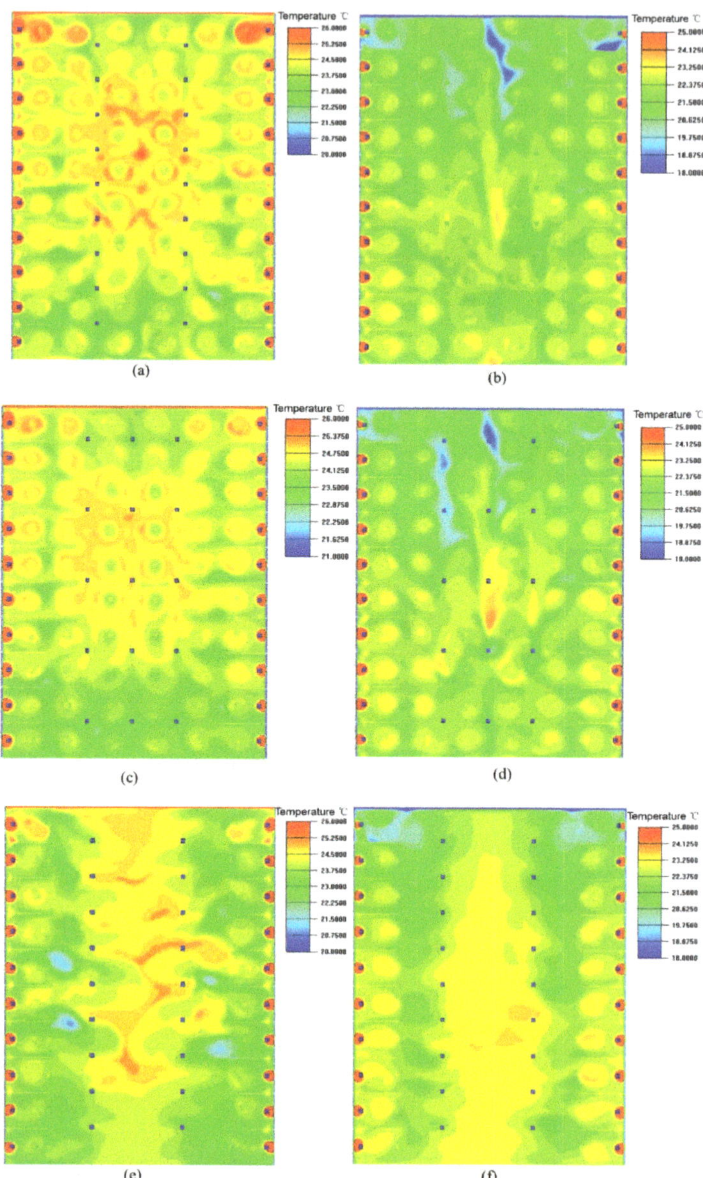

Figure 10. Temperature distribution at Y = 1.5 m: Scheme 1 in summer (**a**), and 1 in winter (**b**); Scheme 2 in summer (**c**), and in winter (**d**); Scheme 3 in summer (**e**), and in winter (**f**).

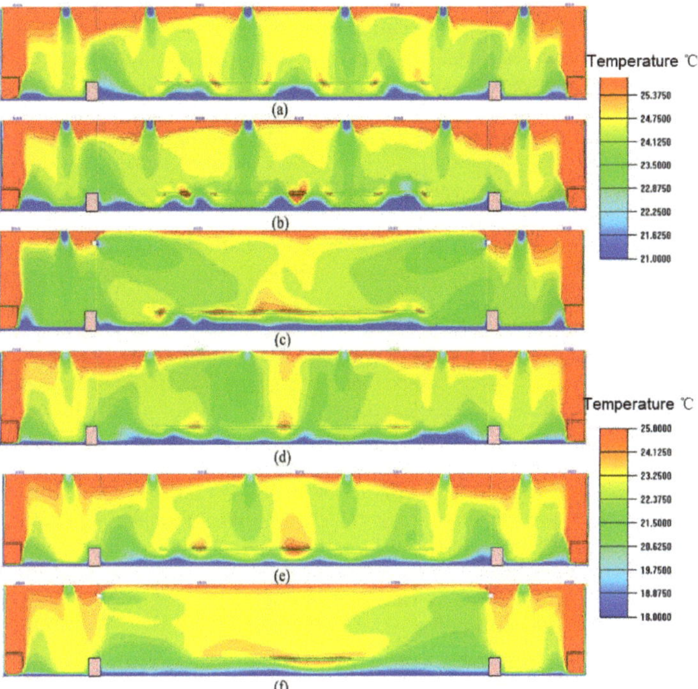

Figure 11. Temperature distribution at Z = 18 m: Scheme 1 in summer (**a**), and in winter (**d**); Scheme 2 in summer (**b**), and in winter (**e**); Scheme 3 in summer (**c**), and in winter (**f**).

(2) Scheme 2: Horizontal Exhaust.

The simulation results in Figure 8c,d show that air velocities in the most occupied regions of the dining area were within 0.3 m/s. The mean air velocity was 0.11 m/s in summer and 0.117 m/s in winter. As shown in Figure 9c,d, the distribution vector diagram of the air velocity shows a trend similar to the actual condition, where the cold air sent from the air supply outlet sank, and the backflow was formed between the two adjacent air supply outlets. Compared with the actual condition, the uniformity of velocity distribution improved after changing the location of the return air outlet, and the phenomenon of excessive wind speed under the air outlets was also reduced.

Figure 10c,d shows the temperature distribution at a height of 1.5 m under the horizontal exhaust scheme. The average indoor temperature in the room was 24.5 °C in summer and 22.3 °C in winter. Compared with the actual conditions, the temperature did not change significantly. In winter, owing to the low outside temperature, the temperature near the entrance was lower, about 18.5 °C, while the maximum temperature in the dining area was 24.4 °C. The temperature difference of 5.9 °C would cause discomfort to customers. Figure 11c,e reveals the longitudinal temperature change in the catering area; it illustrates that the temperature in the middle of the activity area was suitable, although there were extreme temperatures near the ground and roof. The longitudinal temperature difference did not exceed 1 °C, which would be comfortable for customers.

(3) Scheme 3: Side supply.

Figures 8e,f and 9e,f depict the velocity distributions under side supply. The air velocity from the air outlet was relatively high, but the velocity at the height of 1.5 m was less than 0.3 m/s, with an average value of 0.118 m/s. In winter, under the side supply scheme, the air velocity gradually decreased with the sinking of hot air. The velocities in most areas of the room ranged between 0.1 and 0.2 m/s, and the mean air velocity was 0.126 m/s. Compared with Scheme 1 and Scheme 2, it was seen that good uniformity of the velocity could be achieved at a height of 1.5 m when adopting the side supply scheme. The average air velocity of side supply was greater than for the other two schemes but lower in the occupied zone, which had better air mobility and no impact on the occupants.

According to Figure 10e,f, the temperature distribution showed that high temperatures were mainly distributed in the dining area during summer, while the temperatures near the air outlet were low due to the sinking of the cold air. The mean temperatures in summer and winter were 24.1 °C and 22.8 °C, respectively. Figure 11e,f shows that compared with the upper supply, the longitudinal temperature distribution under the side supply was more uniform, indicating that the side supply had better effects on indoor temperature regulation.

3.3.2. Model Validation

The measured air temperatures in the catering area were used to evaluate the simulation performance. The actual conditions included the upper supply upper return air-conditioning scheme with a supply temperature of 23 °C. There were two measuring points in the catering area, one located at the center of the dining area and the other located at the counter. The simulated outdoor temperature was 35 °C, the preset indoor temperature was 24 °C, and the heat of the top was 7.7 W/m^2. All the kitchen stoves and other cooling equipment were concentrated as one heat source, and the calculated heat dissipation was 6 kW. The personnel and food loads in the dining area were uniformly distributed in the plane heat source at a height of about 0.8 m, with a load of 72.4 W/m^2, the lighting load was distributed evenly in the ceiling, and was calculated to be 14 W/m^2. The simulated temperature was compared with the measured temperature during the mealtime (11:00–13:00), as shown in Figure 12, with a 4.02% margin of error. The model was considered reasonable with simplification and uncertain factors, including customer flow, in the actual situation.

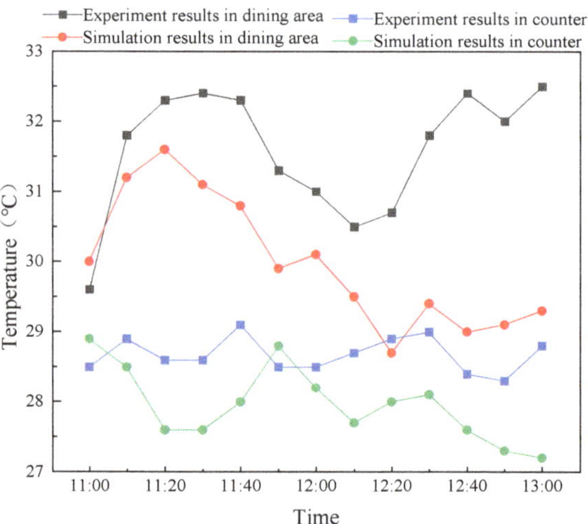

Figure 12. Simulation vs. measured results for measuring points.

4. Discussion

4.1. Thermal Comfort Analysis of Underground Shopping Mall

Thermal comfort analyses in existing studies mostly focus on aboveground buildings, such as office buildings, shopping malls, and residential buildings, while underground shopping malls are underground projects, and the thermal comfort of the personnel may be different from those aboveground. When evaluating the thermal environment comfort of underground shopping malls, four indicators are considered, namely, temperature, relative humidity, airflow speed, and average radiation temperature of the environment. Different combinations of these indicators can result in different thermal environments in underground shopping malls.

4.1.1. Model Validation

The measured air-state points were plotted on a psychrometric chart, and the comfort zone required by ASHRAE Standard 55-2017 was taken as the evaluation basis. As seen from Figure 13, most of the air-state points of the underground shopping mall were not within the comfort zone requirements, due to the high humidity in summer and low temperature in winter. Owing to poor ventilation in the underground shopping mall, the indoor airflow speed was about 0–0.2 m/s during measurement, but 0 m/s at other times. Therefore, when analyzing the thermal comfort of underground shopping malls, the influences of indoor temperature, relative humidity, and average radiant temperature on thermal comfort were mainly considered.

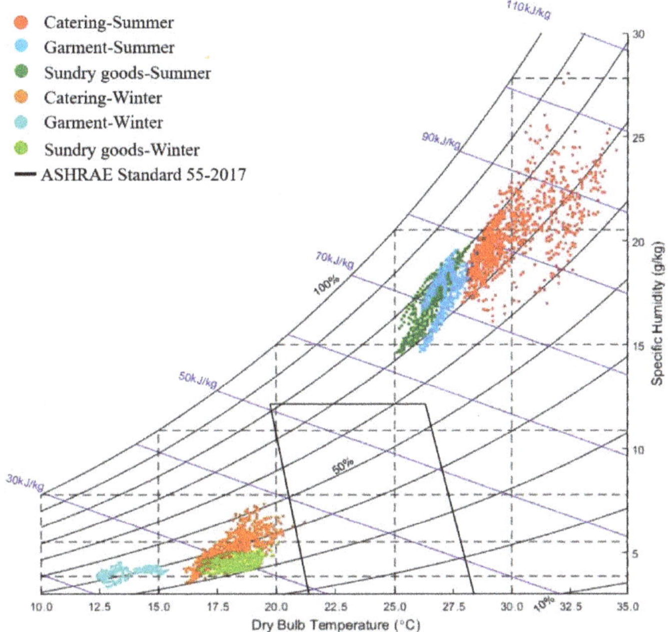

Figure 13. Distribution of air-state points in a psychrometric chart.

To maintain basic physiological processes, the human body often performs energy balance processes including heat production, heat exchange, and heat storage. The indoor mean radiation temperature of an underground space will be affected by the earth temperature of the project, as shown in Figure 14. For people living in underground buildings, human body heat dissipation is mainly through convection and radiation. According to the analysis of indoor heat balance of the human body, air temperature affects convection heat dissipation, and the mean radiation temperature has a strong influence on the radiation heat dissipation. In addition, the different proportions of convection and radiation in sensible heat dissipation result in changes in the thermal physiological state of the human body. Consequently, the variation of indoor air and wall temperatures in different functional areas affect the heat dissipation of the human body.

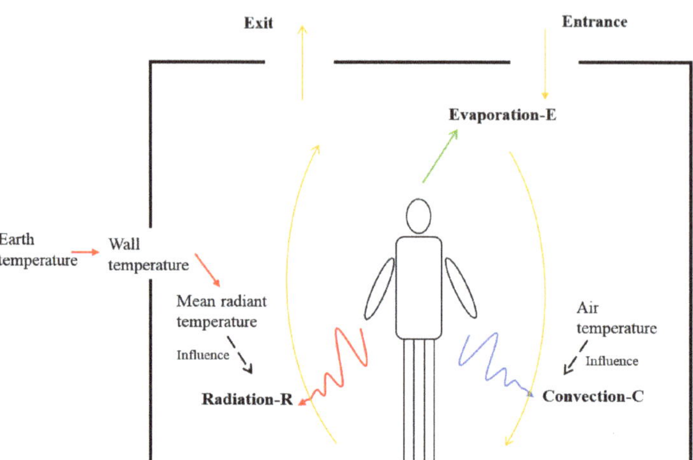

Figure 14. Influence of the underground space thermal environment on human body heat dissipation.

The difference between indoor air and wall temperatures is called indoor temperature difference (Δt = (indoor air temperature) − (wall temperature)). To better reflect the influence of the indoor temperature difference (Δt) on thermal comfort, the human body convection–radiation heat dissipation ratio (C/R) was introduced. The relation between indoor temperature difference and convection–radiation heat dissipation ratio (C/R) is shown in Figure 15.

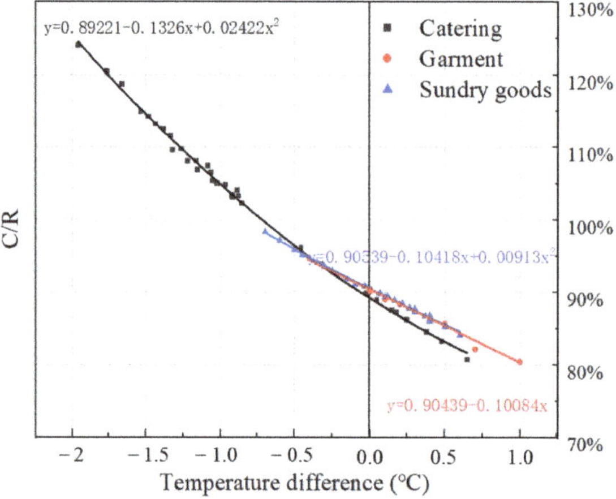

Figure 15. Convection–radiation heat dissipation ratio in different functional areas.

The ASHRAE Manual [56] describes the energy balance process in the human body as follows:

$$M - W - C\text{-}R - E = TL \tag{3}$$

where M is the energy metabolism rate of human body, depending on the amount of activity, W/m^2; W is the mechanical work performed by the human body, W/m^2; C is the convective heat radiating from the outer surface of the human body to the surrounding environment, W/m^2; R is the heat radiating from the outer surface of the human body to the surrounding environment, W/m^2; E is the heat emitted from the outer surface of the human body in the form of sweat evaporation or exhalation to the surrounding environment, W/m^2; and *TL* is the difference between the amount of heat produced and dissipated by the human body, similarly, it is the heat load of the human body, W/m^2.

The calculations [56] for C and R are as follows:

$$C = f_{cl}h_c(t_{cl} - t_a) \tag{4}$$

$$R = f_{cl}h_r(t_{cl} - \overline{t_r}) \tag{5}$$

where f_{cl} is the convective heat transfer coefficient, W/(m^2·K); h_c is the garment surface area factor, ND; t_{cl} is the average temperature of the outer surface of the garment, °C; t_a is the air temperature around the human body, °C; h_r is the radiant heat transfer coefficient, W/(m^2·K); and $\overline{t_r}$ is the mean radiation temperature of the environment, °C.

It was observed that the indoor temperature difference in the catering area was large due to the high wall temperature during the food preparation process, and the proportion of convection heat dissipation was higher. However, the indoor temperature differences in the garment and sundry goods areas were small, and the proportions of convection and radiation heat dissipations were similar.

4.1.2. Convection–Radiation Heat Dissipation Ratio and Human Thermal Comfort

In different functional areas, the indoor temperature differences in an underground shopping mall vary, causing significant differences in the convection–radiation heat dissipation ratio, thereby affecting the thermal comfort of the human body. Consequently, C/R and humidity were used as variables to represent the thermal comfort of each region, as shown in Figure 16. The heat transfer ratio of indoor convective radiation was deemed as the X axis, the relative humidity was deemed as the Y axis, the PMV value was deemed as the Z axis, and the heat transfer ratio of convective radiation and its relationship with human thermal comfort under the same relative humidity were compared.

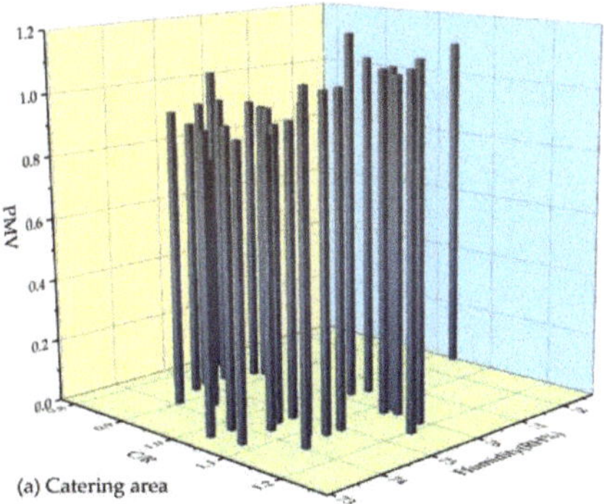

Figure 16. *Cont.*

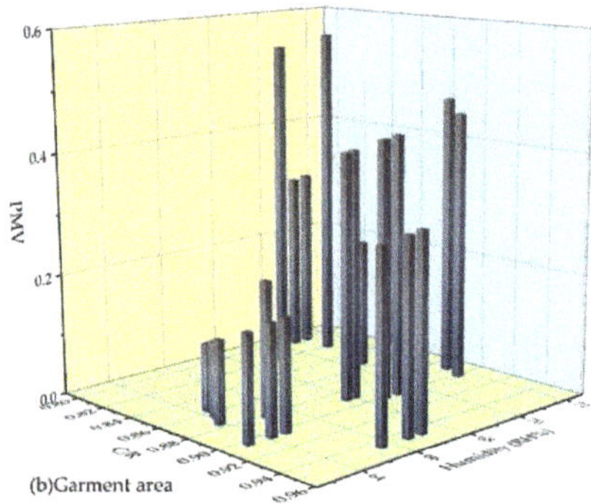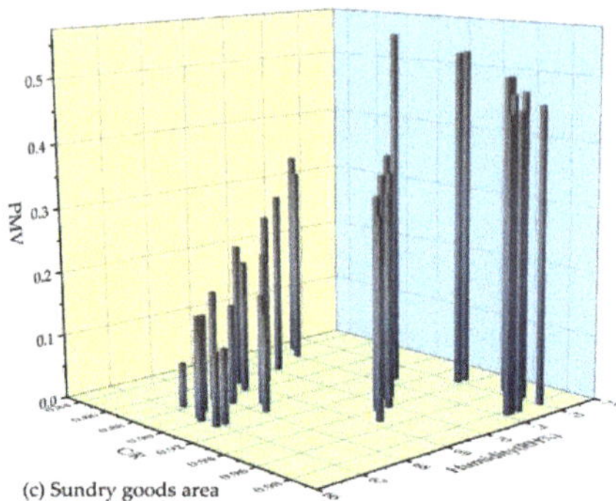

Figure 16. C/R and thermal comfort in catering area (**a**), garment area (**b**) and sundry goods area (**c**).

The catering area in the underground shopping mall had a higher proportion of convection heat dissipation. In contrast, the proportion of radiation heat dissipation in the clothing and sundry goods areas was higher. During measurements, the relative humidity of the underground commercial building was generally greater than 60%. Figure 16 illustrates that under the condition of a constant C/R ratio, the thermal comfort of the human body decreases with an increase in relative humidity. Owing to the high temperature in the catering area, the PMV value was higher than the standard value, as well as increasing with the increase in C/R for the same relative humidity. However, in the garment and sundry goods areas, the PMV values decreased with the increase in C/R under identical humidity conditions. To further analyze the relationships between convection–radiation ratio and PMV, the catering area state points were divided into C/R > 1 and C/R < 1.

Figure 17 shows that in the catering area, when C/R < 1 and air temperature was greater than the mean radiant temperature, the PMV value decreased with the increase in C/R value, showing a negative correlation. When C/R > 1, that is, the air temperature was less than the mean radiant temperature, the PMV value increased with the increase in the C/R value, showing a positive correlation. Low air velocity in an underground building causes a reduction of convection heat dissipation from the human body. To maintain the basic heat dissipation requirements of the human body, the proportion of radiation heat dissipation must be elevated and the C/R ratio must be smaller. In the garment and sundry goods areas, the radiation heat dissipation was of greater account, since there were fewer heat sources. However, the wall temperature in the catering area was considerably higher than the air temperature, owing to the existence of heat sources such as stoves, and the human convection heat dissipation was higher. This illustrates that in underground commercial buildings, there are differences in the wall and air temperatures in different functional areas, hence, indoor temperatures in underground buildings should be adjusted reasonably to reduce differences between the air and average radiation temperatures, to improve indoor thermal comfort.

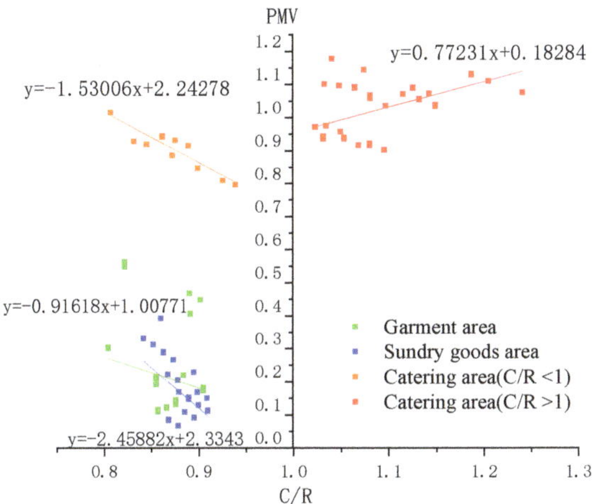

Figure 17. The relationship between C/R and PMV.

4.2. Comparison of the Three Air Supply Schemes

Underground space has attracted much attention due to its energy saving and heat storage advantages, but its internal thermal performance and air quality cannot meet the comfort requirements of personnel due to its particular location. According to the results during the measurements, due to the large flow of people and various functions, underground commercial buildings are in a high state of temperature and humidity in summer, which makes it hard to meet the requirements of thermal comfort of the human body. The concentration of CO_2 and $PM_{2.5}$ in the catering area exceeded the standard and the air quality was worrying, due to the different functional divisions of each region and the different pollutant emission characteristics. This was consistent with the research conclusion of Shang et al. [55] on air quality in four different aboveground shopping malls. Therefore, based on the measured results of thermal characteristics and air quality of multifunctional underground buildings, this study simulated and optimized air distribution in the catering area.

To improve the comfort level of the catering area, the factors that strongly influence air distribution and thermal comfort were compared. Four evaluation parameters, including mean temperature (T_{ave}), mean velocity (V_{ave}), and the non-uniform coefficients of temperature distribution (K_T) and velocity distribution (K_v), were considered under the three different air supply schemes in this study. The results were based on a height of 1.5 m above the ground and are presented in Table 3.

Table 3. Results comparison among the three air supply schemes.

Supply Scheme	Condition	T_{ave} (°C)	V_{ave} (m/s)	K_T	K_v
Actual condition	Summer:	24.5	0.109	0.0183	0.137
	Winter:	22.2	0.119	0.0242	0.158
Horizontal exhaust	Summer:	24.5	0.11	0.0171	0.131
	Winter:	22.3	0.117	0.0231	0.156
Side supply	Summer:	24.1	0.118	0.0173	0.128
	Winter:	22.7	0.122	0.0233	0.144

The non-uniform coefficients of temperature distribution (K_T) and velocity distribution (K_V) were used to evaluate the temperature and wind velocity fields in the working area. Their calculation processes and formulas are as follows.

N sampling points were selected at a height of 1.5 m above the ground. Temperature and wind speed were measured at each point and their arithmetic means were calculated.

$$\bar{t} = \frac{\sum t_i}{n} \qquad \bar{v} = \frac{\sum v_i}{n} \tag{6}$$

The standard error between air temperature and velocity at 1.5 m was root-mean-square error.

$$\sigma_t = \sqrt{\frac{\sum(\bar{t} - t_i)^2}{n-1}} \qquad \sigma_v = \sqrt{\frac{\sum(\bar{v} - v_i)^2}{n-1}} \tag{7}$$

The non-uniform coefficients of temperature distribution (K_T) and velocity distribution (K_v) were obtained:

$$K_T = \frac{\sigma_t}{\bar{t}} \qquad K_V = \frac{\sigma_v}{\bar{v}} \tag{8}$$

where t is the air temperature, $°C$; and v is the wind velocity, m/s. The smaller the K_T and K_V values are, the better the uniformity of temperature and wind velocity distribution is, and the better the air distribution is.

Table 3 shows that under extreme climate conditions, the air-conditioning scheme of side supply in the catering area was obviously better than the upper supply upper return, which not only effectively reduced the indoor temperature but also caused the indoor air distribution to be more uniform. The side supply scheme reduced the temperature by about 0.4 °C in summer and increased it by 0.5 °C in winter. Although the average velocity was higher than those of other schemes, it was still less than 0.2 m/s, which cannot produce draught sensations. The temperature in the room was evenly distributed, and the front side of the room did not experience overcooling. Compared with the actual condition, the temperature uniformities in summer and winter were improved by 5.4% and 3.7%, while the velocity uniformities were improved by 6.5% and 8.8%, respectively. As a large space with large heat dissipation, the temperature of the dining area could effectively be improved by the side supply scheme. The upper supply and upper return scheme, such as in the actual condition and horizontal exhaust, would cause excess wind speed, while the side supply scheme can avoid this situation.

4.3. Limitations

This research was carried out only in one typical underground shopping mall in Zhengzhou, China. More studies conducted in different climate areas may give a better understanding of the thermal environment in underground commercial buildings.

The indoor thermal environment of the underground commercial building was actively regulated by the air conditioning system, and the passive regulation method, such as the atrium, was not analyzed. These passive regulation methods can be combined with the air-conditioning system to achieve energy-savings.

5. Conclusions

This research presented an experiment study of an underground multifunctional commercial building, including its thermal environment, indoor air quality, and simulated air distribution. The key conclusions are summarized as follows.

The thermal environments and air qualities of different functional areas of an underground shopping mall in Zhengzhou were measured and evaluated. It was observed that the air temperature and humidity of the underground shopping mall was more stable than observed outdoors. The air temperature distribution in the underground shopping mall was uneven, with a significant effect on the function and location. The summer temperature in the catering area was too high, with an average temperature of 29.1 °C. The temperature in the sundry goods area during winter was generally below 16 °C. The relative humidity in the underground shopping mall was above 60% in each zone in summer, and more comfortable in winter.

The air quality survey inside the underground shopping mall showed that CO_2 concentrations in the clothing and catering areas rose sharply during the peak business season in summer, resulting in CO_2 concentrations exceeding the standard. The phenomenon of excess pollutant concentration in winter was mainly caused by the high concentration of $PM_{2.5}$ in the dining area. These results showed that the ventilation in the underground shopping mall needs to be improved.

The convection–radiation heat dissipation ratio (C/R) was introduced to analyze the influence of air temperature and mean radiation temperature on human body heat dissipation and thermal comfort. The results showed that the proportions of convection and radiation heat dissipations in the underground commercial buildings were similar, but the C/R in the catering area was higher due to heat sources such as stoves. There is a certain correlation between the C/R and PMV value, so the indoor temperatures of underground buildings should be adjusted reasonably to reduce differences between air and mean radiation temperatures, to improve indoor thermal comfort.

To solve the problem of high temperature, the air distribution in the catering area was optimized. The simulation results showed that the side supply scheme was more suitable for the catering area. Under this air distribution, the air temperature of catering area can effectively be reduced about 0.4 °C in summer, and be increased 0.5 °C in winter. A higher air velocity also contributes to indoor ventilation. Compared with other air conditioning schemes, the indoor air temperature uniformities of the side supply scheme in summer and winter increased by 5.4% and 3.7%, while the velocity uniformities increased by 6.5% and 8.8%, respectively.

Author Contributions: Conceptualization, methodology, data curation, X.Z.; formal analysis, writing—original draft preparation, C.L.; visualization, investigation, J.Z.; supervision, J.L.; writing—review and editing, Y.C. All authors have read and agreed to the published version of the manuscript.

Funding: This research was supported by grants from the Science and Technology Department of Henan Province (grant number 212102310573, 222102320113), Key R&D and Promotion Project of the Department of Science and Technology of Henan Province, China (grant number 222102320113), and China Construction Seventh Bureau Technology Research Project (grant number CSCEC7b-2015-Z-24).

Data Availability Statement: Not applicable.

References

1. Wang, H.; Chen, D.; Duan, H.; Yin, F.; Niu, Y. Characterizing urban building metabolism with a 4D-GIS model: A case study in China. *J. Clean. Prod.* **2019**, *228*, 1446–1454. [CrossRef]
2. P.R.C. Ministry of Housing and Urban-Rural Development. *China Urban Construction Statistical Yearbook*; China Planning Press: Beijing, China, 2021.
3. Deng, X.; Huang, J.; Rozelle, S.; Uchida, E. Growth, population and industrialization, and urban land expansion of China. *J. Urban Econ.* **2008**, *63*, 96–115. [CrossRef]
4. Franco, M.A.J.Q.; Pawar, P.; Wu, X. Green building policies in cities: A comparative assessment and analysis. *Energy Build.* **2020**, *231*, 110561. [CrossRef]
5. Wang, J.; Zhai, T.; Lin, Y.; Kong, X.; He, T. Spatial imbalance and changes in supply and demand of ecosystem services in China. *Sci. Total Environ.* **2018**, *657*, 781–791. [CrossRef] [PubMed]
6. Mat, S.; Shamila, H.; Francesco, F.; Paul, O.; Ding, L.; Deo, P.; Zhai, X.; Wang, R. Urban heat island and overheating characteristics in sydney, australia. an analysis of multiyear measurements. *Sustainability* **2017**, *9*, 712.
7. He, B.J.; Zhao, D.X.; Xiong, K.; Qi, J.; Prasad, D.K. A framework for addressing urban heat challenges and asso-ciated adaptive behavior by the public and the issue of willingness to pay for heat resilient infrastructure in Chongqing, China. *Sustain. Cities Soc.* **2021**, *75*, 103361. [CrossRef]
8. Yla, B.; Fei, L.A.; Cx, A.; Zo, A. Urban development and resource endowments shape natural resource utilization efficiency in Chinese cities. *J. Environ. Sci.* **2022**, *126*, 806–816.
9. Li, C.; Wang, Z.; Li, B.; Peng, Z.-R.; Fu, Q. Investigating the relationship between air pollution variation and urban form. *Build. Environ.* **2019**, *147*, 559–568. [CrossRef]

10. Crippa, M.; Guizzardi, D.; Pisoni, E.; Solazzo, E.; Guion, A.; Muntean, M.; Florczyk, A.; Schiavina, M.; Melchiorri, M.; Hutfilter, A.F. Global anthropogenic emissions in urban areas: Patterns, trends, and challenges. *Environ. Res. Lett.* **2021**, *16*, 074033. [CrossRef]
11. Petroleum, B. *BP Statistical Review of World Energy*; British Petroleum: London, UK, 2021.
12. Mele, C. Human settlements and sustainability: A crucial and open issue. *E3S Web Conf.* **2019**, *119*, 00012. [CrossRef]
13. Wang, J.; Sun, K.; Ni, J.; Xie, D. Evaluation and Factor Analysis of the Intensive Use of Urban Land Based on Technical Efficiency Measurement—A Case Study of 38 Districts and Counties in Chongqing, China. *Sustainability* **2020**, *12*, 8623. [CrossRef]
14. Yao, Y. *Key Technologies for Environmental Quality Assurance of Urban Underground Space*; China Architecture & Building Press: Beijing, China, 2010; p. 392.
15. Zhao, J.-W.; Peng, F.-L.; Wang, T.-Q.; Zhang, X.-Y.; Jiang, B.-N. Advances in master planning of urban underground space (UUS) in China. *Tunn. Undergr. Space Technol.* **2016**, *55*, 290–307. [CrossRef]
16. Bobylev, N. Mainstreaming sustainable development into a city's Master plan: A case of Urban Underground Space use. *Land Use Policy* **2009**, *26*, 1128–1137. [CrossRef]
17. Broere, W. Urban underground space: Solving the problems of today's cities. *Tunn. Undergr. Space Technol.* **2016**, *55*, 245–248. [CrossRef]
18. Peng, F.-L.; Qiao, Y.-K.; Sabri, S.; Atazadeh, B.; Rajabifard, A. A collaborative approach for urban underground space development toward sustainable development goals: Critical dimensions and future directions. *Front. Struct. Civ. Eng.* **2021**, *15*, 20–45. [CrossRef]
19. Strategic Consulting Center of Chinese Academy of Engineering. *Blue Book of Urban Underground Space Development in China*; Strategic Consulting Center of Chinese Academy of Engineering: Beijing, China, 2021.
20. Li, X.; Xu, H.; Li, C.; Sun, L.; Wang, R. Study on the demand and driving factors of urban underground space use. *Tunn. Undergr. Space Technol.* **2016**, *55*, 52–58. [CrossRef]
21. Chen, Y.; Chen, Z.; Guo, D.; Zhao, Z.; Lin, T.; Zhang, C. Underground space use of urban built-up areas in the central city of Nanjing: Insight based on a dynamic population distribution. *Undergr. Space* **2022**, *7*, 748–766. [CrossRef]
22. Jian, P.; Flp, A.; Ny, B.; Tf, B. Factors in the development of urban underground space surrounding metro stations: A case study of Osaka, Japan. *Tunn. Undergr.* **2019**, *91*, 103009.
23. Tan, Z.; Roberts, A.C.; Christopoulos, G.I.; Kwok, K.-W.; Car, J.; Li, X.; Soh, C.-K. Working in underground spaces: Architectural parameters, perceptions and thermal comfort measurements. *Tunn. Undergr. Space Technol.* **2018**, *71*, 428–439. [CrossRef]
24. Li, W.J.; Liao, X.X.; Miao, G.Y.; Zhou, X.Z. Analysis on Built Environment in Underground Shopping Malls in Chongqing in P.R. China. *Adv. Mater. Res.* **2012**, *599*, 233–236. [CrossRef]
25. Xie, Y.; Liao, J.; Guo, C.; Zhang, K.; Zhang, J.; Yang, K. Lighting environment analysis of Chengdu underground commercial spaces. *IOP Conf. Series Mater. Sci. Eng.* **2020**, *741*. [CrossRef]
26. Chun, C.; Tamura, A. Thermal environment and human responses in underground shopping malls vs department stores in Japan. *Build. Environ.* **1998**, *33*, 151–158. [CrossRef]
27. Ganesh, G.A.; Sinha, S.L.; Verma, T.N.; Dewangan, S.K. Numerical simulation for energy consumption and thermal comfort in a naturally ventilated indoor environment under different orientations of inlet diffuser. *Build. Environ.* **2022**, *217*, 109071. [CrossRef]
28. Li, Y.; Geng, S.; Chen, F.; Li, C.; Zhang, X.; Dong, X. Evaluation of thermal sensation among customers: Results from field investigations in underground malls during summer in Nanjing, China. *Build. Environ.* **2018**, *136*, 28–37. [CrossRef]
29. Abbaspour, M.; Jafari, M.J.; Mansouri, N.; Moattar, F.; Nouri, N.; Allahyari, M. Thermal comfort evaluation in Tehran metro using Relative Warmth Index. *Int. J. Environ. Sci. Technol.* **2008**, *5*, 297–304. [CrossRef]
30. Kim, J.; Hong, T.; Lee, M.; Jeong, K. Analyzing the real-time indoor environmental quality factors considering the influence of the building occupants' behaviors and the ventilation. *Build. Environ.* **2019**, *156*, 99–109. [CrossRef]
31. Wen, Y.; Leng, J.; Shen, X.; Han, G.; Sun, L.; Yu, F. Environmental and Health Effects of Ventilation in Subway Stations: A Literature Review. *Int. J. Environ. Res. Public Heal.* **2020**, *17*, 1084. [CrossRef] [PubMed]
32. Tagiyeva, N.; Sheikh, A. Domestic exposure to volatile organic compounds in relation to asthma and allergy in children and adults. *Expert Rev. Clin. Immunol.* **2014**, *10*, 1611–1639. [CrossRef]
33. Luo, F.; Guo, H.; Yu, H.; Li, Y.; Feng, Y.; Wang, Y. PM2.5 organic extract mediates inflammation through the ERβ pathway to contribute to lung carcinogenesis in vitro and vivo. *Chemosphere* **2020**, *263*, 127867. [CrossRef] [PubMed]
34. Darby, S.; Hill, D.; Deo, H.; Auvinen, A.; Barros-Dios, J.M.; Baysson, H.; Bochicchio, F.; Falk, R.; Farchi, S.; Figueiras, A.; et al. Resi-dential radon and lung cancer—Detailed results of a collaborative analysis of individual data on 7148 persons with lung cancer and 14 208 persons without lung cancer from 13 epidemiologic studies in Europe. *Scand. J. Work. Environ. Health* **2006**, *32*, 1–84. [PubMed]
35. Gao, J.M.; Chen, L.; Zhang, Y.; Jin, F.; Li, B.Z. Occurrence and pollution source of TVOC in underground stores in Chongqing. *J. Cent. South Univ. (Sci. Technol.)* **2012**, *43*, 4554–4558.
36. Yang, J.; Wu, X.; Chan, K.T.; Yang, X. Investigation of Indoor Air Quality in the Underground Shopping Mall. In Proceedings of the 2010 4th International Conference on Bioinformatics and Biomedical Engineering, Chengdu, China, 18–20 June 2010; pp. 1–4. [CrossRef]
37. Tao, H.; Fan, Y.; Li, X.; Zhang, Z.; Hou, W. Investigation of formaldehyde and TVOC in underground malls in Xi'an, China: Concentrations, sources, and affecting factors. *Build. Environ.* **2015**, *85*, 85–93. [CrossRef]

38. Wen, Y.; Leng, J.; Yu, F.; Yu, C.W. Integrated design for underground space environment control of subway stations with atriums using piston ventilation. *Indoor Built Environ.* **2020**, *29*, 1300–1315. [CrossRef]

39. Dong, L.; He, Y.; Qi, Q.; Wang, W. Optimization of daylight in atrium in underground commercial spaces: A case study in Chongqing, China. *Energy Build.* **2021**, *256*, 111739. [CrossRef]

40. Aflaki, A.; Esfandiari, M.; Mohammadi, S. A Review of Numerical Simulation as a Precedence Method for Prediction and Evaluation of Building Ventilation Performance. *Sustainability* **2021**, *13*, 12721. [CrossRef]

41. da Graça, G.C.; Martins, N.R.; Horta, C.S. Thermal and airflow simulation of a naturally ventilated shopping mall. *Energy Build.* **2012**, *50*, 177–188. [CrossRef]

42. Kim, G.; Schaefer, L.; Lim, T.S.; Kim, J.T. Thermal comfort prediction of an underfloor air distribution system in a large indoor environment. *Energy Build.* **2013**, *64*, 323–331. [CrossRef]

43. Liu, C.; Li, A.; Yang, C.; Zhang, W. Simulating air distribution and occupants' thermal comfort of three ventilation schemes for subway platform. *Build. Environ.* **2017**, *125*, 15–25. [CrossRef]

44. Shen, H.; Zhang, M.; Wang, J. Air purification of urban underground space. *Build. Energy Effic.* **2015**, *43*, 27–29.

45. Guo, H.X.; Deng, M.R.; Li, Y. Effect of sunken plaza on ventilation performance of underground commercial buildings. *J. South China Univ. Technol. Nat. Sci.* **2014**, *42*, 114–120.

46. Chen, Z.; Xin, J.; Liu, P. Air quality and thermal comfort analysis of kitchen environment with CFD simulation and experimental calibration. *Build. Environ.* **2020**, *172*, 106691. [CrossRef]

47. Li, Q.; Yoshino, H.; Mochida, A.; Lei, B.; Meng, Q.; Zhao, L.; Lun, Y. CFD study of the thermal environment in an air-conditioned train station building. *Build. Environ.* **2009**, *44*, 1452–1465. [CrossRef]

48. *CN-CECS402-2015*; Operation and Management Standard of Urban Underground Space. China Planning Press: Beijing, China, 2015.

49. Cheng, L.; Li, B.; Cheng, Q.; Baldwin, A.N.; Shang, Y. Investigations of indoor air quality of large department store buildings in China based on field measurements. *Build. Environ.* **2017**, *118*, 128–143. [CrossRef]

50. *GB/T18883-2002*; Indoor Air Quality Standard, (In Chinese). Health Department: Beijing, China, 2002.

51. Zhang, K.; Zhang, X.; Li, S.; Wang, G. Numerical Study on the Thermal Environment of UFAD System with Solar Chimney for the Data Center. *Energy Procedia* **2014**, *48*, 1047–1054. [CrossRef]

52. Amodio, M.; Dambruoso, P.R.; de Gennaro, G.; de Gennaro, L.; Loiotile, A.D.; Marzocca, A.; Stasi, F.; Trizio, L.; Tutino, M. Indoor air quality (IAQ) assessment in a multistorey shopping mall by high-spatial-resolution monitoring of volatile organic compounds (VOC). *Environ. Sci. Pollut. Res.* **2014**, *21*, 13186–13195. [CrossRef] [PubMed]

53. Brohus, H.; Nielsen, P.V. Dispersal of exhaled air and personal exposure in displacement ventilated rooms. *Indoor Air* **2010**, *12*, 147–164.

54. Cheng, Y.H.; Yan, J.W. Comparisons of particulate matter, co, and co 2 levels in underground and ground-level stations in the taipei mass rapid transit system. *Atmos. Environ.* **2011**, *45*, 4882–4891. [CrossRef]

55. Shang, Y.; Li, B.; Baldwin, A.; Ding, Y.; Yu, W.; Cheng, L. Investigation of indoor air quality in shopping malls during summer in Western China using subjective survey and field measurement. *Build. Environ.* **2016**, *108*, 1–11. [CrossRef]

56. American Society of Heating, Refrigerating and Air Conditioning Engineers, Inc. *Ashrae Handbook of Fundamentals*; American Society of Heating, Refrigerating and Air Conditioning Engineers, Inc.: Atlanta, GA, USA, 1972.

 *processes*

Article

Pollution Dispersion and Predicting Infection Risks in Mobile Public Toilets Based on Measurement and Simulation Data of Indoor Environment

Ruixin Li [1], Gaoyi Liu [1,2], Yuanli Xia [1,*], Olga L. Bantserova [3], Weilin Li [1] and Jiayin Zhu [1]

[1] School of Water Conservancy and Civil Engineering, Zhengzhou University, Zhengzhou 450001, China
[2] MOE Key Laboratory of Thermo-Fluid Science and Engineering, School of Energy and Power Engineering, Xi'an Jiaotong University, Xi'an 710049, China
[3] Institute of Construction and Architecture, Moscow State University of Civil Engineering, 129337 Moscow, Russia
[*] Correspondence: summer13592218725@yeah.net

Abstract: Since the 21st century, in several public health emergencies that have occurred across the world, the humid enclosed environment of the toilet has become one of the places where bacteria, viruses, and microorganisms breed and spread. Mobile public toilets, as a supplement of urban fixed public toilets, are also widely used in densely populated areas. According to statistics, since the outbreak of COVID-19 in 2019, multiple incidents of people being infected by the COVID-19 virus due to aerosol proliferation in public toilets have been confirmed. It is an urgent issue to resolve the internal environmental pollution of mobile public health and reduce the risk of virus transmission in public spaces under the global epidemic prevention. This paper utilized a typical combined mobile public toilet as the research object and measured and evaluated the indoor thermal environment in real time over a short period of time. The diffusion mode and concentration change law of pollutants in mobile public toilets were predicted and analyzed based on CFD. Regression analysis was also used to clarify the relationship between indoor thermal environment variables and aerosol diffusion paths, and a ventilation optimization scheme was proposed to reduce the risk of virus transmission.

Keywords: mobile public toilets; thermal environment; public health security; mechanical ventilation; aerosol transmission risk

Citation: Li, R.; Liu, G.; Xia, Y.; Bantserova, O.L.; Li, W.; Zhu, J. Pollution Dispersion and Predicting Infection Risks in Mobile Public Toilets Based on Measurement and Simulation Data of Indoor Environment. *Processes* **2022**, *10*, 2466. https://doi.org/10.3390/pr10112466

Academic Editor: Avelino Núñez-Delgado

Received: 19 October 2022
Accepted: 14 November 2022
Published: 21 November 2022

Publisher's Note: MDPI stays neutral with regard to jurisdictional claims in published maps and institutional affiliations.

1. Introduction

Urban public toilets (UPTs), as urban infrastructures, are related to the basic physiological needs of people; they play an important role in the construction of modern civilization in cities, and also demonstrate the social and economic development and the government's attention to people's livelihood. Around the world, countries are paying more and more attention to the environmental conditions of public toilets. This is happening in response to the World Public Health Summit scheduled every year to discuss the practical hygiene improvement measures. Since the "toilet revolution" proposed by UNICEF, developing countries, such as China, the Philippines, and India, have made extensive efforts on the issue of public health construction. China first set off a vigorous "toilet revolution" in the countryside by continuously improving the quality requirements for toilets and the supporting facilities of public toilets. The Philippine government encouraged the construction of toilets with septic tanks, which India made a requirement for any village office to run. In 2011, Bill Gates also launched his "toilet reconstruction plan" to provide low-cost, hygienic, and sustainable toilet services to approximately 260 million people without sewers around the world. Developed countries, including Singapore, Canada, and Germany, also have extremely high requirements regarding the appearance and internal facilities of UPTs to create a healthy and comfortable toilet environment for people. Based on diversified toilet needs, mobile public toilets have been widely used and popularized in recent years. They are environmentally friendly and energy-saving. Further, in recent times, artificial-intelligence-based mobile public toilets have emerged. The mobile public restroom has the advantages of being movable, combinable, and convenient for transportation. It is almost not restricted by any working environment. It can be used as a good supplement to the traditional fixed public toilets, in the case a sudden increase in the flow of people. Recent research efforts have mainly focused on the design of the internal environment and the disposal methods of excrement within them [1,2]. However, there is little research on the thermal environment and sanitation issues in mobile toilets.

Public toilet facilities possess both hygienic, cultural, and economic values [3]. The main pollutant gases in the toilet are hydrogen sulfide, methyl mercaptan, ammonia, and indole. Ammonia is a colorless gas with a strong pungent odor [4]. Because of its high solubility, ammonia is often adsorbed to the skin, mucous membranes, and conjunctiva of the eyes, resulting in irritation and inflammation. Exposure to a large amount of ammonia in a short period of time in public toilets can cause tears, a sore throat, and a cough, accompanied by dizziness, headache, nausea, and other symptoms. Severe cases may be subject to pulmonary edema, adult respiratory distress syndrome, and respiratory irritation symptoms. Among them, the volatile "ammonia", with a strong offensive odor, is one of the most important pollutants in public toilets. Its pollution to air quality in small enclosed spaces is affected by the amount of use and seasons. Ao Yongan et al. [5] analyzed the diffusion law of toilet pollutants and found that the concentration of heavy gas is greater when the heavy gas is closer to the pollution source, while the concentration of light gas is the opposite. Compared with fixed public toilets, mobile public toilets are confined and narrow, making it is easier for them to become a highly dangerous place for various pathogens and bacteria. Because of human excretion, our environment will be forced to accept a large influx of pollutants [6]. Many scholars have analyzed public toilets, e.g., in hospitals and schools, and found that the surfaces of frequently contacted objects, such as internal door handles, faucets, floors, and flush buttons, are contaminated by bacteria and pathogens [7,8]. In the process of using public toilets, 31.6% of people are unwilling to flush the toilets, which intensifies the pollution of the toilets. Because of human excretion, our environment will be forced to accept a large influx of pollutants. In the 21st century, there have been three relatively serious public health incidents in the world, namely SARS in 2003, influenza H_1N_1, and the new coronavirus in 2019. In particular, SARS and new coronaviruses have occurred in narrow confined spaces and crowd gathering events in large spaces. The direct transmission of infectious viruses, such as the new coronavirus, is closely related to the methods and effects of indoor natural ventilation and mechanical ventilation. Effective urban ventilation prevents virus aerosols and other pollutants from accumulating in indoor space and promotes their dissipation.

Poor indoor air quality reduces the comfort experienced in the environment and can also harm our physical health [9]; this long-distance air transmission method is called aerosol transmission, which is one of the main forms of disease transmission [10–13]. The aerosol is a gaseous suspended dispersion system. From the perspective of fluid mechanics, the aerosol is essentially a multi-phase fluid with a continuous gas state and dispersed solid and liquid phases [14]. Among them, those that carry pathogens or bacteria are called microbial aerosols [15]. Viruses are the smallest micro-organisms with a diameter of 0.02–0.3 μm. Although viruses only survive in host cells, they can still attach to the respiratory secretions and other droplets to form virus aerosols without host cells and spread through the air, leading to infectious diseases, such as chickenpox, tuberculosis, measles, etc. [16]. The cleanliness of water closets not only has a visual effect, but the systematic cleaning of surfaces in any locations significantly reduces the risks of infection and the spread of various diseases [17]. In Hong Kong, following the Amoy Garden incident during the SARS outbreak in 2003, the virus entered the sewage pipe through feces and then entered the bathroom of the household from the floor drain, infecting more than 300 people within just ten days. The outbreak of the new coronavirus in 2019 proved the possibility of viruses and other microorganisms spreading through aerosols again. The aerosol transmission of the new coronavirus is regarded as droplets mixed in the air to form aerosols and cause infection after inhalation. "Stool-oral transmission" is most likely to be transmitted through breathing and aerosols. The sixth edition of the "New Corona-virus Diagnosis and Treatment Plan" was issued by the National Health Commission of China. This plan also proposed that "The new corona-virus can be isolated in feces and urine" and that "attention should be paid to the aerosol and contagious transmission by feces and urine". In addition, viruses such as SARS and PV-1 have also been proven to survive in the water environment. It takes 10 days for PV-1 to decrease in wastewater. Therefore, the water environment that contains the virus source in the bathroom is more dangerous [18].

In summary, improving the internal environment and sanitation of mobile public toilets is the main goal. Many scholars have conducted experiments on reducing the concentration of indoor pollutants and found that airflow can affect the escape probability of microorganisms and affect the diffusion of pollutants [19]. Li Cheng et al. [20] simulated the toilets in the World Expo and found that the use of mechanical ventilation and airflow organization with up and down distributed exhausts can ensure that the pollutant concentration and thermal comfort in the toilet meet the requirements. Li et al. [21] proved that optimizing the form of ventilation in hospitals can reduce the aerosol transmission of pathogens. Yue Gaowei [22] proposed that the use of same-side return air can effectively reduce the indoor formaldehyde concentration. Ji Shaojie [23] proposed that the H_2S concentration gradually decreases when the height of the partition increases, and the toilet that is located in the building with a high wind pressure can effectively improve the air quality in the bathroom. Liu Haoran [24] observed that when the wind pressure is 25 Pa and above, determined as the best wind pressure value, the indoor pollutants will not flow back. Tung et al. [25,26] concluded that the removal efficiency of pollutants was higher with the shorter distance between the toilet and the exhaust vent, and the ceiling exhaust can effectively remove the polluted gas from the toilet by gas tracing. Chung et al [27]. predicted the airflow rate and the characteristics of aerosol particle transmission in the toilet and analyzed that the use of floor air supplies can create a comfortable and appropriate internal environment for the next user. To improve the sanitary environment inside the toilet, Seo et al. proposed an exhaust method that directly sucks odor and bacteria from the inside of the toilet [28].

Although many studies have been carried out on indoor odor intensities, most have only focused on improving indoor ventilation efficiency. In previous reports, many efforts have been dedicated to urban fixed toilets. However, there are few studies on the internal hygiene of mobile public toilets and residential toilets. This paper is based on measuring the indoor thermal environment of mobile public toilets, using CFD to simulate the indoor thermal environment distribution and pollutant diffusion of mobile public toilets, find the main influencing factors of indoor pollutants in mobile public toilets, predict the spread of viruses (such as COVID-19) in mobile public toilets in emergencies to reduce the further transmission of viral aerosols, and provide an important reference for virus transmission risk and environmental quality control in mobile public toilets.

2. Field Test Method for the Thermal Environment

The most common types of mobile public toilets are single mobile toilets and composite mobile toilets. In densely populated urban areas, composite mobile toilets are used as a supplement to urban fixed mobile toilets, which is the same as the single mobile public toilets used in cabin hospitals during the global COVID-19 outbreak.

This study conducted a field test of three UPTs in Zhengzhou in August 2019. Zhengzhou is located in the south of the warm temperate zone and experiences a continental monsoon humid climate. The average daytime temperatures in Zhengzhou are 33.4 °C in August. The permanent resident population is 10.352 million.

2.1. Research Subjects

The main objective of this study was to investigate UPTs in Zhengzhou. Considering the current wide application of mobile public toilets in the city and the risk to public health security caused by using the frequency of mobile public toilets, this paper selected three groups of composite mobile public toilets located in Zhengzhou, China. The three groups of measured objects were all located in the high-density population area. Object A was located near the city's university town, while B was located in Zhengzhou Railway Station. Zhengzhou Railway Station is a comprehensive transportation hub integrating a high-speed railway, an inter-city railway, EMU trains, general-speed trains, and other passenger transportation services. It sends more than 40,000 passengers a day, and the number of people passing by is close to 150,000 every day. Object C was located in a local industrial park. The three groups of measured objects had the same structure and building materials, but different forms of ventilation, as shown in Table 1.

Table 1. Details for testing objects.

Object	Structure	Material	Towards	Shelter	Ventilation	Opening Size (mm)	Structure Size (mm) External	Structure Size (mm) Internal
A			North 350°	No	Natural ventilation	20 × 20	1100 × 1100 × 2300	970 × 970 × 2300
B	Composite	Metal Carved Board	Southeast 149°	Yes	Mechanical ventilation	29 × 29	1420 × 1540 × 2630	1120 × 1370 × 2000
C			West 285°	No	Intermittent mechanical ventilation	14.4 × 14.4	1250 × 1500 × 2250	1200 × 1400 × 2000

2.2. Test Principles

According to typical meteorological year (TMY) data, the high-temperature period for actual measurements was acquired from 11 August 2019 to 17 August 2019. Considering the effect of cloudy and sunny days on the indoor environment of the mobile public toilets and the accuracy of the test, the 72 h continuous test data were selected for analysis, with an interval of 5 min. The indoor and outdoor temperature and humidity; the temperature of the inner roof, the inner wall, and the floor; and the exhaust air velocity were measured. The actual measurement process was divided into three parts. The first measurement was temperature and humidity (indoor and outdoor). The second was the temperature of the internal roof, wall, and ground. The third was the air velocity at the exhaust outlet.

The indoor temperature was selected as the same value satisfying more than 80 percent of a person from the following equation performed by Ashrae [29]:

$$t_{in} = 0.31\overline{t_{pma(out)}} + 21.3 \tag{1}$$

The high-comfort indoor environment elements and control values are defined in ISO 7730 "Ergonomics of the thermal environment—Analytical determination and interpretation of thermal comfort using calculation of the PMV and PPD indices and local thermal comfort criteria". It is instructed that the indoor temperature should be no lower than 20 °C in winter and no higher than 26 °C in summer, and the indoor relative humidity is maintained between 40% and 60%.

According to "Urban Public Toilet Hygiene Standards" [30], for II UPTs, the indoor temperature should not be lower than 10 °C, the relative humidity should not be higher than 80%, and the odor intensity should not exceed level 2. It is required that the indoor temperature is maintained at 22–28 °C in summer, and the relative humidity at 40–80% in the standard. According to the "Healthy Building Evaluation Standard" (T/ASC 02) [31], "Mold can grow and spread fungal diseases when it is kept at a temperature of 25–30 °C, a humidity above 80% with sufficient oxygen conditions". It is recommended that the relative humidity is maintained in the range of 30% to 70% to reduce the cutaneous and visual stimulation of damp and dryness, as well as avoid the harm of static electricity, bacterial growth, and respiratory diseases, in order to help create a comfortable and healthy indoor air environment for the human body [32–34].

2.3. The Layout of the Test Point

The data collection interval was carried out in a shaded place to avoid direct sunlight. The indoor air temperature and humidity were measured at a height of 1100 mm from the ground (ASHRAE55-2013); the indoor roof, wall, and ground temperatures were measured in the middle area where the value was relatively stable. In addition, it was important to close the gate to reduce the impact on the outdoor environment. The wind speed of the exhaust outlet was selected at the center of the exhaust outlet, allowing the data to be recorded once the instrument was stable. During the test, the mobile public toilets were in normal use to achieve a more realistic evaluation and comparison of the indoor temperature, humidity, and thermal sensation of the mobile public toilets.

2.4. Test Instruments

Table 2 lists the instrument types and related errors for the test.

Table 2. Measurement equipment.

Equipment	Measuring Parameter	Equipment Parameter	Equipment Photos
FLUKE 971 temperature and humidity recorder	The indoor temperature and humidity	Temperature range: $-20\sim60$ °C, accuracy: ± 0.5 °C ($0\sim45$ °C); relative humidity range: $10\sim90\%$, accuracy: $\pm 2.5\%$ (23 °C)	
FLUKE VT02 infrared thermometer	The temperature of the wall	Temperature range: $-10\sim250$ °C, accuracy: ± 2 °C (14$\sim$482 °F); infrared band: 6.5 μm to 14 μm	
FLUKE 923 hot-wire anemometer	The airflow rate of the outlet	Wind speed range: $0.2\sim20$ m/s, accuracy: $5\% \pm 0.05$ m/s; temperature range $-20\sim60$ °C, accuracy ± 0.1 °C	

2.5. Aerosol Transmission Risk Assessment Approach Using Gas Tracers

The risk of aerosol transmission during the COVID-19 pandemic cannot be underestimated. Ge Zi-Yu et al. [35] argued that the routes of transmission are direct contact, as well as droplet and possible aerosol transmissions. Gorbunov [36] found that aerosol particles generated by coughing and sneezing can travel over 30 m. Based on the coupling effect between the CFD modelling of bioaerosol dispersion and the calculation of probability of contact events, Maria Portarapillo et al. [37] proposed a methodology to perform risk analysis of the virus spread. The use of tracer gas to simulate and discuss aerosol propagation risk is one of the main research methods. Rencken Gerhard K et al. [38] used tracer gases to approximate the flow and dispersion of aerosol–air mixtures.

Tracer gases are gases that can be mixed with air; they do not undergo chemical changes and can be detected at low concentrations [39]. Common tracer gases are carbon dioxide, hydrogen, and so on. C.J. Ghazi et al. [40] locally measured air leakage rates using carbon dioxide as a tracer gas. Aditya Togi et al. [41] proposed a new method of estimating the mass of total molecular gas in galaxies directly from pure mid-infrared rotational radiation using hydrogen as a tracer gas. Hydrogen sulfide and ammonia can be used as tracers for aerosols. Zhao J. et al. [42] used hydrogen sulfide as a tracer gas to investigate the effectiveness of forced ventilation strategies for eliminating the toxic and oxygen deficient atmospheres in confined-space manure storages. Kirk Helen et al. [43] considered that ammonia is an excellent tracer. Furthermore, the use of hydrogen sulfide and ammonia can not only simulate the aerosol propagation, but can also fully illustrate the distribution law and level of pollutants in the mobile public health. It is a novel topic worthy of discussion in this study.

3. Results and Discussion

3.1. Field Test Results for the Thermal Environment

3.1.1. Indoor Temperature and Humidity Distribution

To find the degree of indoor temperature and humidity affected by the outdoor environment, the curves of indoor and outdoor temperature and humidity were measured, as shown in Figures 1 and 2. The indoor temperature and humidity in summer were greatly affected by the outdoor temperature and humidity, and the indoor temperature and humidity were higher than the outdoor temperature and humidity, which shows that the material had poor heat storage properties. The indoor and outdoor temperatures showed the same changing trend. The indoor temperature ranged from 28.5 °C to 35.5 °C. The highest indoor temperature was 35.5 °C when the solar radiation was the strongest, while the lowest was 28.5 °C when the solar radiation was the weakest. Moreover, it is shown that the indoor temperature of samples A and B both exceeded the ASHRAE and indoor air quality standards, and thus only the temperature of sample C in the initial period meets the ASHRAE standard.

In summer, the maximum indoor humidity of the mobile public toilets could reach 60% in the peak off-duty period with a large flow of people. The changing trend of the indoor and outdoor relative humidity of A was the same because A was in a sparsely crowded area and indoor natural ventilation was adopted; the relative humidity of C suddenly rose at 16:30 with the largest flow of people. Among the three subjects, only UPT B used continuous mechanical ventilation; thus, B had the lowest relative humidity. It can be shown that the indoor humidity meets the highest ISO standard, but C does not meet the lowest ISO standard.

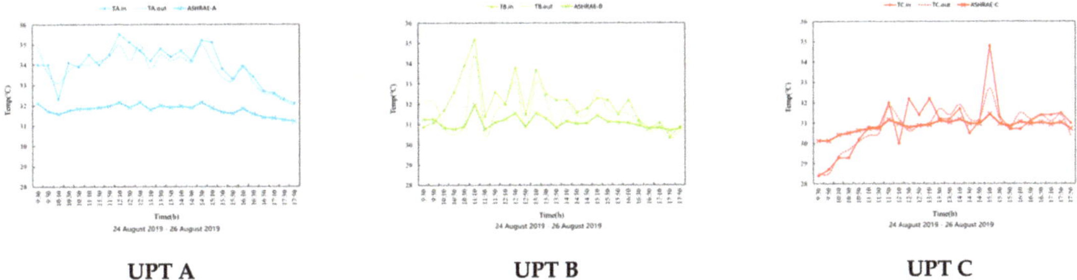

UPT A   **UPT B**   **UPT C**

Figure 1. Air temperature and relative humidity changes in tested UPTs.

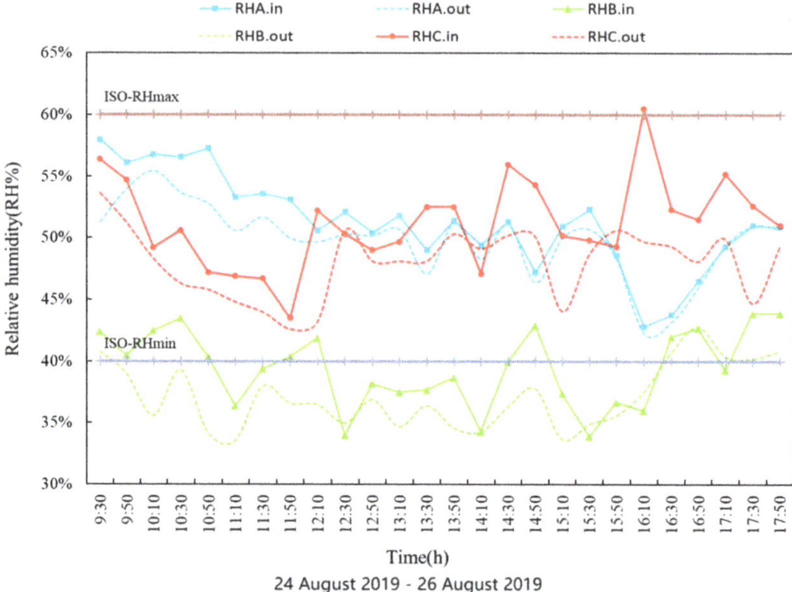

Figure 2. Combination of relative humidity changes in tested UPTs.

Living conditions of fungi were 22–36 °C and relative humidity was 95–100%. From the measurement data, the temperature in the mobile public toilets was just suitable for fungi growth. Some fungi, such as Cryptococcus neoformans and Aspergillus, etc., could infect the lungs of the human and cause harm to the body.

The pollutant gas molecules inside the toilet were always in constant motion. As the temperature increased, the movement speed of the pollutant molecules accelerated, which increased the concentration of air pollutants where the personnel breathes in the toilet. The increase in humidity was predicted to affect people's comfort, and the effective control of humidity could alleviate the growth of fungi in the building, but a proper increase in indoor humidity could reduce the survival period of the virus.

It has also been found that too low or too high humidity could increase the survival time of virus-carrying aerosols. The survival and inactivation time of viruses depended more on the humidity. When the relative humidity was close to 80%, the virus was the most active and had the longest survival time. Therefore, if the temperature and humidity inside the public toilets were effectively controlled, people's comfort could be improved and the growth of pathogens in the public toilets could be reduced.

3.1.2. Indoor Air Temperature Vertical Stratification

The temperature measurement of the inner roof, inner wall, and ground inside the mobile public toilets are shown in Figure 3. The temperature of the three groups of objects had obvious vertical stratifications. The temperature change trend of the inner roof and inner wall surface was the same, while that of the floor temperature was relatively large. The temperature decreased from the top to the bottom in the vertical direction. The inner roof had the highest temperature. This is mainly because the outer roof was exposed to strong solar radiation and was then transferred to the inner roof through heat conduction; the ground temperature was the lowest, but it was strongly affected by the outside environment. The floor temperature of C rose sharply at 11:50 and exceeded the temperature of the inner wall surface after using the toilet.

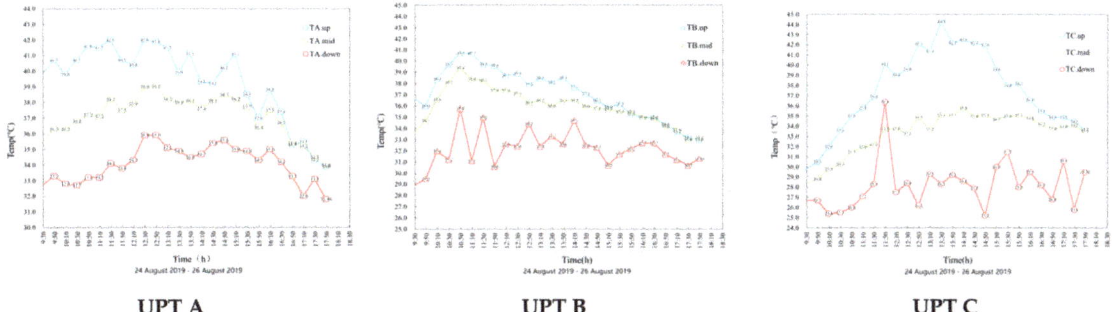

UPT A   **UPT B**   **UPT C**

Figure 3. Indoor air temperature vertical stratification of tested UPTs.

The pollutant gases inside the mobile public toilets were mainly ammonia and hydrogen sulfide. The density of ammonia gas was lower than air and mainly gathered in the upper layer of the public toilets, while the density of hydrogen sulfide was higher and mainly gathered in the lower layer of the public toilets. The movement rate of the ammonia molecules was faster with higher temperature, which was predicted to increase the concentration of ammonia in the breathing height of the personnel in toilets. The internal temperature ranged between 25 and 45 °C in the mobile public toilets. For the coronavirus, 40 °C was more likely to be inactivated than 20 °C under the same humidity. Therefore, viruses were mostly distributed in the lower space in the mobile public toilets. In addition, virus aerosols were also more stable at lower temperatures.

3.1.3. Evaluation and Analysis of Odor Intensity

An evaluation of the indoor odor intensity of mobile public toilets was mainly based on the subjective judgment of the human sense of smell. Eight people involved in scoring were all college students. The number of men and women was equal and they were in good health and had a sensitive sense of smell. The evaluation process was completed synchronously with the measured time node. The odor intensity adopted a six-level intensity representation: level 1 is odorless; level 2 is a weaker odor that can determine the properties of the odor; level 3 is an obvious odor that is easy to smell; level 4 is a strong odor; and level 5 is very strong odor. When the odor intensity exceeded level 3, it was considered that the atmosphere was foul-smelling. Figure 4 shows that the odor intensity of the three groups had the same changing trend. UPT A had the highest odor intensity, and was not in the odor pollution stage only 7.69% of the time. UPT A used natural ventilation and hot indoor conditions, resulting in a difficulty to discharge the odor, while UPTs B and C were not in the odor pollution stage 19.23% and 26.92% of the time, respectively. UPTs B and C both used mechanical ventilation. UPT B was located in the railway station with a large flow of people, while C was located in the suburbs far away from the city with a low flow of people, and the number of users was small.

The main pollutants in the toilet were ammonia, hydrogen sulfide, etc. The pollutant that caused bad smells was mainly hydrogen sulfide. When the concentration of hydrogen sulfide reached 0.00041 ppm, people already started to smell the odor. Therefore, when the room was polluted by bad odors, the hydrogen sulfide content in the mobile public toilets was already high. This shows that, in summer, indoor aerosol pollutants are easy to accumulate inside mobile public toilets, causing the rapid deterioration of the internal environment and posing a threat to people's health.

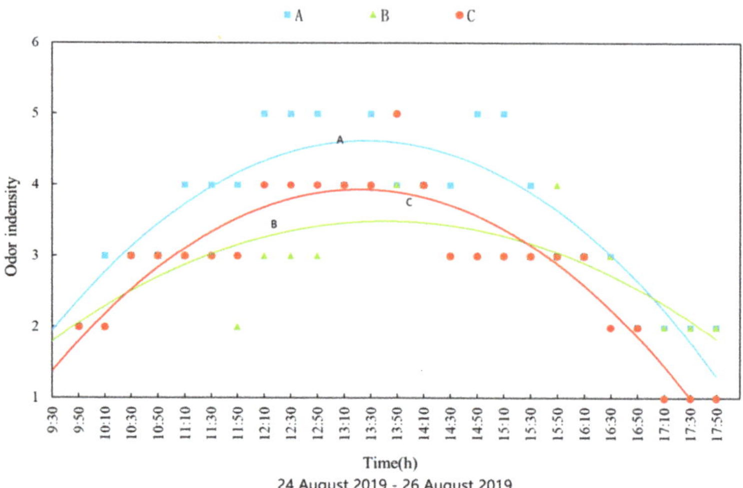

Figure 4. Subjective evaluation of indoor odor intensity in tested UPTs.

4. Numerical Simulation

4.1. Simulation Model

A numerical model was built based on the actual mobile public toilets size and room layout. The three-dimensional numerical model used using an Airpak design modeler, as shown in Figure 5. A mobile toilet with x × y × z = 1100 mm × 1100 mm × 2300 mm dimensions was also used, and the mobile toilets were modeled with a down-supply and up-supply ventilation mode. A 1730 mm male standing model, a 200 mm × 100 mm air inlet, a 200 mm × 200 mm air outlet, and a 420 mm × 220 mm pollutant release source were placed in the center of the ground. The main pollutant was released as ammonia and hydrogen sulfide.

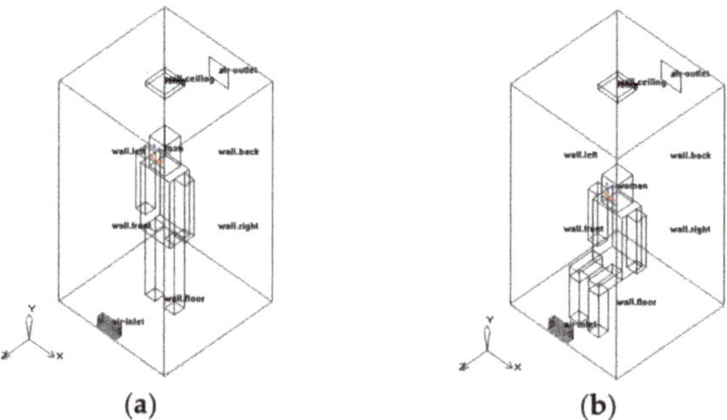

Figure 5. The three-dimensional numerical model: (**a**) standing position; (**b**) sitting position.

4.2. Model Equations

The air in the toilet was assumed to have low velocity, impressible, and turbulent fluid characteristics. Therefore, airflow can be regarded as continuous fluid and can be represented by the followed equation.

Ideal gas law:

$$P = \rho RT \tag{2}$$

Continuity equation:

$$\frac{\partial u}{\partial x} + \frac{\partial v}{\partial y} + \frac{\partial w}{\partial z} = 0 \tag{3}$$

Energy conservation equation:

$$u\frac{\partial T}{\partial x} + v\frac{\partial T}{\partial y} + w\frac{\partial T}{\partial z} = a\left(\frac{\partial^2 T}{\partial x^2} + \frac{\partial^2 T}{\partial y^2} + \frac{\partial^2 T}{\partial z^2}\right) \tag{4}$$

After a comprehensive consideration of simulation reliability, a standard zero equation was used to generate the turbulence model.

Zero Equation

This is a model that does not apply differential equations but uses algebraic relations to connect the vortex viscosity coefficient to the time means value [44]. The zero-equation model can accurately determine the indoor air distribution, air temperature, and pollutant concentration [45,46], and the turbulent viscosity can be determined from

$$\mu_t = \rho l^2 S \tag{5}$$

where ρ, l, and S are the fluid density, mixed length, and modulus of the average strain rate tensor, respectively. l and S are given by $l = \min(kd, 0.09d_{\max})$ and

$$S = \sqrt{2S_{ij}S_{ij}}$$

where k = 0.0417 is the von Kármán constant; d, d_{max}, and S_{ij} are a respective distances from the wall, the maximum distance, and average strain rate tensor, given by:

$$S_{ij} = \frac{1}{2}(\frac{\partial u_j}{\partial x_i} + \frac{\partial u_i}{\partial x_j}) \tag{6}$$

where u is the velocity.

4.3. Boundary Conditions

The isothermal boundary condition was employed in the walls of mobile toilets in this simulation when λ was 0.4. The air temperature outside the wall was based on the experimental data. The measured data were used in the temperature of the floor. The solar radiation intensity (W/m^2) on a typical day at 12 a.m. had a strong influence on the wall heating process. The inlet boundary conditions of the domain, including a free inlet, experimental velocity, and air temperature, were used in the outlet boundary with a standard intensity of pressure of 101.325 kPa. In addition, the following assumptions were made in the present simulation:

(1)   It could be assumed that the air is uniformly in compressible;
(2)   In the steady state, the 3D flow could be fully developed;
(3)   The influence of cold air penetration could be ignored on the heat transfer process;
(4)   All thermo-physical fluid properties could be considered constant.

4.4. Meshing and Model Validation

Considering the room size, the maximum size of the fine grid in the X, Y, and Z directions was set to 0.2 m. The grid at the exhaust outlet, air inlet, and pollution source was finely processed. After calculation, the parameters of the fine grid were reported as follows: NODES: 25134, HEXAS: 21761, QUADS: 6634, Faces + solids: 43. The meshing of the field was carried out using structural mesh, as shown in Figure 6.

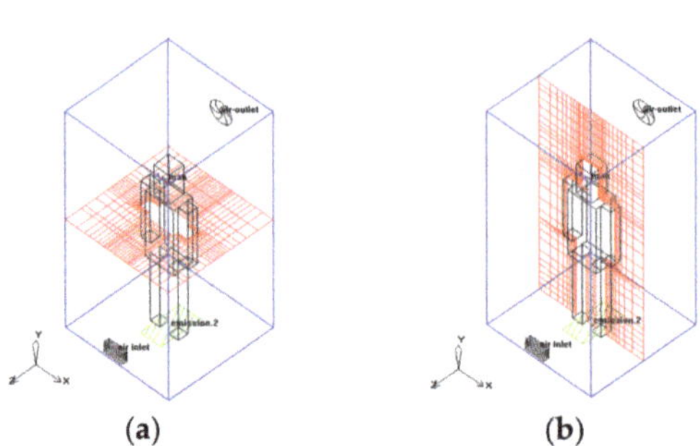

(a)   (b)

Figure 6. Structured mesh of computational domain: (**a**) XZ plane view; (**b**) XY plane view.

The parameters representing the quality of the mesh were as follows: Face alignment = 0.986928 > 1; Quality = 0.882239 > 1 min (quality (volume)) = 1.10135×10^{-6} max (quality (volume)) = 0.986928 > 1 when the mesh quality is good [47].

Correlation analysis was performed by combining the measured data of indoor air temperature and humidity with the simulated results. The coefficient of determination $R^2 = 0.6987 > 0.5$ and $R^2 = 0.891 > 0.5$ was obtained by regression analysis (as shown in Figure 7). This indicates that the simulated results have a good correlation with the measured results and that the constructed physical model is reliable.

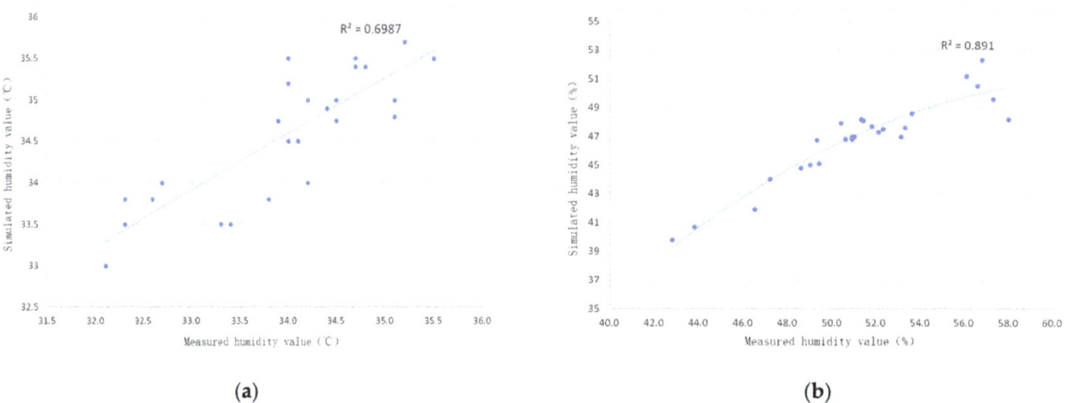

(a)   (b)

Figure 7. Correlation analysis between experiment and simulation: (**a**) air temperature; (**b**) relative humidity.

4.5. Simulation Results

4.5.1. Pollution Distribution

According to the World Health Organization's "Air Quality Guidelines", when the stay interval is within ten minutes, the average sulfide content should not exceed 500 ug/m^3. According to the "Urban Public Toilet Hygiene Standards", for class II public toilets, the indoor ammonia concentration should not exceed 1 mg/m^3, and the hydrogen sulfide concentration should not exceed 0.01 mg/m^3. According to the "Indoor Air Quality Standard" [48], indoor ammonia should not exceed 0.2 mg/m^3.

To explore the distribution of indoor pollutants in mobile public toilets, the distribution of indoor pollutants is shown in Figure 5. NH_3 was released from the pollution source. Due to the effect of the air inlet, NH_3 diffused along the ground surface at the rear of the mobile toilets and then spread upwards along the back wall, so the ammonia concentration at the bottom of the mobile toilets was lower. The ammonia concentration around the human body was 2.8 ppmv (2.16 mg/m^3). Thus, this meets the sanitary standards of three types of UPTs, but the indoor air quality does not meet the standard requirements. The NH_3 concentration was higher near the wall.

As shown in Figure 8, after H_2S was released from the pollution source, outdoor air entered the room from the air inlet. H_2S diffused toward the rear of the mobile toilets and then spread upward along the wall. At the two corners of the rear of the mobile toilets, the H_2S concentration was higher. This is probably because the pollutant H_2S tended to accumulate in the corners. The H_2S concentration around the human body was 0.03 ppmv (0.046 mg/m^3). The H_2S in the mobile public toilets does not meet the standards, and when the H_2S concentration exceeded 0.0041 ppmv, the odor could be smelled. If 0.41 ppmv was reached, a strong odor could be smelled.

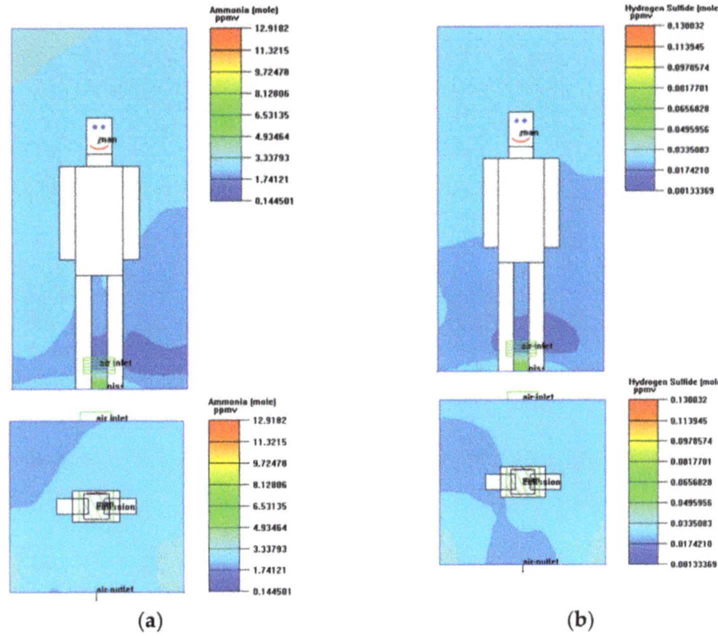

(a)   (b)

Figure 8. Concentration distribution of (**a**) NH_3 and (**b**) H_2S in the UPTs.

4.5.2. Effects of Indoor Pollutant Concentration

To study the effect of indoor temperature on the concentration of pollutants, correlation analysis between temperature and the concentration of pollutants was carried out (Figure 9a), and the determination coefficient of ammonia was $R^2 = 0.6876 > 0.5$, showing that the concentration of ammonia has a positive and strong correlation with temperature. As the ambient temperature increased, the ammonia concentration increased, but the determination coefficient of hydrogen sulfide was $R^2 = 0.027 < 0.09$, indicating that there is a scarce correlation between the hydrogen sulfide concentration and temperature.

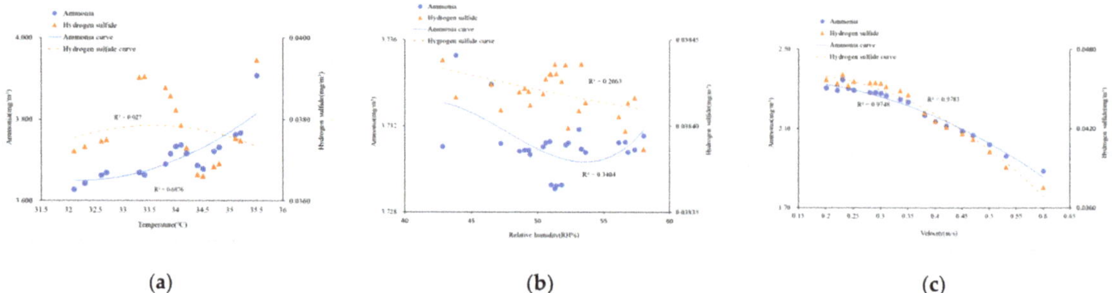

(a)   (b)   (c)

Figure 9. (**a**) Regression analysis of pollutant concentration with indoor air temperature; (**b**) regression analysis of pollutant concentration with relative humidity; (**c**) regression analysis of pollutant concentration with airflow.

To study the effect of indoor humidity on the concentration of pollutants, correlation analysis between humidity and the concentration of pollutants was carried out (Figure 9b), and the determination coefficient of ammonia gas was $R^2 = 0.3404 < 0.5$, while the determination coefficient of hydrogen sulfide was $R^2 = 0.2063 < 0.5$ Therefore, the concentration of ammonia and hydrogen sulfide has a weak negative correlation with humidity.

To study the effect of the wind speed at the exhaust outlet on the concentration of pollutants, correlation analysis between the wind speed and the concentration of pollutants was carried out (Figure 9c), and the determination coefficient of ammonia was $R^2 = 0.9783 > 0.8$, while the determination coefficient of hydrogen sulfide was $R^2 = 0.9748 > 0.8$. Therefore, the concentration of ammonia and hydrogen sulfide has a strong negative correlation with the wind speed. As the wind speed increased, the exhaust air volume from the exhaust vent increased and the indoor pollutant concentration decreased.

4.5.3. Indoor Air Velocity Distribution

It can be seen that the concentration of pollutants has a strong correlation with wind speed. The higher the wind speed is, the lower the concentration of indoor pollutants. The air velocity around the human body was 0.06 m/s, as shown in Figure 10, which does not meet the quality standards [48]. The air velocity on the left side of the human body was high, with low pollutant concentration on the same side of the human body. When fresh air entered the room, the wind speed gradually decreased from the direction of the air outlet, and the wind speed close to the human body reached 0.71 m/s.

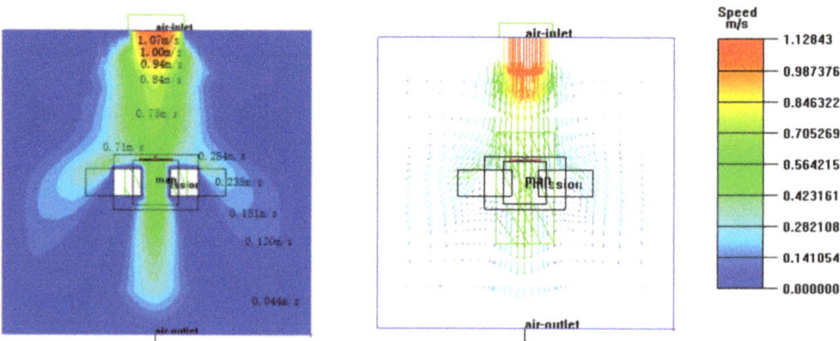

Figure 10. Cloud chart at Y = 0.15 m.

4.5.4. Indoor Air Age Distribution

The age of air reflected the freshness of indoor air, making it an important indicator for evaluating indoor air quality. The air quality in the lower part of the mobile public toilets is clearly shown in Figure 11. With the increase in pollutants, the air quality in the upper part worsened. The air quality was the worst behind the head of the users where bacteria were most likely to grow. In addition, the air quality at the breathing height of the human body was also poor, which is the main factor leading to the high odor intensity.

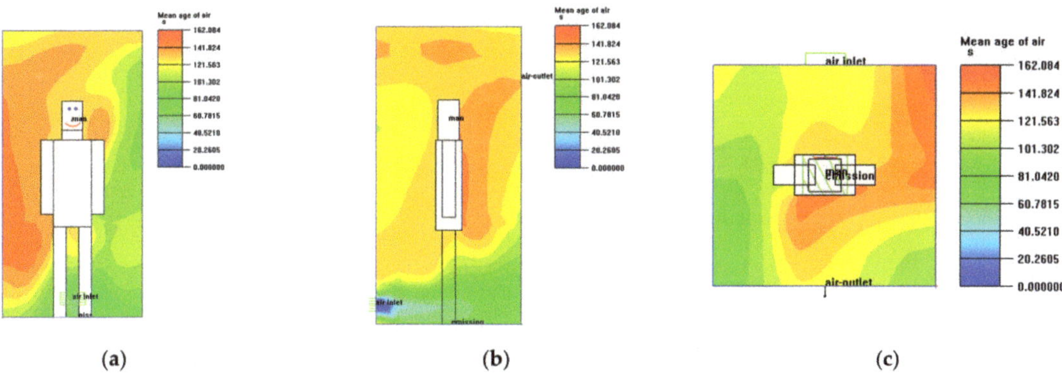

(a)   (b)   (c)

Figure 11. Air age distribution at (**a**) Z = 0.55 m, (**b**) X = 0.55 m, and (**c**) Y = 1.1 m.

5. Ventilation Methods for Reducing Public Health Safety Risks

5.1. H_2S and NH_3 Mass Concentrations in Different Ventilation Modes

It was shown that the concentration of pollutants has the strongest correlation with wind speed, so the airflow was reorganized by changing the indoor ventilation form (Figures 12 and 13). In Figure 12, the indoor NH_3 concentration decreased when the downward ventilation mode was adopted, and that of the human breathing height was 2 ppmv (1.5 mg/m^3), which meets the hygienic standards of the second-class UPTs. Heavy ammonia was discharged outdoors after it was released from the pollution source, and part of it was diluted by indoor fresh air. The indoor concentration of H_2S was also reduced, and the concentration of hydrogen sulfide around the human body was 0.015 ppmv (0.023 mg/m^3), which still does not meet the sanitary standards of UPTs.

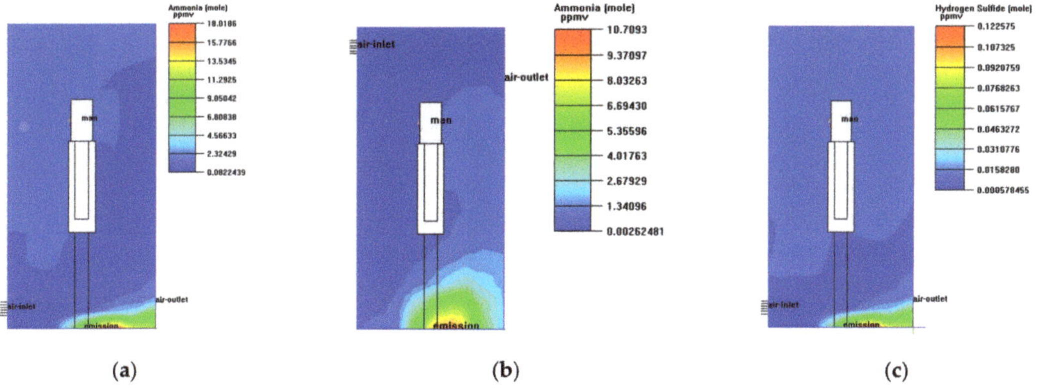

(a)   (b)   (c)

Figure 12. Concentration distribution of NH_3 in different ventilations: down-supply and down-exhaust, up-supply and up-exhaust, up-supply and down-exhaust, respectively. (**a**–**c**): Pollution distribution at X = 0.55 m.

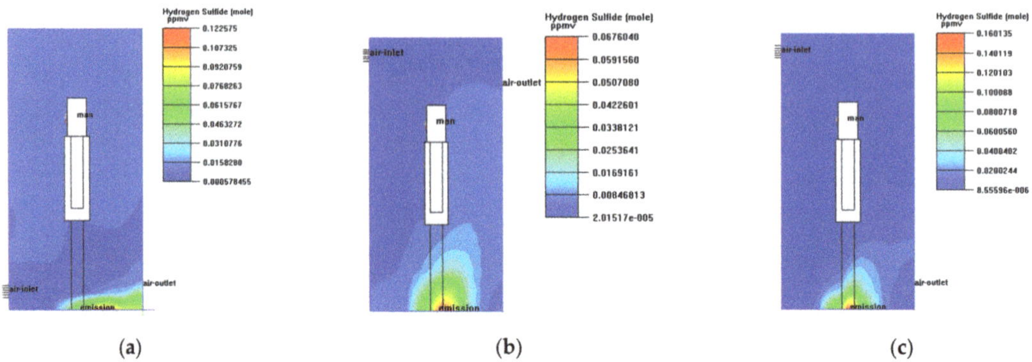

(a)   (b)   (c)

Figure 13. Concentration distribution of H_2S in different ventilations: down-supply and down-exhaust, up-supply and up-exhaust, up-supply and down-exhaust, respectively. (**a–c**): Pollution distribution at X = 0.55 m.

Figure 12 shows that most of the indoor ammonia gas accumulated in the lower part of the mobile public toilets and the back wall when the up-supply and up-discharge ventilation mode was adopted. The concentration of ammonia gas around the user was about 0.67 ppmv (0.52 mg/m^3), which conforms to the hygienic standards for second-class UPTs. The distribution of H_2S concentration was similar to NH_3, and the concentration in the breathing area of users was about 0.004 ppmv (0.007 mg/m^3), which meets the norms for first-class urban toilets.

Figure 13 shows that the ammonia concentration around the indoor user was about 0.0017 ppmv (0.0013 mg/m^3) when the up-supply and down-exhaust ventilation mode was adopted. The concentration of H_2S at the breathing height of the toilet users was about 8.6×10^{-6} ppmv (1.3×10^{-5} mg/m^3); thus, this pollutant concentration meets the standards for first-class UPT sanitation. The toilet users in the mobile public toilets could hardly smell the odor. The concentration of pollutants in mobile public toilets was drastically reduced, and the health threat to the human body also decreased. The aerosol concentration at the height of breathing under different ventilation modes is shown in Figure 14.

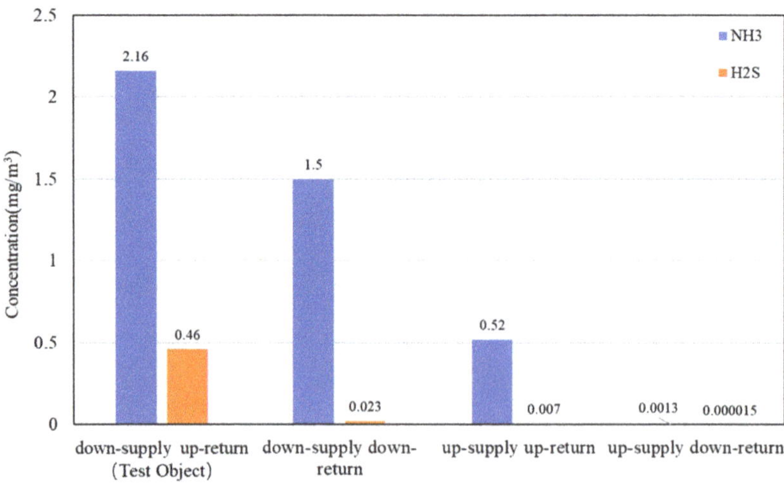

Figure 14. The aerosol concentration at the height of breathing under different ventilation modes.

5.2. Indoor Air Age in Different Ventilation Modes

Figure 15 shows that when the downward ventilation mode was adopted, although it could expel pollutants quickly, the indoor airflow organization was poor and the fresh air entering the room could hardly reach the breathing height of toilet users.

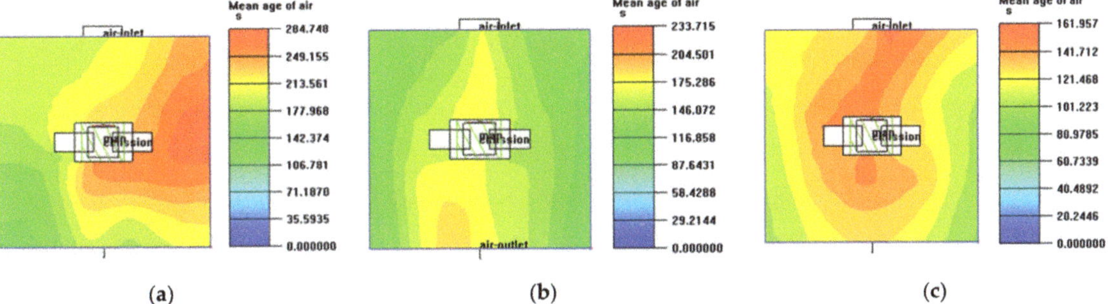

Figure 15. Age of air in the UPTs at Z = 1.15 m with (**a**) down-supply and down-exhaust, (**b**) up-supply and up-exhaust, and (**c**) up-supply and down-exhaust ventilation modes.

When the up-supply and down-exhaust were adopted, the fresh air entered the room from the top inlet and then the fresh air sank, resulting in better air quality in the toilet. When the up-supply and down-exhaust ventilation mode was used, the air quality was good and the indoor airflow organization was suitable.

6. Conclusions

Unlike the outdoor open environment which is usually cleaned automatically by wind, mobile toilets are continuously closed with the pollution released from its source; thus, contamination may severely accumulate and eventually raise the risk of infection and lower the comfortability. In the present review, the design of humanization was discussed. The innovation of the study was to utilize CFD simulation to analyze the problem of tracer gas to reflect the danger of aerosol transmission, and the pollution distribution of urban mobile public health was studied for the first time. The results can be summarized as follows.

(1) There is no obvious difference in the change rules of thermal environment in the three mobile public toilets. Temperature and humidity have the same large amplitude and short delay, the values of which are acceptable for people based on GB/T 17217-1998. However, odor intensity indicates that the environment in the mobile public toilets is unadopted but suitable for bacteria.

(2) Different characteristics in the thermal environments can affect pollution dispersion. An increase in the temperature increases the intensity of ammonia. The temperature increased by 1 °C and the ammonia concentration increased by 0.036 mg/m^3. While the hydrogen sulfide concentration does not have a significant correlation with temperature, and an increase in humidity does not drastically change the pollution concentration in mobile public toilets. This shows that this aerosol pollution dispersion will be stronger when the temperature is varied.

(3) The pollution concentration level increases linearly with air velocity. Exhaust air volume and pollutant concentration show an obvious inverse relationship. With an increase of 0.1 m/s in the wind speed at the exhaust air outlet, the indoor ammonia concentration and hydrogen sulfide concentration were reduced by 0.1 mg/m^3 and 0.002 mg/m^3, respectively. Thus, it is possible to reduce the aerosol pollution concentration with enough ventilation.

(4) It is best to use up-supply and down-exhaust for pollution discharge in mobile public toilets. Compared to the measured object, ammonia and hydrogen sulfide concentration decreased by 2.2 mg/m^3 and 0.046 mg/m^3, respectively, indicating that

there is a higher potential for improving the indoor environment in mobile public toilets. Therefore, the air distribution mode of mobile toilets is extremely important.

Author Contributions: Conceptualization, R.L., G.L. and Y.X.; methodology, R.L. and G.L.; software, R.L. and G.L.; validation, R.L., G.L. and O.L.B.; formal analysis, G.L., W.L. and O.L.B.; investigation, G.L. and Y.X.; resources, W.L. and J.Z.; data curation, R.L. and G.L.; writing—original draft preparation, G.L.; writing—review and editing, R.L., G.L. and Y.X.; visualization, G.L. and Y.X.; supervision, R.L., J.Z. and G.L.; project administration, R.L. and O.L.B.; funding acquisition, R.L. and J.Z. All authors have read and agreed to the published version of the manuscript.

Funding: This research was supported by a grant from the National Natural Science Foundation of China (grant number: 51808506) and Science and Technology Department of Henan Province (grant number: 212102310573).

Institutional Review Board Statement: Not applicable.

Informed Consent Statement: Informed consent was obtained from all subjects involved in the study.

Data Availability Statement: Not applicable.

Acknowledgments: The authors would like to acknowledge and sincerely thank Yaling He, Yang Bai, and Jingshu Men for their valuable help in the experimental process of this study.

Conflicts of Interest: The authors declare no conflict of interest. The funders had no role in the design of the study; in the collection, analyses, or interpretation of the data; in the writing of the manuscript; or in the decision to publish the results.

Nomenclature

Parameters

t_{in}	The indoor temperature of the object, $^\circ$C
μ_t	The turbulent viscosity, $\mathrm{N \cdot s/m^2}$
ρ	The fluid density, $\mathrm{kg/m^3}$
l	The mixed length of the fluid, m
S	The modulus of the average strain rate tensor of the fluid, $\mathrm{s^{-1}}$
k	The von Kármán constant
d	The respective distance from the wall, m
u	The fluid velocity, m/s
P	The fluid pressure, Pa
R	The molar gas constant
T	The thermodynamic temperature of the fluid, K
Abbreviations	
UPT	Urban public toilet
EMU	Electric multiple unit
TMY	Typical meteorological year
A, B, C	Object A, object B, and object C
HVAC	Heating, ventilation, and air conditioning
CFD	Computational fluid dynamics
Subscripts	
max	Maximum of data
min	Minimum of data

References

1. Piao, S. Research on design of movable simple public toilets by the street. In Proceedings of the 2006 International Conference on Industrial Design & The 11th China Industrial Design Annual Meeting, Hangzhou, China, 18–20 November 2006.
2. Alepu, O.E. Integrated Technology Development of Feces Disinfection, Odor Control and Resource Oriented Treatment for UDDT. Ph.D. Thesis, University of Science and Technology Beijing, Beijing, China, 2019. Available online: https://kns.cnki.net/KCMS/detail/detail.aspx?dbname=CDFDLAST2019&filename=1019146321.nh (accessed on 30 May 2019).
3. Fu, B.; Xiao, X.; Li, J. Big Data-Driven Measurement of the Service Capacity of Public Toilet Facilities in China. *Appl. Sci.* **2022**, *12*, 4659. [CrossRef]

4. Mawumenyo, S.B.; Patrick, B.; Sampson, O.K.; Damertey, S.D.; Esi, A.; Appiah, O.P.; Kobina, A. Inhibition of ammonia and hydrogen sulphide as faecal sludge odour control in dry sanitation toilet facilities using plant waste materials. *Sci. Rep.* **2021**, *11*, 17803.

5. Ao, Y.; Wang, L.; Jia, X. Numerical simulation of pollutant diffusion law in bathroom and optimization of the locations and ways of exhaust outlet and makeup air. *J. Shenyang Jianzhu Univ. (Nat. Sci.)* **2011**, *27*, 720–724.

6. Lee, H.K. Designing a Waterless Toilet Prototype for Reusable Energy Using a User-Centered Approach and Interviews. *Appl. Sci.* **2019**, *9*, 919. [CrossRef]

7. Zhang, W.; Liang, J.; Lin, J. Contamination state of Staphylococcus aureus from toilets of primary schools in Guangzhou. *Chin. J. Sch. Health* **2020**, *41*, 295–297.

8. Zhang, L.; Liu, W. Investigation on the pollution status of toilet door handles on passenger trains of Xi'an Railway Administration. *Pract. Prev. Med.* **2016**, *23*, 97–98.

9. Zhao, L.; Zhang, H.; Wang, Q.; Sun, B.; Liu, W.; Qu, K.; Shen, X. Digital Twin Evaluation of Environment and Health of Public Toilet Ventilation Design Based on Building Information Modeling. *Buildings* **2022**, *12*, 470. [CrossRef]

10. Barker, J.; Stevens, D.; Bloomfield, S.F. Spread and prevention of some common viral infections in community facilities and domestic homes. *J. Appl. Microbiol.* **2001**, *91*, 7–21. [CrossRef]

11. Mendell, M.J.; Fisk, W.J.; Kreiss, K.; Levin, H.; Alexander, D.; Cain, W.S.; Girman, J.R.; Hines Cynthia, J.; Jensen Paul, A.; Milton, D.K.; et al. Improving the health of workers in indoor environments: Priority research needs for a national occupational research agenda. *Am. J. Public Health* **2002**, *92*, 1430–1440. [CrossRef]

12. Nicas, M.; Nazaroff, W.W.; Hubbard, A. Toward understanding the risk of secondary airborne infection: Emission of respirable pathogens. *J. Occup. Environ. Hyg.* **2005**, *2*, 143–154. [CrossRef]

13. Tellier, R. Review of aerosol transmission of influenza A virus. *Emerg. Infect. Dis.* **2006**, *12*, 1657–1662. [CrossRef]

14. Wang, X.; Wang, Y.; Liu, Y. *Hygienics Dictionary*; Huaxia Publishing House: Beijing, China, 2000; p. 554.

15. *Environmental Science Dictionary*; China Environmental Press: Beijing, China, 1991; p. 589.

16. Tellier, R.; Li, Y.; Cowling, B.J.; Tang, J.W. Recognition of aerosol transmission of infectious agents: A commentary. *BMC Infect. Dis.* **2019**, *19*, 101. [CrossRef]

17. Berik, T.; Zhanerke, B.; Martin, S. Assessing Access to WASH in Urban Schools during COVID-19 in Kazakhstan: Case Study of Central Kazakhstan. *Int. J. Environ. Res. Public Health* **2022**, *19*, 6438.

18. Gundy, P.M.; Gerba, C.P.; Pepper, I.L. Survival of Coronaviruses in Water and Wastewater. *Food Environ. Virol.* **2008**, *1*, 294–299. [CrossRef]

19. Liu, Z.; Wang, L.; Rong, R.; Fu, S.; Cao, G.; Hao, C. Full-scale experimental and numerical study of bioaerosol characteristics against cross-infection in a two-bed hospital ward. *Build. Environ.* **2020**, *186*, 107373. [CrossRef]

20. Li, C.; Wang, R.; Lu, Z.; Zhai, X. Research on ventilation airflow organization and indoor environment for public lavatory. In Proceedings of the 2009 Annual Conference of the Chinese Society of Refrigeration, Tianjing, China, 1 November 2009.

21. Li, Y. An evaluation of the ventilation performance of new SARS isolation wards in nine Hospitals in Hong Kong. *Indoor Built Environ.* **2007**, *16*, 400–410. [CrossRef]

22. Yue, G.; Lu, M.; Ji, H. Numerical simulation of indoor pollutant diffusion on ventilation optimization. *Fluid Mach.* **2014**, *42*, 81–85.

23. Ji, S.; Mao, J. Pollutant concentration research of toilet in natural ventilation conditions. *Contam. Control Air Cond. Technol.* **2012**, *4*, 22–24.

24. Liu, H.; Wei, X.; Chou, X. Simulation calculation of the distribution law of pollutant diffusion concentration in office based on FLUENT. *J. Jilin Jianzhu Univ.* **2019**, *36*, 48–52.

25. Tung, Y.; Hu, S.; Tsai, T. Influence of bathroom ventilation rates and toilet location on odor removal. *Build. Environ.* **2009**, *44*, 1810–1817. [CrossRef]

26. Tung, Y.; Shih, Y.; Hu, S.; Chang, Y. Experimental performance investigation of ventilation schemes in a private bathroom. *Build. Environ.* **2010**, *45*, 243–251. [CrossRef]

27. Chung, K.-C.; Chiang, C.-M.; Wang, W.-A. Predicting contaminant particle distributions to evaluate the environment of lavatories with floor exhaust ventilation. *Build. Environ.* **1997**, *32*, 149–159. [CrossRef]

28. Xu, M. Improvement of ventilation in residential toilets. *Hous. Sci.* **1994**, *5*, 39.

29. ASHRAE. Thermal environmental conditions for human occupancy. In *ASHRAE-55*; American Society of Heating, Refrigerating and Air-Conditioning Engineers: Atlanta, GA, USA, 2013.

30. *GB/T 17217—1998*; Hygienic Standard for Communal Toilet in City. Standards Press of China: Beijing, China, 1998.

31. *T/ASC 02-2016*; Assessment Standard for Healthy Building. Standards Press of China: Beijing, China, 2016.

32. Wang, H.-H.; Lin, K.-J.; Chu, S.-H.; Chen, H.-W.; Chiang, Y.-J.; Lin, P.-H.; Wang, T.-M.; Liu, K.-L. The impact of climate factors on the prevalence of urolithiasis in Northern Taiwan. *Biomed. J.* **2014**, *37*, 24–30. [CrossRef] [PubMed]

33. Sirohi, M.; Katz, B.F.; Moreira, D.; Dinlenc, C. Monthly variations in urolithiasis presentations and their association with meteorologic factors in New York City. *J. Endourol.* **2014**, *28*, 599–604. [CrossRef] [PubMed]

34. Akers, T.G.; Bond, S.; Goldberg, L.J. Effect of temperature and relative humidity on survival of airborne Columbia SK group viruses. *Appl. Microbiol.* **1966**, *14*, 361–364. [CrossRef]

35. Ge, Z.-Y.; Yang, L.-M.; Xia, J.-J.; Fu, X.-H.; Zhang, Y.-Z. Possible aerosol transmission of COVID-19 and special precautions in dentistry. *J. Zhejiang Univ.* **2020**, *21*, 361–368. [CrossRef]

36. Gorbunov, B. Aerosol particles laden with viruses that cause COVID-19 travel over 30m distance. *J. Prepr.* **2020**, *546*, v2. [CrossRef]
37. Portarapillo, M.; di Benedetto, A. Methodology for risk assessment of COVID-19 pandemic propagation. *J. Loss Prev. Process Ind.* **2021**, *72*, 104584. [CrossRef]
38. Rencken, G.K.; Rutherford, E.K.; Ghanta, N.; Kongoletos, J.; Glicksman, L. Patterns of SARS-CoV-2 aerosol spread in typical classrooms. *Build. Environ.* **2021**, *204*, 108167. [CrossRef]
39. Sherman, M.H. Tracer-gas techniques for measuring ventilation in a single zone. *Build. Environ.* **1990**, *25*, 365–374. [CrossRef]
40. Ghazi, C.J.; Marshall, J.S. A CO_2 tracer-gas method for local air leakage detection and characterization. *Flow Meas. Instrum.* **2014**, *38*, 72–81. [CrossRef]
41. Togi, A.; Smith, J.D.T. Lighting the dark molecular gas: H2 as a direct tracer. *Astrophys. J.* **2016**, *830*, 18. [CrossRef]
42. Zhao, J. Simulation and Validation of Hydrogen Sulfide Removal from Fan Ventilated Confined-Space Manure Storages. Ph.D. Thesis, The Pennsylvania State University, Stetter Colledge, PA, USA, 2006. Available online: https://elibrary.asabe.org/abstract.asp?aid=23278 (accessed on 3 October 2006).
43. Kirk, H.; Friesen, R.K.; Pineda, J.E.; Rosolowsky, E.; Offner, S.S.R.; Matzner, C.D.; Myers, P.C.; di Francesco, J.; Caselli, P.; Alves, F.O.; et al. The Green Bank Ammonia Survey: Dense Cores under Pressure in Orion A. *Astrophys. J.* **2017**, *846*, 144. [CrossRef]
44. Wang, F. *Computational Fluid Dynamics-Principle and Application of CFD Software*; Tsinghua University Press: Beijing, China, 2004.
45. Chen, Q.; Xu, W. A zero-equation turbulence model for indoor airflow simulation. *Energy Build.* **1998**, *28*, 137–144. [CrossRef]
46. Zhao, B.; Li, X.; Yan, Q. Simulation of indoor air flow in ventilated room by zero-equation turbulence model. *J. Tsinghua Univ. (Sci. Technol.)* **2001**, *41*, 109–113.
47. *Airpak 3.0 User's Guide*; Fluent Inc.: Lebanon, NH, USA, 2007.
48. *GB/T 18883-2002*; Indoor Air Quality Standard. Standards Press of China: Beijing, China, 2002.

Article

Thermal Performance and Energy Conservation Effect of Grain Bin Walls Incorporating PCM in Different Ecological Areas of China

Yan Wang [1], Changnv Zeng [2],* and Chaoxin Hu [2]

[1] School of Energy and Environment Engineering, Zhongyuan University of Technology, Zhengzhou 450007, China
[2] College of Civil Engineering, Henan University of Technology, Zhengzhou 450000, China
*   Correspondence: zengchangnv@163.com

Abstract: China, as one of the largest grain production countries, is faced with a storage loss of at least 20 billion kilograms each year. The energy consumption from grain bin buildings has been rising due to the preferred environmental demand for the long-term storage of grain in China. A prefabricated phase change material (PCM) plate was incorporated into the bin walls to reduce energy consumption. The physical model of PCM bin walls was numerically simulated to optimize the latent heat and phase change temperature of PCMs for ecological grain storage area. The thermal regulating performance of the prefabricated PCM plate on the grain bin wall was optimized. It was indicated that a higher value of latent heat of the PCM is more suitable for the hotter region for storing grain in bins in this paper. The energy saving did not increase in the same proportion as the increase in latent heat, suggesting a diminishing return. In this study, the optimal latent heat ranged from 180 to 250 kJ/kg. The values of phase change temperature were selected as 31 °C, 28 °C, and 28 °C for Guangzhou, Zhengzhou, and Harbin cities, respectively, corresponding to hot, warm, and cold climates. The percentages of energy saving were 12.5%, 14.8%, and 17.5% with the corresponding phase change temperatures, which showed an advantage of the PCM used in grain bin walls.

Keywords: grain bin; PCM; phase change temperature; energy saving; PCM latent heat

Citation: Wang, Y.; Zeng, C.; Hu, C. Thermal Performance and Energy Conservation Effect of Grain Bin Walls Incorporating PCM in Different Ecological Areas of China. *Processes* **2022**, *10*, 2360. https://doi.org/10.3390/pr10112360

Academic Editors: Yabin Guo, Zhanwei Wang, Yunpeng Hu and João M. M. Gomes

Received: 25 September 2022
Accepted: 9 November 2022
Published: 11 November 2022

Publisher's Note: MDPI stays neutral with regard to jurisdictional claims in published maps and institutional affiliations.

1. Introduction

The energy consumption of buildings accounts for 40% of the total world's primary energy consumption [1]. The major portion of the energy consumption in buildings is to maintain a comfortable interior environment, which is mainly related to the requirement of heating or cooling [2]. The thermal performance of building walls greatly affects the energy consumption to provide a required indoor temperature. Commonly, lightweight heat insulation materials, such as expanded polystyrene (EPS) and extruded polystyrene (XPS), have been widely used to isolate the effect of outdoor conditions on the interior environment of buildings [3,4]. However, the disadvantages of these insulation materials are insufficient durability, inflammability, and complicated construction, etc.

Nowadays, PCMs as a renewable energy technology have been increasingly applied in many fields, such as buildings [5,6], foundations [7,8], and highways [9,10]. Thus, PCM walls are suitably used to construct low-energy-consumption buildings [11,12]. The desired properties of PCMs are high thermal storage capacity, suitable phase change temperature (T_m), chemical stability, low cost, small volume change during solidification, and availability without flammability, toxic distortion, and crystallization problems [13]. During the phase change process, the PCMs can absorb (or release) a great amount of energy at almost isothermal temperature or in a very narrow temperature range [14].

In order to select a suitable PCM for the specified requirement, the thermal storage capacity and phase change temperature of the PCM should be first considered. The thermal storage capacity of PCMs is mostly affected by their latent heat [15]. Paraffin has a high thermal storage capacity, which is extensively used as a PCM [16]. It can be directly incorporated into the building materials, e.g., concrete or plaster, although it might cause the leakage of the PCM [17]. To solve the leakage problem, the composite PCM, for example, high-density polyethylene complexed with paraffin, can effectively prevent the leakage of the PCM during the phase melting process [18]. Thus, different composite PCMs are developed and used in different fields.

The selection of the phase change temperature of PCMs plays a role in greatly utilizing the PCMs, which depends on both the targeted demand of the indoor temperature and the effect of solar radiation [19]. The indoor temperature can be decreased and smoothed in a narrow temperature amplitude range due to the use of PCM walls because the exterior heat transition is successfully delayed by the PCMs [20]. Therefore, the usage of the PCM walls can enhance the energy saving as well as decrease the carbon emissions during the grain storage for the grain bins. To evaluate the energy saving, different analysis methods were applied in terms of the reduction in cooling energy consumption or the total heat gain [21]. In most studies from the literature, the energy saving is represented by comparing the total heat gain reduction of building walls with and without PCMs [22].

The previous studies reported that the PCMs have been capable of regulating the indoor temperature by application in civil buildings [23,24]. The PCMs were usually mixed into cement mortar and concrete to construct the phase change boards or walls [25,26]. Thus, the incorporation of the PCM may enhance the heat storage capacity of these walls. Meanwhile, the PCMs were numerically studied for the PCM-based walls or roofs [27,28]. The high heat storage mass of the PCM was proven to enhance the thermal performance of building walls and roofs, maintaining a comfortable indoor environment.

As for grain bins in China, the technologies of air-conditioning and ventilation cooling have been widely applied to maintain the grain storage temperature in bins, especially in summer, which should be below 25 °C for indoor temperature [29,30]. However, the previous cooling technologies generate high energy consumption and even induce environmental pollution. In this paper, the PCMs were proven to be useful in the grain bin wall to enhance the energy savings during the long-term grain storage.

China is one of the largest grain production countries, with an annual yield of more than 0.65 trillion kg. However, improper grain storage can bring in a loss of more than 20 billion kg each year, equivalent to the food of 100 million people for one year. One of the most efficient ways to reduce grain storage loss is to maintain a low temperature within grain bins. In this paper, the thermal performance of bin walls with PCMs was studied in order to provide a suitable temperature for the long-term stored grain in different climate regions around China. As known, the required temperature of stored grain should be below 25 °C according to the Chinese code [31], at which grain insects will not be produced to decrease the quality of grains or even damage grains. The PCM bin wall can effectively regulate the temperature of the stored grain; thus, it can greatly decrease the energy consumption by air-conditioner cooling, especially in summer.

In this paper, the prefabricated PCM plate was integrated into the concrete wall to construct the PCM bin wall, which was composed of a 40 mm thick PCM layer and a 330 mm thick concrete layer. The PCM layer was located on the exterior side of the concrete wall, which aimed to absorb the exterior heat flux and regulate the temperature of storage grain stored in bins. The physical model of PCM bin walls was numerically simulated to optimize the latent heat and phase change temperature of PCMs for different ecological grain storage areas in China. Due to the different climate effects in different grain storage regions, three typical cities were taken into consideration, representing the hot climate region in Guangzhou city, the warm climate region in Zhengzhou city, and the cold climate in Harbin city around China, respectively. The thermal regulating performance of the PCM bin wall was optimized. Finally, the corresponding energy savings of PCM walls were presented by comparing with the common concrete bin wall.

2. Methodology

2.1. Physical Model

Each typical week in Guangzhou, Zhengzhou, and Harbin city was selected to analyze the thermal performance of grain bin walls, which represent the hot, warm, and cold climates for grain storage in China, respectively.

The typical week was determined by the hottest summer week in July in these three cities. Climatic conditions of each typical week were represented by the hourly outdoor solar air temperature T_{eq}, as shown in Figure 1 and Equation (1). It combines the outdoor dry bulb temperature T_{ex}, solar radiation equivalent temperature $\rho \cdot I / h_{ex}$, and the external surface temperature of longwave radiation t_γ [32].

$$T_{eq} = T_{ex} + \frac{\rho \cdot I}{h_{ex}} - t_\gamma \qquad (1)$$

where ρ is the absorption coefficient of the external surface, I is the solar radiation intensity (W/m^2), and h_{ex} is the convective heat transfer coefficient of the exterior surface (W/m·K). Here, the value of t_γ was 1.8 W/m^2 and ρ was 0.48 W/m·K.

Two types of walls were modeled in the simulations. One was named the common bin wall, which was purely a concrete layer with a total width of 370 mm. The other was called a PCM bin wall consisting of a 330 mm thick concrete layer and a 40 mm thick PCM layer presented in Figure 2. The height of the two walls was 2000 mm. The thermal and physical properties of the PCM and concrete are detailed in Table 1. The initial temperature of the wall was 25 °C. Here, T_{eq} was used as the exterior surface temperature of the wall.

Table 1. Thermal properties of materials used in the model [33–35].

Materials	Density (kg/m^3)	Specific Heat (kJ/kg·K)	Thermal Conductivity (W/m^2·K)	Latent Heat (kJ/kg)
Concrete	2400	1.030	1.23	0
PCM	890	1.500	0.8	60–250

2.2. Governing Equations and Boundary Conditions

The assumptions were as follows:

(1) A one-dimensional heat transfer was assumed.
(2) The PCM layer was a pure, homogeneous, and isotropic material.
(3) The thermal contract resistance between the PCM layer and concrete layer was negligible.
(4) The surrounding radiations except solar were ignored.
(5) Heat generation, radiation heat transfer, and natural convection in materials were not considered.
(6) The top and bottom boundary of the wall were assumed in adiabatic conditions.

According to the assumptions given above, the concrete heat conduction equation can be expressed by Equation (2) [12]:

$$\rho_1 c_1 \frac{\partial T}{\partial t} = \frac{\partial}{\partial x}\left(\lambda_1 \frac{\partial T}{\partial x}\right) \qquad (2)$$

where ρ_1 is the density of concrete (kg/m^3), c_1 is the specific heat of concrete (J/kg·K), λ_1 is the thermal conductivity of concrete (W/m^2·K), T denotes temperature (°C), and t is time. The apparent heat capacity method was used to simulate the phase change process of the PCM. The heat equation of the PCM layer, considering the variation in the latent heat with temperature, is described by Equation (3) [22]:

$$\rho_2 c_2 \frac{\partial T}{\partial t} = \frac{\partial}{\partial x}\left(\lambda_2 \frac{\partial T}{\partial x}\right) \qquad (3)$$

where ρ_2 is the density of the PCM, c_2 is the specific heat of the PCM, and λ_2 is the thermal conductivity of the PCM (W/m^2·K).

The exterior surface of the PCM layer was simulated by the combined effects of convective and solar radiation. Meanwhile, the interior surface of the concrete layer was exposed by convective heat transfer. Thus, the mathematical boundary conditions of the exterior surface can be expressed by Equation (4) [32]:

$$-\lambda_2 \frac{\partial T}{\partial x} = h_{ex}\left(T_{eq} - T_{ex}\right) + \rho I + \varepsilon\sigma\left(T_{eq}^4 - T_{ex}^4\right) \tag{4}$$

where h_{ex} is the heat transfer coefficient of the exterior surface with a value of 18.3 W/m^2·K [32], ρ is the absorption coefficient of the external surface, I is the solar radiation intensity (W/m^2), ρI represents the absorption of solar radiation, and ε is the surface emissivity. σ is the Stefan–Boltzmann constant with a value of 5.67×10^{-8} W/m^2·K^4 [32]. The heat transfer in the interior surface of the wall is represented as Equation (5):

$$-\lambda_1 \frac{\partial T}{\partial x} = h_{in}\left(T_1 - T_{in}\right) \tag{5}$$

where h_{in} is the convective heat transfer coefficient of the interior surface (8.7 W/m^2·K) [32], and T_1 and T_{in} are the indoor temperature and indoor air temperature, respectively.

The heat flux was investigated to analyze the energy saving. The instantaneous heat flux $q_{(t)}$ was defined as the heat flux going through the wall, which can be expressed by Equation (6):

$$q_{(t)} = \begin{cases} h_{in}\left(T_{out} - T_{in}\right) \text{ for heat storage cycle} \\ h_{in}\left(T_{in} - T_{out}\right) \text{ for heat release cycle} \end{cases} \tag{6}$$

where T_{out} is the outdoor air temperature. Heat storage cycles occurred during solar radiation after sunrise in the daytime, corresponding to heat flux increasing through the wall. In addition, the heat release cycle was identified after sunset at night, corresponding with heat flux decreasing through the wall.

The weekly heat gain Q_{gain} can be calculated by Equation (7), which was integrated by $q_{(t)}$ in a week including the heat storage cycle and the heat release cycle. The weekly heat gain difference represented as ΔQ_{gain} in Equation (8) was expressed by the difference between the common bin wall Q_{gain-0} and PCM bin wall $Q_{gain-PCM}$, as shown in Figure 3. Then, the energy saving can be obtained by Equation (9).

$$Q_{gain} = \int_0^{168} q(t)dt \tag{7}$$

$$\Delta Q_{gain} = Q_{gain-0} - Q_{gain-PCM} \tag{8}$$

$$E_{saving} = \frac{\Delta Q_{gain}}{Q_{gain-0}} \tag{9}$$

The economy analysis of the PCM bin wall was analyzed based on the energy saving analysis. Q_{season} was the electricity saving of the PCM bin wall in whole cooling season, as shown in Equation (10) [36].

$$Q_{season} = \Delta Q \cdot A_1 \cdot \frac{\tau}{168} \tag{10}$$

where A_1 is the total area of exterior walls, m^2; τ is the total time in the cooling season, 2160 h [37]; 168 is the time of one week, h. P is the cost saving of electricity in the whole cooling season, which can be expressed as Equation (11) [36].

$$P = \frac{Q_{season}}{EER} \cdot a \tag{11}$$

where EER is the air-conditioning cooling energy efficient ratio, 3.2; a is the unit price of electricity, 0.56 CNY/kW·h [37].

The thermal resistance, heat storage coefficient, and thermal inertia index were used to evaluate the thermal properties of grain bin walls. Thermal resistance (R_{wall}) was calculated by Equation (12), consisting in the sum of the thermal resistance of every grain bin wall. The heat storage coefficient (S_{168}) represented the heat storage capacity of the concrete layer and PCM layer, which was calculated by Equation (13). The thermal inertia index D calculated by Equation (14) indicates the periodic attenuated temperature waves within the wall [38].

$$R_{wall} = \sum R_i = \sum \frac{\delta_i}{\lambda_i} \tag{12}$$

$$S_{168} = \sqrt{\frac{2\pi}{168} c_1 \lambda_1 \rho_1} = \sqrt{\frac{2\pi}{168} \cdot \lambda_2 \rho_2 \frac{\triangle h}{\triangle t}} \tag{13}$$

$$D = \sum D_i = \sum R_i \cdot S_{168} \tag{14}$$

where δ_i (i = 1,2) is the thickness of the concrete layer or PCM layer, $\triangle h$ is the enthalpy of the PCM, and $\triangle t$ is the phase change diameter of the PCM. D_i (i = 1, 2) is the thermal inertia index of the concrete layer or PCM layer.

2.3. The Validation and Mesh Independence of Physical Model

The experiment conducted by Pasupathy et al. [39] was used to verify the reasonability of the numerical method. The experimental room (1.22 m × 1.22 m × 2.44 m) was constructed to assess the effect of the PCM layer incorporated in the roof on regulating the room temperature. The indoor temperature was set-up at a constant temperature of 27 °C; in addition, the outdoor and indoor heat transfer coefficients were 5 W/m²·K and 1 W/m²·K, respectively.

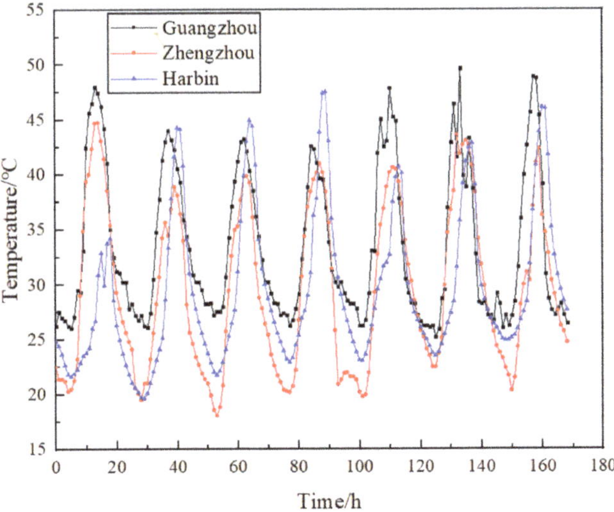

Figure 1. The temperature distribution in the three cities in July [40].

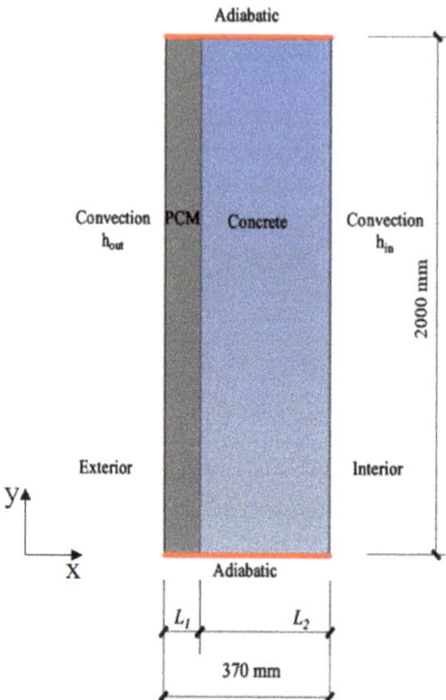

Figure 2. Physical model of grain bin wall (L1: 40 mm; L2: 330 mm).

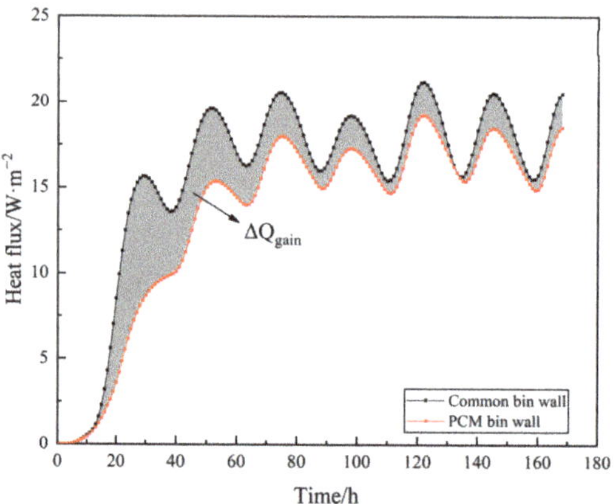

Figure 3. Weekly heat gain description of bin wall.

Figure 4a represents the ambient temperature, the experimental and the numerical results of the indoor surface temperatures are presented, as well as a maximum difference of 0.85 °C and a relative error of 3.26% between the experimental and numerical values. The difference was mainly caused by the variation in the outdoor environment, which may influence the heat transfer in the roof. Thus, the numerical results showed a good agreement with the experimental results, which provided greater reasonableness for the numerical method.

In Figure 4b, the physical model of the grain bin wall was divided into four kinds of element numbers, corresponding to 450, 1900, 7400, and 29,600. The outdoor temperature was selected by the Zhengzhou region. It can be seen that the four temperature curves presented a great uniformity, which proved the mesh independence of the physical model.

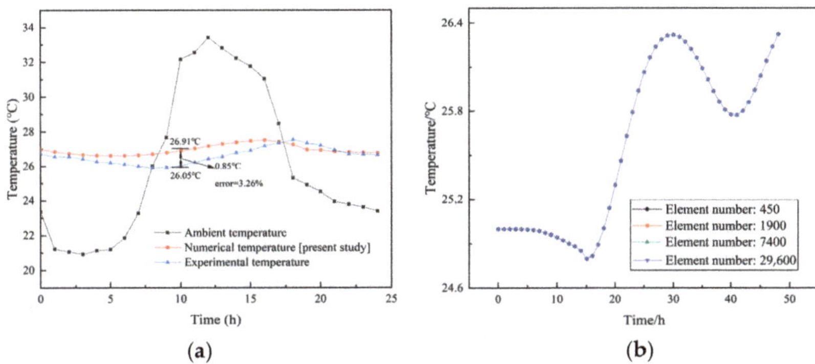

Figure 4. The numerical validation and mesh independence of physical model [39]. (**a**) Numerical validation. (**b**) Mesh independence.

3. Results

In the current work, the thermal performance of different PCM bin walls was evaluated in every typical week of the three cities mentioned above. Two parameters of the PCM were mainly taken into consideration, including the latent heat (60, 90, 120, 180, and 250 kJ/kg) and phase change temperature (26 °C to 33 °C). The thermal performance of the bin wall with or without the PCM was evaluated by its heat flux, heat gain, and temperature amplitude.

3.1. Effects of Latent Heat of PCM

The latent heat of the PCM had a significant effect on the thermal performance of the wall, which primarily affected the regulation capacity of the indoor temperature of the grain bin. In this paper, the latent heat of the PCM ranged from 60 to 250 kJ/kg, considering the common value used in the literature and also the potential maximum value obtained from the tests in the future.

3.1.1. Effects of Latent Heat on T_{in} and Heat Flux of the Bin Wall

Obviously, the maximum latent heat can provide better regulation capacity of the building. However, the maximum utilization rate of latent heat may differ. Simultaneously, different climates would not affect the use of latent heat. Therefore, Guangzhou was typically selected to evaluate the effect of the latent heat of the PCM on the thermal performance of the bin wall because the difference is very little for different climates.

In Guangzhou, there is a great demand to control the temperature condition of the bin for safely storing grain because of its hot climate. The variations in the indoor temperature T_{in} with different latent heats of the PCM in the PCM bin wall are described in Figure 5. The peak indoor temperature of the PCM bin wall reduced compared to that of the common bin wall. As shown in Figure 5 during the fifth day, the highest reduction of 0.62 °C was shown by the PCM layer with a latent heat of 250 kJ/kg, while the smallest reduction of 0.3 °C was presented by the PCM layer with a latent heat of 60 kJ/kg. As shown in Figure 6, similar periodic changes of the interior heat flux in the PCM bin wall were also presented during the typical week. The peak instantaneous heat flux of the PCM bin wall gradually reduced by 3 to 5.5 W·h·m^{-2} compared to the common bin wall. The reason is that the heat energy was effectively absorbed by the PCM layer because of its state from solid to liquid. The heat transition can be profitably decreased by the PCM layer of the bin wall. It is worth noticing that the indoor heat flux was also reduced with increasing latent heat of the PCM.

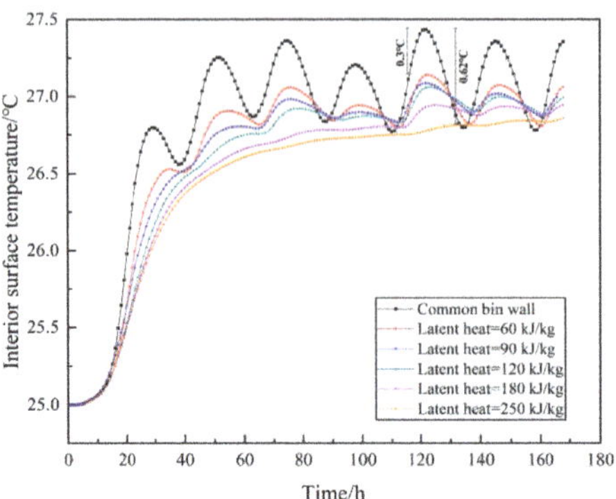

Figure 5. T_{in} distribution with different latent heats of PCM.

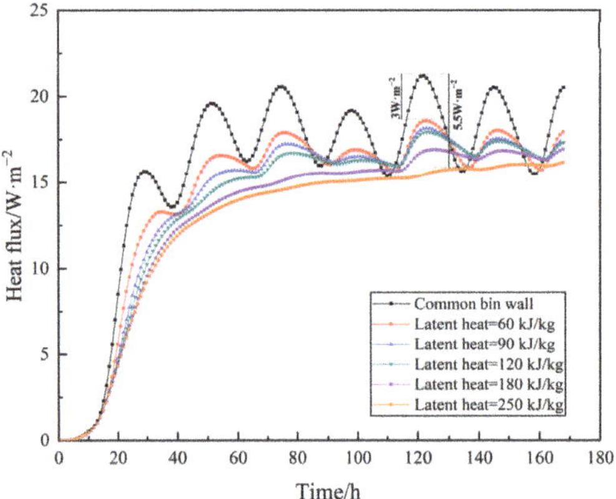

Figure 6. The interior heat flux of PCM bin wall.

3.1.2. Effects of Latent Heat on Heat Gain of the Bin Wall

To evaluate the energy saving associated with the effect of latent heat of the PCM, the weekly heat gain Q_{gain} was investigated in Guangzhou city. As described in Figure 7, Q_{gain} gradually decreased with increasing latent heat of the PCM. The weekly heat gains Q_{gain} were 2621, 2390.5, 2317.7, 2293.3, 2196.4, and 2125.1 W·h·m^{-2}, corresponding to the common bin wall and PCM bin walls with latent heats of 60, 90, 120, 180, and 250 kJ/kg. Thus, the energy saving E_{saving} was 18.9% for a latent heat of 250 kJ/kg, while it was 16.2%, 12.5%, 11.6%, and 8.8% for latent heats of 180, 120, 90, and 60 kJ/kg, respectively. It can be concluded that the energy saving was promoted with increasing latent heat of the PCM.

As shown in Figure 6, when latent heat changed from 60 to 180 kJ/kg, the energy saving of the PCM bin wall enhanced by 3.7% with every latent heat increment of 60 kJ/kg. For the case from 180 to 250 kJ/kg, the energy saving enhanced by 2.7%. Thus, the energy saving of the PCM bin wall optimized by increasing latent heat, but the energy saving did not increase in the same proportion as an increase in latent heat, suggesting a diminishing return.

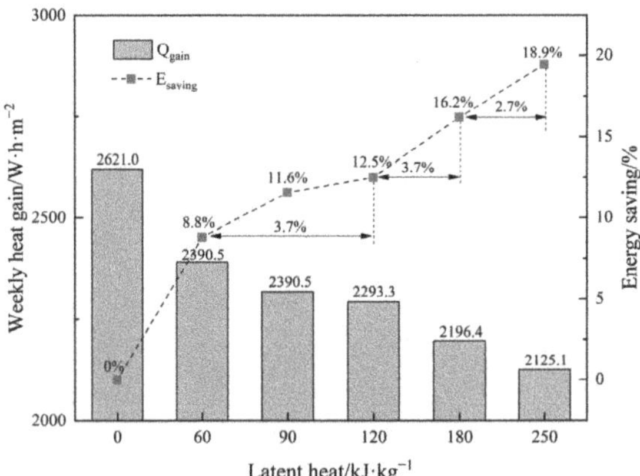

Figure 7. Heat gain and energy saving of PCM (Guangzhou).

3.2. Effects of Phase Change Temperature of PCM

The outdoor temperature will greatly affect the phase change temperature in different climate regions. Thus, all cases in the three cities were studied. The hourly indoor temperatures of the grain bin wall at different phase change temperatures are shown in Figures 8 and 9 for these three cities. For all the cases in one typical week, there were slight fluctuations in the indoor temperature T_{in} during the initial time. After that, the T_{in} rapidly increased to its peak value, and then it decreased gradually to the valley value for each case. A heat storage cycle occurred during solar radiation after sunrise, which showed an increasing T_{in}, while the heat release cycle was identified after sunset when the T_{in} began to decrease. As shown in Figure 8, the phase change temperature of the PCM had a significant effect on the thermal performance of the grain bin wall. The distribution of T_{in} of the PCM wall was compared to that of the common bin wall without the PCM layer. It was observed that T_{in} obviously varied for the common bin wall and the PCM bin wall. The fluctuation in temperature was effectively smoothed with the application of the PCM in the bin wall.

3.2.1. Effect of Phase Change Temperature on T_{in}

The phase change temperature T_m was investigated during one typical week in summer to evaluate its effect on the bin wall in the three cities mentioned above.

As shown in Figure 8a, the distributions of T_{in} of the PCM bin wall and the common bin wall in Guangzhou are presented. It can be demonstrated that the T_{in} curve linearly increased from 0 to 27 h. Then, the amplitude of T_{in} attained different peak and valley values with the outdoor temperature oscillating.

With the application of the PCM layer in the bin wall, the peak T_{in} of the PCM bin wall was reduced by 0.2 to 0.4 °C on the sixth day. The highest T_{in} reduction of 0.4 °C was shown in the PCM bin wall with T_m changing from 31 °C to 33 °C, whereas the PCM bin wall with T_m of 29 °C presented a smaller T_{in} reduction of 0.2 °C. The phenomenon was due to a suitable T_m profitably decreasing the T_{in} of the PCM bin wall.

The amplitude of T_{in} of the PCM bin wall was smoothed in comparison to that of the common bin wall. When T_m was 31 °C, the amplitude of T_{in} of the PCM bin wall was lower than 0.2 °C. It was indicated that the PCM with suitable T_m can effectively absorb and release the heat energy during the heat storage and release cycle, which can produce more energy saving. A T_m of 31 °C was identified as the optimal phase change temperature of the PCM in Guangzhou city.

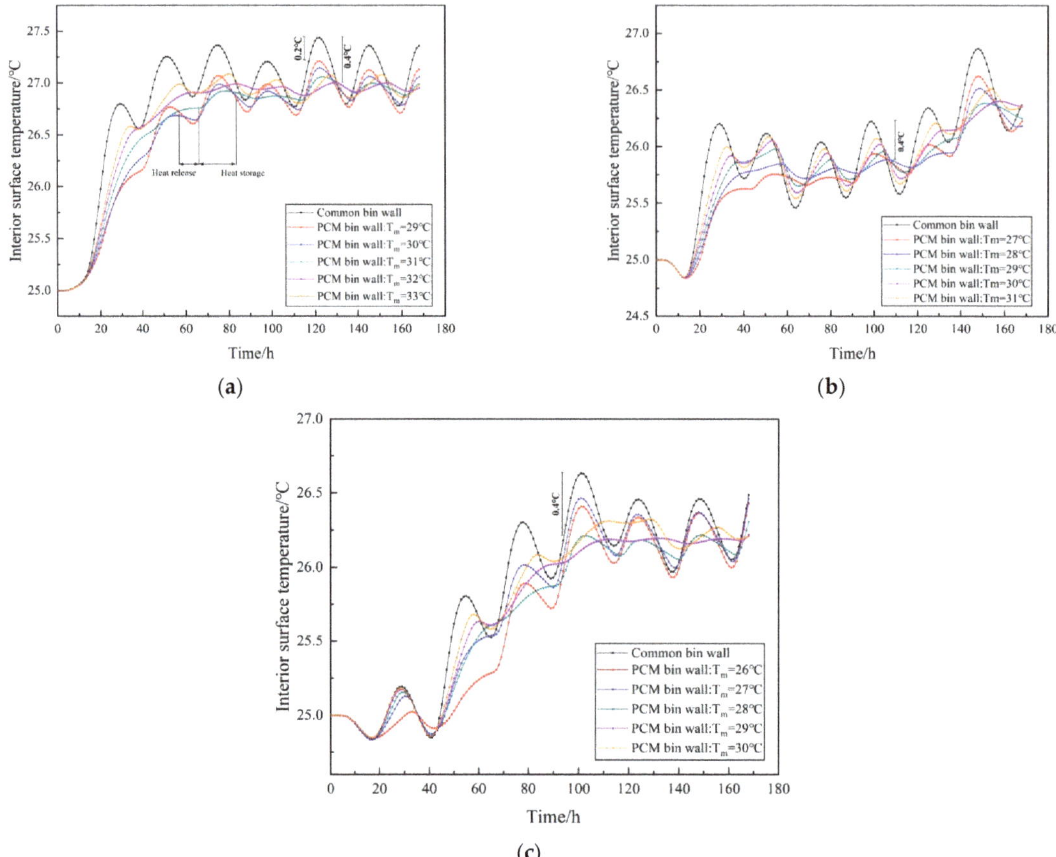

Figure 8. The T_{in} distribution with different phase change temperatures. (**a**) Guangzhou. (**b**) Zhengzhou. (**c**) Harbin.

Figure 8b displays the indoor temperature profiles of the bin wall with and without the PCM in Zhengzhou, corresponding to the warm climate region. It can be seen that the PCM with T_{m} of 27 °C can maintain a lower T_{in} than the PCM bin wall with T_{m} of 28 °C during the initial 4 days. Then, a higher fluctuation in T_{in} was presented by the PCM bin wall with T_{m} of 27 °C. This was because the PCM had completely melted, which could not prevent the heat interaction with the outdoor temperature. It is obvious that a stabilized amplitude was presented by the PCM bin wall with T_{m} of 28 °C, and the maximum reduction in T_{in} was 0.4 °C. Thus, the PCM bin wall with T_{m} of 28 °C could maintain a stabler temperature for the grain storage. The optimal phase change temperature was 28 °C in Zhengzhou city.

For the case of a cold climate region as Harbin, its variation in T_{in} is represented in Figure 8c for the common bin wall and different PCM bin walls. In this investigation, the PCM bin wall with T_{m} of 28 °C was optimal to regulate the T_{in} of the grain bin. Taking the common bin wall as a reference, the largest T_{in} reduction of 0.4 °C was presented by the PCM bin wall with T_{m} of 28 °C and it also tends to maintain a T_{in} below 26 °C. Obviously, a higher T_{m}, for example, 29 or 30 °C, used in the cold regions could not be effective. Meanwhile, a PCM with a lower T_{m}, for example, 26 or 27 °C, may release less of the stored heat energy to the outdoor environment during the heat release cycle, which is low efficiency for the heat cycle. Thus, the thermal performance of the PCM bin wall with T_{m} of 28 °C was especially optimal in this region.

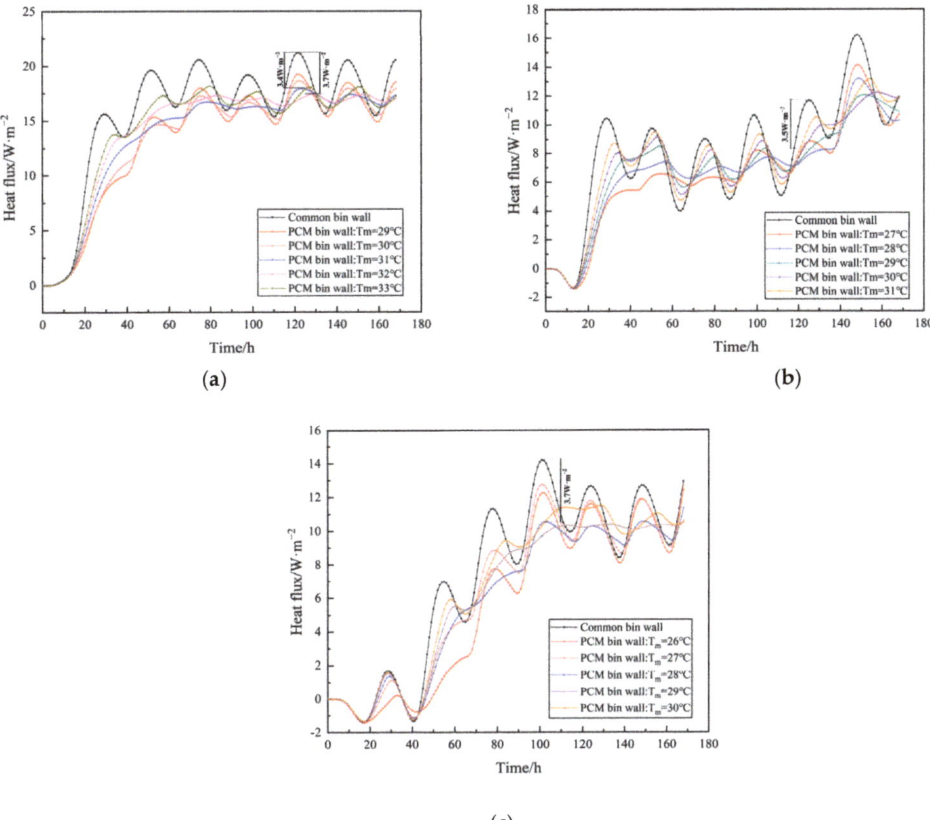

Figure 9. The indoor heat fluxes distribution of PCM bin wall. (**a**) Guangzhou. (**b**) Zhengzhou. (**c**) Harbin.

3.2.2. Effect of T_m on the Heat Flux

Figure 9 displays the indoor heat flux of PCM bin walls with different T_m in the three cities. As depicted in Figure 9a, a uniform downtrend of peak heat flux was presented by the instantaneous heat flux profiles in Guangzhou. The peak heat flux of the common bin wall was 21.2 W·h·m^{-2}, which appeared on the sixth day. Taking the common bin wall as a reference, the indoor heat flux of the PCM bin wall with T_m of 32 °C was reduced by 3.7 W·h·m^{-2}. It was followed by the PCM bin wall with T_m of 31 °C, which was reduced by 3.3 W·h·m^{-2}. However, the maximum heat flux reduction of the PCM bin wall was presented by the PCM with T_m of 31 °C during the other days. The reason was that the outdoor temperature on the sixth day was higher than those on the other days, which resulted in the different reduction in peak heat flux.

As shown in Figure 9b,c, the indoor heat flux clearly decreased in Zhengzhou and Harbin. As mentioned above, the thermal performance of the PCM with T_m of 28 °C was optimal in both areas. In comparison to the common bin wall, the indoor heat flux of the PCM bin wall with T_m of 28 °C was reduced by 3.5 W·m^{-2} and 3.7 W·m^{-2} in Zhengzhou and Harbin, respectively. The PCM with an optimal phase change temperature can easily absorb the outdoor heat flux during the heat storage cycle. Furthermore, the heat release capacity can be enhanced during the heat release cycle. Therefore, the PCM wall with an optimal phase change temperature had the greatest capacity for heat storage and release.

3.2.3. Effect of T_m on Energy Saving

To obtain a preferred energy saving result, the effect of T_m on the energy saving of PCM bin walls was analyzed in the three cities.

As shown in Table 2, Table 3, Table 4, the maximum heat gain Q_{gain} was 2621.03 W·h·m^{-2}, 1324.23 W·h·m^{-2}, and 1189.6 W·h·m^{-2} for the common bin wall in Guangzhou, Zhengzhou, and Harbin, respectively. Taking the corresponding common bin wall as a reference, the maximum heat gain differences ΔQ_{gain} were 378.82 W·h·m^{-2}, 228.88 W·h·m^{-2}, and 244.59 W·h·m^{-2}, with energy savings E_{saving} of 14.5%, 17.3%, and 20.6%, respectively. It was concluded that the higher energy saving can be provided by the PCM layer with a lower T_m during the typical week for all the cases.

Table 2. Weekly heat gain and energy saving results with different Tm for hot grain storage region.

City	Parameters	Common Bin Wall	PCM Bin Wall Phase Change Temperature of PCM (°C)				
			29	30	31	32	33
Guangzhou	Q (W·h·m^{-2})	2621.0	2244.6	2242.2	2293.1	2372.8	2403.7
	ΔQ (W·h·m^{-2})	/	376.5	378.8	327.4	248.3	217.3
	E_{saving} (%)	/	14.4	14.5	12.5	9.5	8.3

Table 3. Weekly heat gain and energy saving results with different T_m for warm grain storage region.

City	Parameters	Common Bin Wall	PCM Bin Wall Phase Change Temperature of PCM (°C)				
			27	28	29	30	31
Zhengzhou	Q (W·h·m^{-2})	1324.3	1095.4	1128.0	1166.0	1209.2	1236.8
	ΔQ (W·h·m^{-2})	/	228.9	196.2	158.2	115.1	87.5
	E_{saving} (%)	/	17.3	14.8	11.9	8.8	6.6

Table 4. Weekly heat gain and energy saving results with different T_m for cold grain storage region.

City	Parameters	Common Bin Wall	PCM Bin Wall Phase Change Temperature of PCM (°C)				
			26	27	28	29	30
Harbin	Q (W·h·m^{-2})	1189.6	945.0	1056.8	981.5	1036.7	1099.7
	ΔQ (W·h·m^{-2})	/	244.6	132.8	208.1	152.9	89.9
	E_{saving} (%)	/	20.6	11.6	17.5	12.9	7.6

In the hot climate regions, for example, in Guangzhou, the exterior climate maintained a continuous high temperature, and the maximum exterior temperature was 51 °C in the summer week. The higher exterior temperature may consume more energy to regulate the grain storage temperature in Guangzhou. As shown in Table 2, it was observed that the energy saving E_{saving} was decreased by 14.5% to 8.3% in the typical week. The energy savings E_{saving} of the PCM bin wall with T_m of 29 °C and 30 °C were practically identical at 14.4% and 14.5%, respectively. Additionally, the energy saving E_{saving} of the PCM bin wall with T_m of 31 °C was decreased by 12.5%, which was followed by the PCM bin wall with T_m of 32 °C and 33 °C.

Obviously, the energy saving E_{saving} of the PCM bin wall with T_m of 30 °C was higher than the other cases. However, referring to the above, the T_m of 31 °C was more optimal than in the other cases. The main reason is that the PCM bin wall with T_m of 30 °C had a higher indoor temperature fluctuation than that with T_m of 31 °C. When the outdoor temperature fell in between the PCM with T_m of 29 °C and 31 °C, more heat energy can be released by the PCM with T_m of 29 °C. Thus, it is reasonable that the maximum energy saving effect may not correspond to the optimal phase change temperature. The energy saving E_{saving} of the PCM bin wall with an optimal T_m of 31 °C attained 12.5%.

In the warm climate region of Zhengzhou and cold region of Harbin, the exterior temperatures presented a lower temperature of 20 °C in the nighttime, while the exterior temperature represented a sharp rise in the daytime. Although the exterior temperatures in Zhengzhou and Harbin were lower than that of Guangzhou, the cooling energy demand was also extensively required to control the grain storage temperature in summer. Thus, it was necessary to use the PCMs in the bin walls to reduce the energy consumption.

Table 3 displays a gradual downtrend of energy saving, which was presented by PCM bin walls with T_m of 27 to 31 °C in Zhengzhou city. The energy savings E_{saving} were 17.3%, 14.8%, 11.9%, 8.8%, and 6.6%, corresponding to PCMs with T_m values of 27 °C, 28 °C, 29 °C, 30 °C, and 31 °C. It was concluded that the energy saving E_{saving} of the PCM bin wall was 14.8% with an optimal T_m of 28 °C.

As detailed in Table 4, the energy savings E_{saving} of PCM bin walls were provided for the case in Harbin, which were 20.6%, 11.6%, 17.5%, 12.9%, and 7.6%, corresponding to PCMs with T_m values of 26 °C, 27 °C, 28 °C, 29 °C, and 30 °C. It can be seen that the energy saving E_{saving} of the PCM bin wall was 17.5% with an optimal T_m of 28 °C. It is worth emphasizing that the highest energy saving with the PCM bin wall may not remain a stable thermal environment during the typical week. In the initial 3 days, the maximum energy saving was provided by the PCM with T_m of 27 °C and the PCM with T_m of 26 °C, corresponding to Zhengzhou and Harbin, respectively. Its phase change temperature was lower than those of other PCMs, which contributed to more heat energy absorbed into the PCM layer. When the PCM with T_m of 26 °C (or with T_m of 27 °C) completely melted, its interaction with the exterior environment was then restricted. Then, the heat flux was only reduced by the thermal resistance of the concrete wall and PCM layer. To sum up, the heat energy had been effectively decreased by the PCM with T_m of 28 °C.

Based on the above investigation, the energy saving of PCM bin walls could attain 12.5%, 14.8%, and 17.5%, corresponding to the optimal phase change temperature in Guangzhou, Zhengzhou, and Harbin.

4. Analysis and Discussion

In general, the thickness of the common concrete grain bin wall is 370 mm. The heat transition is resisted by the thermal resistance of the concrete wall, which cannot remain a stable interior environment for the stored grain. Thus, it is necessary to optimize the structure of the grain bin wall. In this study, a 40 mm thick PCM layer followed by a 330 mm thick concrete layer are modeled as the PCM bin wall. The PCM layer is set on the exterior surface of the wall, which mainly aims to reduce the heat transiting into the indoor environment [39].

As shown in Table 5, the thermal inertia index and heat storage coefficient of the PCM bin wall are approximately four times higher than those of the concrete bin wall. It is indicated that the inward heat transition can be effectively reduced by the PCM incorporated in the bin wall.

Table 5. The thermophysical properties of bin wall with and without PCM [38].

	PCM Thickness (mm)	Thermal Resistance ($m^2 \cdot K/W$)	Regenerative Indicator ($W \cdot m^{-2} \cdot K$)	Thermal Inertia Index
Concrete bin wall	-	0.3	10.66	3.21
PCM bin wall	40	0.32	39.97	15.63

With the outdoor heat going through the PCM layer, the PCM absorbs and releases heat energy with a solid–liquid phase change process. In the heat storage cycle, the inward heat energy continuously charges the PCM bin wall. Much heat energy is resisted by the thermal inertia of the bin wall and absorbed by the PCM as latent heat. It also can maintain a relevant constant PCM temperature as long as both solid and liquid phases coexist. After the PCM completely melts, its interaction with the exterior environment is restricted. Then, the heat energy only reduces by the thermal resistance of concrete. On the other hand, a portion of heat energy will transfer to the interior of the grain bin, resulting in an increasing indoor temperature. During the heat release cycle, the stored energy in the PCM primarily releases to the outdoor environment. Then, the indoor temperature reduces with the heat energy transferring to the outdoor environment. In total, the PCM changes its state between solid and liquid with the outdoor temperature changing; therefore, it can automatically regulate the indoor temperature by the PCM itself.

Moreover, the application of the PCM in the walls is desired for maintaining the thermal stability in the thermal cycling process. In our previous work, the leakage test was examined to validate the thermal stability of the composite PCM. There was no leakage of the composite PCM when the heating source rose from 20 °C to 110 °C [35]. The previous studies had heated and cooled the composite PCM from 20 °C to 80 °C to prove its thermal stability by the accelerated thermal cycling analysis [41,42]. In this work, the PCM bin wall displays a great regulating temperature effect on the interior surface temperature in the whole week. The PCM layer can effectively absorb the heat source in the daytime, while releasing the stored energy to the outdoor environment in the nighttime. Thus, the PCMs may provide a great thermal stability in the thermal cycling process.

A suitable PCM latent heat is essential to effectively regulate indoor temperature. The effect of latent heat on the thermal performance is improved with the increase in latent heat [27], whereas the energy saving does not show a continuous increase with the same increment in latent heat. Zhou et al. [43] stated that high latent heat has an important effect on the daily energy storage when the PCM can complete the melting–solidification process during a day. Kishore et al. [21] reported a rise in the cumulative heat gain with the increase in latent heat, while the heat gain increase was reduced. Thus, it can be demonstrated that the maximum utilization rate of latent heat might appear as an increase in latent heat.

The phase change temperature is crucial for the thermal performance of the PCM bin wall for different climate regions. The optimal phase change temperatures are 31 °C, 28 °C, and 28 °C, corresponding to Guangzhou, Zhengzhou, and Harbin. Li et al. [44] reported that the optimal phase change temperature was 36 °C in Las Vegas, NV, USA. It is primarily affected by outdoor inward heat energy. However, some prior studies reported that the optimal phase change temperature is close to the indoor temperature, such as 22 to 24 °C [28]. The difference is mainly caused by the difference of building constructions and climates.

The optimal phase change temperature varies with climate change [44]. In a hot climate region such as in Guangzhou, the PCM phase change temperature is higher than that in the warm climate region of Zhengzhou and cold climate region of Harbin. Mechouet et al. [45] demonstrated that both the thickness and phase change temperature of the PCM used in different climate regions need to be selected, to achieve different energy saving optimization results. Yu et al. [46] reported that compared with the reference wall, the optimum phase change temperature of PCMs varied with different climatic regions. Obviously, climates play a decisive role in the selection of PCMs for different climatic regions.

In this study, the energy saving differs with different PCM latent heats and phase change temperatures in different climate regions. The energy saving is positively promoted by suitable latent heats and phase change temperatures of the PCM, and the energy saving effect can be profitably realized by PCMs in different climates.

According to the physical model of the grain bin (length 36 m × width 24 m × height 8.3 m) [47], the PCM bin walls with optimal phase change temperatures were used to analyze the economy analysis of the grain bin in Table 6. The Q_{season} values of three grain storage regions were 2095.7 kW·h, 1256.1 kW·h, and 1337.1 kW·h, corresponding to Guangzhou, Zhengzhou, and Harbin. Furthermore, the cost of electricity can be saved by 366.7 CNY, 219.8 CNY, and 234.0 CNY for three grain storage regions in the cooling season.

It can be seen that the application of the PCM layer in the grain bin is an effective method to decrease the electricity consumption in different grain storage regions.

Table 6. The economy analysis of PCM bin walls in different grain storage regions [36].

Grain Storage Regions	ΔQ_{gain}/W·h·m^{-2}	Q_{season}/kW·h	P/CNY
Guangzhou-T_m—31 °C	327.4	2095.7	366.7
Zhengzhou-T_m—28 °C	196.2	1256.1	219.8
Harbin-T_m—28 °C	208.8	1337.1	234.0

5. Conclusions

In this study, the thermal performance of the PCM bin wall was simulated, considering the heat storage and release cycle process in three cities for safe grain storage. The main goal was to analyze the effect of latent heat and phase change temperature of the PCM on the thermal performance of the PCM grain bin wall. The obtained results are as follows.

(1)   The latent heat of the PCM is vital to its heat storage capacity to maximally regulate the indoor temperature of the grain bin. The cumulative heat gain presents an uptrend with the rise in latent heat. By contrast, a downtrend is observed for the heat gain increment. The selection of latent heat of the PCM should consider the effective utilization rate of the PCM, rather than the direct selection of high PCM latent heat. The maximum utilization rate of latent heat might appear when the latent heat falls within 180 to 250 kJ/kg.

(2)   The optimal phase change temperature is highly dependent on the climate conditions. The hot solar-air temperature directly requires a higher phase change temperature of the PCM. Proper phase change temperatures are selected by 31 °C, 28 °C, and 28 °C in Guangzhou, Zhengzhou, and Harbin city, respectively.

(3)   The energy saving of the PCM bin wall mainly depends on the phase change temperature and latent heat of the PCM. Based on the above investigation, the energy saving of the PCM bin wall can attain 12.5%, 14.8%, and 17.5%, corresponding with the optimal phase change temperature in Guangzhou, Zhengzhou, and Harbin.

Author Contributions: Conceptualization, C.Z. and Y.W.; methodology, C.Z. and Y.W.; software, C.H.; validation, C.Z. and C.H.; investigation, C.Z. and C.H.; writing—original draft preparation, C.Z. and C.H.; writing—review and editing, Y.W. and C.H.; supervision, C.Z.; funding acquisition, C.Z. All authors have read and agreed to the published version of the manuscript.

Funding: This research is funded by the Henan Province Key Specialized Research and Development Breakthrough Plan (212102110027), Henan Province Joint Fund Project of Science and Technology (222103810075).

Data Availability Statement: Not applicable.

Conflicts of Interest: The author declares no potential conflict of interest regarding the publication of this article.

References

1.   Bhamare, D.K.; Rathod, M.K.; Banerjee, J. Numerical model for evaluating thermal performance of residential building roof integrated with inclined phase change material (PCM) layer. *J. Build. Eng.* **2020**, *28*, 101018. [CrossRef]

2.   Yan, T.; Sun, Z.W.; Gao, J.J.; Xu, X.H.; Yu, J.J.; Gang, W.J. Simulation study of a pipe-encapsulated PCM wall system with self-activated heat removal by nocturnal sky radiation. *Renew. Energy* **2019**, *146*, 1451–1464. [CrossRef]

3.   Fard, P.M.; Alkhansari, M.G. Innovative fire and water insulation foam using recycled plastic bags and expanded polystyrene (EPS). *Constr. Build. Mater.* **2011**, *305*, 124785. [CrossRef]

4.   Priyanka, E.; Dhanya, S.; Mini, K.M. Functional and strength characteristics of EPS beads incorporated foam concrete wall panels. *Mater. Today: Proc.* **2021**, *46*, 5167–5170. [CrossRef]

5.   Duan, S.P.; Wang, L.; Zhao, Z.Q.; Zhang, C.W. Experimental study on thermal performance of an integrated PCM Trombe wall. *Renew. Energy* **2021**, *163*, 1932–1941. [CrossRef]

6. Al-Yasiri, Q.; Szabó, M. Incorporation of phase change materials into building envelope for thermal comfort and energy saving: A comprehensive analysis. *J. Build. Eng.* **2021**, *36*, 102122. [CrossRef]

7. Yang, W.B.; Sun, T.F.; Yang, B.B.; Wang, F. Laboratory study on the thermo-mechanical behaviour of a phase change concrete energy pile in summer mode. *J. Energy Storage* **2021**, *41*, 102875. [CrossRef]

8. Mousa, M.M.; Bayomy, A.M.; Saghir, M.Z. Phase change materials effect on the thermal radius and energy storage capacity of energy piles: Experimental and numerical study. *Int. J. Thermofluids.* **2021**, *10*, 100094. [CrossRef]

9. Mahedi, M.; Cetin, B.; Cetin, K.S. Freeze-thaw performance of phase change material (PCM) incorporated pavement subgrade soil. *Constr. Build. Mater.* **2019**, *202*, 449–464. [CrossRef]

10. Ryms, M.; Denda, H.; Jaskuła, P. Thermal stabilization and permanent deformation resistance of LWA/PCM-modified asphalt road surfaces. *Constr. Build. Mater.* **2017**, *142*, 328–341. [CrossRef]

11. Kishore, R.A.; Bianchi, M.V.A.; Booten, C.; Vidal, J.; Jackson, R. Enhancing building energy performance by effectively using phase change material and dynamic insulation in walls. *Appl. Energy* **2021**, *283*, 116306. [CrossRef]

12. Sharma, V.; Rai, A.C. Performance assessment of residential building envelopes enhanced with phase change materials. *Energ. Build.* **2020**, *208*, 109664. [CrossRef]

13. Cabeza, L.F.; Navarro, L.; Pisello, A.L.; Olivieri, L.; Bartolomé, C.; Sánchez, J.; Álvarez, S.; Álvarez, J.S. Behaviour of a concrete wall containing micro-encapsulated PCM after a decade of its construction. *Sol. Energy* **2020**, *200*, 108–113. [CrossRef]

14. Lamrani, B.; Johannes, K.; Kuznik, F. Phase change materials integrated into building walls: An updated review. *Renew. Sustain. Energy Rev.* **2021**, *140*, 110751. [CrossRef]

15. Lee, K.O.; Medina, M.A.; Sun, X.Q.; Jin, X. Thermal performance of phase change materials (PCM)-enhanced cellulose insulation in passive solar residential building walls. *Sol. Energy* **2018**, *163*, 113–121. [CrossRef]

16. Liu, Z.Q.; Huang, J.H.; Cao, M.; Jiang, G.W.; Yan, Q.H.; Hu, J. Experimental study on the thermal management of batteries based on the coupling of composite phase change materials and liquid cooling. *Appl. Therm. Eng.* **2021**, *185*, 116415. [CrossRef]

17. Torres-Rodríguez, A.; Morillón-Gálvez, D.; Aldama-Ávalos, D.; Hernández-Gómez, V.H.; Kerdan, I.G. Thermal performance evaluation of a passive building wall with CO2-filled transparent thermal insulation and paraffin-based PCM. *Sol. Energy* **2020**, *205*, 1–11. [CrossRef]

18. Qu, Y.; Wang, S.; Tian, Y.; Zhou, D. Comprehensive evaluation of Paraffin-HDPE shape stabilized PCM with hybrid carbon nano-additives. *Appl. Therm. Eng.* **2019**, *163*, 114404. [CrossRef]

19. Tunçbilek, E.; Arıcı, M.; Krajčík, M.; Nižetić, S.; Karabay, H. Thermal performance based optimization of an office wall containing PCM under intermittent cooling operation. *Appl. Therm. Eng.* **2020**, *179*, 115750. [CrossRef]

20. Bhamare, D.K.; Rathod, M.K.; Banerjee, J. Proposal of a unique index for selection of optimum phase change material for effective thermal performance of a building envelope. *Sol. Energy* **2021**, *218*, 129–141. [CrossRef]

21. Kishore, R.A.; Bianchi, M.V.A.; Booten, C.; Vidal, J.; Jackson, R. Parametric and sensitivity analysis of a PCM-integrated wall for optimal thermal load modulation in lightweight buildings. *Appl. Therm. Eng.* **2021**, *187*, 116568. [CrossRef]

22. Kishore, R.A.; Bianchi, M.V.A.; Booten, C.; Vidal, J.; Jackson, R. Optimizing PCM-integrated walls for potential energy savings in U.S. Buildings. *Energ. Build.* **2020**, *226*, 110335. [CrossRef]

23. Li, Z.X.; Al-Rashed, A.A.A.A.; Rostamzadeh, M.; Kalbasi, R.; Shahsavar, A.; Afrand, M. Heat transfer reduction in buildings by embedding phase change material in multi-layer walls: Effects of repositioning, thermophysical properties and thickness of PCM. *Energy Convers. Manag.* **2019**, *195*, 43–56. [CrossRef]

24. Pirasaci, T. Investigation of phase state and heat storage form of the phase change material (PCM) layer integrated into the exterior walls of the residential-apartment during heating season. *Energy* **2020**, *207*, 118176. [CrossRef]

25. Yuan, X.S.; Wang, B.M.; Chen, P.; Luo, T. Study on the frost resistance of concrete modified with steel balls containing phase change material (PCM). *Materials* **2021**, *14*, 4497. [CrossRef] [PubMed]

26. Tian, Y.; Lai, Y.M.; Pei, W.S.; Qin, Z.P.; Li, H.W. Study on the physical mechanical properties and freeze-thaw resistance of artificial phase change aggregates. *Constr. Build. Mater.* **2022**, *329*, 127225. [CrossRef]

27. Lqbal, S.; Tang, J.W.; Raza, G.; Cheema, I.I.; Kazmi, M.A.; Li, Z.R.; Wang, B.M.; Liu, Y. Experimental and numerical analyses of thermal storage tile-bricks for efficient thermal management of buildings. *Buildings* **2021**, *11*, 357.

28. Al-Yasiri, Q.; Szabó, M. Case study on the optimal thickness of phase change material incorporated composite roof under hot climate conditions. *Case Stud. Constr. Mater.* **2021**, *14*, e00522. [CrossRef]

29. Yang, W.Q.; Li, X.H.; Liu, X.; Zhang, Y.B.; Gao, K.; Lv, J.H. Improvement of soybean quality by ground source heat pump (GSHP) cooling system. *J. Stored Prod. Res.* **2015**, *64*, 113–119. [CrossRef]

30. Li, X.J.; Han, Z.Q.; Lin, Q.; Wu, Z.D.; Chen, L.; Zhang, Q. Smart cooling-aeration guided by aeration window model for paddy stored in concrete silos in a depot of Guangzhou, China. *Comput. Electron. Agr.* **2020**, *173*, 105452. [CrossRef]

31. *GB/T29890-2013, Technical Criterion for Grain and Oil-Seeds Storage.* China Standards Press: Beijing, China, 2013.

32. Yan, Q.S.; Zhao, Q.Z. *Building Thermal Process*; China Architecture & Building Press: Beijing, China, 1986; pp. 86–88.

33. Qiu, L.; Ouyang, Y.X.; Feng, Y.H.; Zhang, X.X. Review on micro/nano phase change materials for solar thermal applications. *Renew. Energy* **2019**, *140*, 513–538. [CrossRef]

34. Lawag, R.A.; Ali, H.M. Phase change materials for thermal management and energy storage: A review. *J. Energy Storage* **2022**, *55*, 105602. [CrossRef]

35. Qu, J.Y. The Construction of a New Type of Granary Phase Change Roof and Its Heat Transfer Characteristics. Master's Thesis, Henan Technology and University, Zhengzhou, China, 2022.
36. Ma, Y.L.; Li, W.W.; Nian, Y.L. Comparison on performance of phase change mortar and board wall with high energy storage. *J. Therm. Sci. Technol.* **2022**, *21*, 337–346.
37. *GB/21455-2019, Minimum Allowable Values of The Energy Efficiency and Energy Efficiency Grades for Room Air Conditioners.* The General Administration of Quality Supervision, Inspection and Quarantine of the People's Republic of China: Beijing, China, 2019.
38. *GB/T50504-2009, Standard for Terminology of Civil Architectural Design.* China Planning Press: Beijing, China, 2009.
39. Al-Absi, Z.A.; Isa, M.H.M.; Ismail, M. Phase change materials (PCMs) and their optimum position in building walls. *Sustainability* **2020**, *12*, 1294. [CrossRef]
40. Energy Plus. Available online: https://energyplus.net/weather (accessed on 5 August 2021).
41. Yang, Y.Y.; Xiao, Y.; Kong, W.B. Reliable and recyclable dynamically combinatorial epoxy networks for thermal energy storage. *Sol. Energy* **2021**, *230*, 825–831. [CrossRef]
42. Xiong, Y.X.; Wang, H.X.; Wu, Y.T.; Zhang, J.H.; Li, H.M.; Xu, Q.; Zhang, X.X.; Li, C.; Ding, Y.L. Carbide slag based shape-stable phase change materials for waste recycling and thermal energy storage. *J. Energy Storage* **2022**, *50*, 104256.
43. Zhou, D.; Shire, G.S.F.; Tian, Y. Parametric analysis of influencing factors in Phase Change Material Wallboard (PCMW). *Appl. Energy* **2014**, *119*, 33–42. [CrossRef]
44. Li, M.L.; Cao, Q.; Pan, H.; Wang, X.Y.; Lin, Z.B. Effect of melting point on thermodynamics of thin PCM reinforced residential frame walls in different climate zones. *Appl. Therm. Eng.* **2021**, *188*, 116615. [CrossRef]
45. Mechouet, A.; Oualim, E.; Hassan, T.M. Effect of mechanical ventilation on the improvement of the thermal performance of PCM-incorporated double external walls: A numerical investigation under different climatic conditions in Morocco. *J. Energy Storage* **2021**, *38*, 102495. [CrossRef]
46. Yu, J.H.; Yang, Q.C.; Ye, H.; Luo, Y.Q.; Huang, J.C.; Xu, X.H.; Gang, W.J.; Wang, J.B. Thermal performance evaluation and optimal design of building roof with outer-layer shape-stabilized PCM. *Renew. Energy* **2020**, *145*, 2538–2549. [CrossRef]
47. Wang, H.T.; Hang, Y.G.; Liu, Y. Integrated economic and environmental assessment-based optimization design method of building roof thermal insulation. *Buildings* **2022**, *12*, 916. [CrossRef]

Article

Real-Time Temperature and Humidity Measurements during the Short-Range Distribution of Perishable Food Products as a Tool for Supply-Chain Energy Improvements

Martim L. Aguiar [1,2], **Pedro D. Gaspar** [1,2,*], **Pedro D. Silva** [1,2], **Luísa C. Domingues** [1] and **David M. Silva** [1]

[1] Department of Electromechanical Engineering, University of Beira Interior, Rua Marquês D'Ávila e Bolama, 6201-001 Covilha, Portugal
[2] C-MAST, Center for Mechanical and Aerospace Science and Technologies, Faculty of Engineering, University of Beira Interior, 6201-001 Covilha, Portugal
* Correspondence: dinis@ubi.pt

Abstract: Food waste results in an increased need for production to compensate for losses. Increased production is directly related to an increase in the environmental impact of agriculture and in the energy needs associated with it. To reduce food waste, the supply chain should maintain ideal preservation conditions. In horticultural products, temperature, and relative humidity are two of the main parameters to be controlled. Monitoring these parameters can help decision-making in logistics and routes management, as well as to diagnose and timely prevent food losses. In the present work, eighteen wireless traceability devices with temperature and relative humidity sensors monitored crates with horticultural products along a short-range distribution route with five stops (4 h 30 m). Sensor data and a location tag were sent via GSM for real-time monitoring. The results showed fluctuations in temperature and relative humidity that reached up to 7.4 °C and 35.3%, respectively. These fluctuations happened mostly due to frequent door opening, operational procedures, and irregular refrigeration conditions. Furthermore, the results brought attention to a procedure that creates unnecessary temperature fluctuations and energy losses. This study highlights the importance of individual monitorization of goods, for quality control and optimization of energy efficiency along the supply chain.

Keywords: intelligent packaging; temperature; humidity; supply chain; food waste; food quality; food preservation; transportation

Citation: Aguiar, M.L.; Gaspar, P.D.; Silva, P.D.; Domingues, L.C.; Silva, D.M. Real-Time Temperature and Humidity Measurements during the Short-Range Distribution of Perishable Food Products as a Tool for Supply-Chain Energy Improvements. *Processes* **2022**, *10*, 2286. https://doi.org/10.3390/pr10112286

Academic Editors: Yabin Guo, Zhanwei Wang, Yunpeng Hu and João M. M. Gomes

Received: 27 September 2022
Accepted: 24 October 2022
Published: 4 November 2022

Publisher's Note: MDPI stays neutral with regard to jurisdictional claims in published maps and institutional affiliations.

1. Introduction

The degradation of food quality is caused by microbial growth [1]. This degradation can lead to health hazards, food waste, and profit losses. The factors that affect microbial growth can be intrinsic (food pH, nutrients, moisture, etc.) or extrinsic (relative humidity, temperature, and atmosphere composition). Extrinsic factors are easier to control to increase food preservation [2–5]. Different food products require different ideal extrinsic conditions for optimum preservation [6,7]. Temperature is the extrinsic factor that affects food preservation the most, allied with the fact that inefficiencies leading to thermal losses are still a problem in the food industry [8–11]; this means that it is crucial to monitor the temperature of perishable foods along their life cycle to diagnose faults and cut losses. The thermal losses can be caused not only by inefficiencies such as bad insulation, system design and frost formation [12], but also by common routines such as doors opening and closing [13] which are unavoidable events during the distribution of refrigerated goods. The impacts of these events should be thoroughly monitored.

Refrigeration is essential for the preservation of horticultural products, but it requires significant amounts of energy [14]. In retail, the energy spent during the commercialization of refrigerated food products is about 50% of the supermarket's total energy consumption [15]. Along with the energy used for storage and transport, refrigeration is one of the main causes of energy consumption along the life cycle of perishable food products [16]. To make matters worse, the amount of food products that are wasted before reaching the final consumer is approximately one-third of the total production; in horticultural products, this waste reaches 45% [17]. This leaves plenty of room for the development of technologies that monitor and help maintain food quality to reduce waste, energy consumption, and associated environmental impact [12,18,19].

Packaging has the objective of facilitating transportation while maintaining the quality of food products. Nowadays, new technologies are being developed to make packaging more efficient and sustainable [20,21]. Packages can incorporate materials to enhance food preservation [22], such as phase change materials (PCM) that help maintain a stable temperature when conditions outside the package vary [23–25]. Traceability is also increasingly being incorporated into packaging along the supply chain [26,27]. The use of sensors for agricultural applications is not a new subject [28] but it is a field that has been through accelerated growth in both scientific and industrial sectors [13]. In the packaging sector, intelligent packaging technologies use sensors to measure and transmit the packaging conditions, with the aim of reducing food waste and meeting new food safety regulations, which are frequently updated to be stricter [26].

Intelligent packaging is the result of technologies that allow for the incorporation of communication and sensing tools in a food packaging system to measure the conditions of the environment in which the system is deployed [13]. Measurement of parameters such as temperature and relative humidity or vibrations, and light can be useful to perform quality control, objectively, which would be otherwise impossible [26]. Of these parameters, temperature is the most associated with energy consumption, and food quality degradation is the most important to monitor [29].

Traceability of food quality leads to increased sustainability and profit [30] while decreasing health hazards [31,32]. This can be magnified if traceability systems are integrated into decision support systems for perishable food products [33,34]. This way strategies such as price rebates can be applied to sell products whose storing and transport conditions monitored can be associated with quality degradation [35], resulting in dynamic pricing according to the quality of the food products [1]. Furthermore, with the emergence of artificial intelligence (AI), the data gathered by intelligent packaging can be fed to AI algorithms for the best decisions to be made to increase sustainability and profitability [36].

In previous works, Morais et al. [37,38] developed a monitoring device to read temperature and relative humidity during transportation. The measurements were transmitted via GSM, along with GPS coordinates. Subsequently, the iTrace platform was developed, allowing real time access to measurements made with an array of temperature and relative humidity sensors [6]. Testing this system in a real case scenario was the aim of the present study. This way, extrinsic factors for food preservation can be monitored along the cargo volume and as close to the food product as possible. This is important as the arrangement in which the products are conditioned can have an impact on temperature and relative humidity distribution and fluctuations, which ultimately, if inadequate, results in quality degradation [39]. Temperature fluctuations among different products have already been studied in supermarkets, within the display cabinets [40,41]. The heterogeneous cargo and produce crate distribution of a real case scenario only accentuate the differences in the fluctuations of temperature and relative humidity. Ultimately, the aim of the study is to highlight the importance of monitoring produce by crate, to track and diagnose flaws in the conservation of extrinsic conditions for food preservation, and energy losses associated with it, that otherwise would go undetected.

2. Materials and Methods

The present work aimed to study the temperature and relative humidity distribution close to the different horticultural products during a distribution route. For this, a system to read and transmit the readings of temperature and relative humidity is needed. Moreover, the cargo must be organized and characterized to facilitate the interpretation of large volumes of data.

2.1. Temperature and Relative Humidity Measurement and Communication System

The sensing module was composed of several parts, shown in Figure 1. The measurements of temperature and relative humidity were taken using an SHT30 sensor. This sensor had an accuracy of $\pm 0.3\ ^\circ$C with a repeatability of 0.06 $^\circ$C for temperatures between 10 $^\circ$C and 55 $^\circ$C, and accuracy of $\pm 3\%$ with a repeatability of 0.13% for relative humidity values between 10% and 90%. The sensor was connected to a LOLIN D1 mini pro V2.0 microcontroller using an LOLIN SHT30 v2.1.0 shield. The sensor, microcontroller, and a 3.7 V 500 mAh lithium battery were conditioned within a 3D printed box for protection. This box had a mesh, visible in Figure 1e, so that the sensor could make direct contact with the air surrounding the food product.

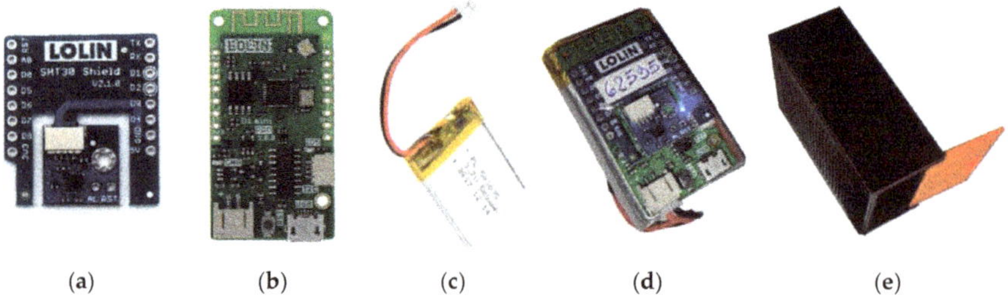

(a)   (b)   (c)   (d)   (e)

Figure 1. Sensing module components: LOLIN SHT30 v2.1.0 shield (**a**), LOLIN D1 mini pro V2.0 microcontroller (**b**), 3.7 V 500 mAh lithium battery (**c**), assembled sensing module (**d**) and 3D-printed case (**e**).

It is worth noting that the measurements of relative humidity have significant errors which required a calibration of the sensors to be made before experimental testing. After calibration, a repeatability of 0.13% relative humidity ensured accurate data, considering these are low-cost sensors. More precise and expensive sensing methods would render the application unfeasible for real case applications, due to higher costs associated.

An array of these sensing modules communicated with the gateway composed of an industruino microcontroller, with a GSM/GPRS module to allow for real time data transmission with GPS location pinpointing. The values were measured and transmitted every five minutes. These data could be accessed at the iTrace platform, developed for this purpose [8]. The gateway and system schematics can be seen in Figure 2.

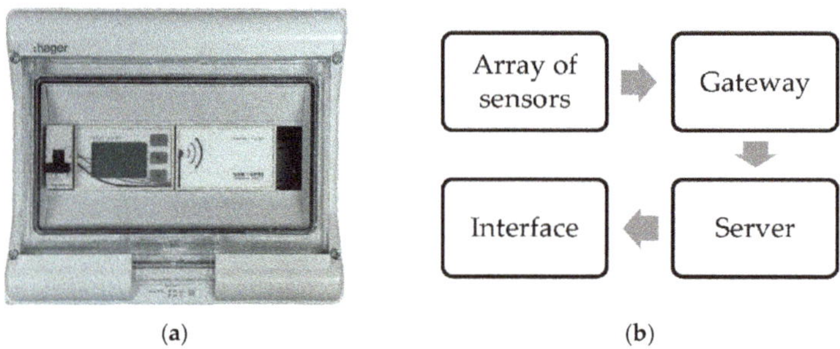

(a)   (b)

Figure 2. Gateway (**a**) and system schematics (**b**).

The system was developed so that the sensing modules could be placed in the crates within the horticultural products. This proximity to the produce allowed measurements of temperature and relative humidity to be taken as close to the food product as possible.

2.2. Distribution Vehicle and Refrigeration System

The fleet of distribution vehicles was composed by IVECO Daily vans, as shown in Figure 3.

Figure 3. Distribution vehicle example.

The cargo area and loading doors of these vans can be characterized by the dimensions shown in Table 1.

Table 1. Van dimensions.

Description		Dimension (mm)
Load compartment	Length	4680
	Width	1800
	Height	1800
Sliding side door	Width	1200
	Height	1800
Rear doors	Width	1530
	Height	1800

The van had a hwasung HT-250RT-ESC refrigeration system installed with a controller in the driver's compartment that displayed the instantaneous temperature value, as shown in Figure 4a. These displayed temperatures were measured at the outlet of the heat exchanger, as shown in Figure 4b, and were compared to the measurements taken by the sensors within the produce crates. As this temperature was only displayed while the van was working, only measurements before and after stops were taken. Between stops, temperature values were manually registered in intervals of 15 min. The temperature setpoint was 3 °C. No relative humidity values are monitored by the refrigeration system; therefore, no van relative humidity measurements were taken for later comparison.

(a)   (b)

Figure 4. Controller (**a**) and heat exchanger with temperature sensor (**b**) of the trucks refrigeration system.

2.3. Cargo and Sensor Distribution

The route consisted of the distribution of an assortment of horticultural products along five stops. The diversity of horticultural products made it necessary to characterize the distribution and content of the crates along the cargo area to facilitate the interpretation of results later. Table 2 displays the abbreviations used to represent the different horticultural products.

Table 2. References for the horticultural products in the crates.

Ref.	Product	Ref.	Product	Ref.	Product	Ref.	Product
AP	Apple	EP	Eggplant	MG	Mango	PO	Potato
AR	Apricot	GA	Garlic	NE	Nectarine	PP	Papaya
BC	Bagged	GR	Grape	ON	Onion	PR	Pear
BN	Banana	KW	Kiwi	OR	Orange	PS	Packaged spinach
CA	Carrot	LE	Lettuce	PA	Cauliflower	TO	Tomato
CB	Cabbage	LK	Leek	PE	Peach	UM	Mushroom
CH	Cherry	LN	Lemon	PI	Pineapple	WM	Watermelon
CT	Cantaloupe	ME	Melon	PL	Parsley	ZU	Zucchini

The distribution of the products was arranged by the drivers, to facilitate deliveries; therefore, the arrangement order was diverse and complex. In the example shown in Figure 5, there were 108 crates of diverse produce and 26 bags of 20 kg of potatoes and onions.

(a)   (b)

Figure 5. Final produce arrangement within the truck (**a**), and sample of the diversity of produce transported (**b**).

The sensors were distributed along the piles (columns) of crates/bags, in the top, middle and bottom. Figure 6 shows a column of crates with a sensor (sensor 3) placed within the tomatoes, in the middle crate.

Figure 6. Placement of sensor 3.

To explain the sensor distribution across such a heterogeneous cargo, the top and section views were schematized as shown in Figure 7. In this representation, the cargo area is divided into groups, each group is divided into columns and each column is layered into lines. As an example, sensor 19 is placed in group 4, column 3, line 3, so G4C3L3 is a short reference that gives exactly the position of the crate the sensor is in. This representation also allows for a sectioned view, that can be helpful to interpret results, considering how the surroundings can influence the environment.

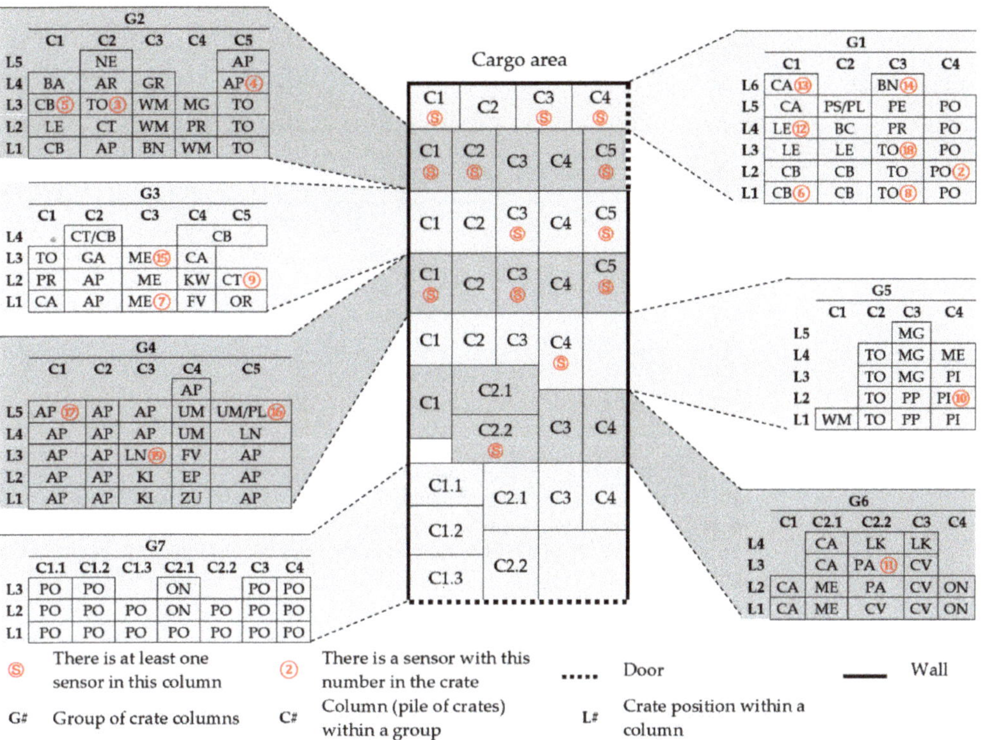

Figure 7. Organization of crates among the cargo area and sensor distribution.

2.4. Route and Stops

The route consisted of the distribution of an assortment of horticultural products, along with five different local retail stores within the northern region of Castelo Branco, Portugal. The measurements were taken during the month of June, in the beginning of the Portuguese summer season, with the temperatures within the minimum average of 15.1 °C and maximum average of 29.4 °C, reaching a maximum absolute value of 38.5 °C [42]. Figure 8 displays a map of the route along with the stops and a graphical representation of the duration of those stops. This graphical representation was overlayed with the sensor data in the results section of the present work, so that the context of the stops could be taken into consideration while analyzing the results.

The route started at a point Ⓞ. At 8:40 h, the van started, and the van's refrigeration system was turned on. After leaving the warehouse, five more stops were made for unloading, before returning. Table 3 displays the details about all the unloading stops made.

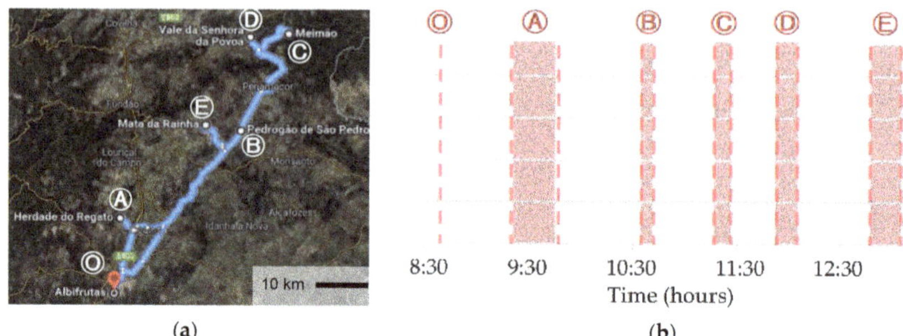

Figure 8. Route stops (**a**) and graphical representation of the duration of those stops (**b**).

Table 3. Stops and relevant information.

Stop	Events			
	Stop Time	**Stop Duration**	**Unloading Door**	**Sensors Turned Off**
Ⓐ	9:20 h	27 min	Back door	10, 11, 19
Ⓑ	10:35 h	7 min	Back door	None
Ⓒ	11:17 h	9 min	Side door	4, 7, 9, 16
Ⓓ	11:53 h	12 min	Side door	3, 5, 17, 15
Ⓔ	12:47 h	18 min	Side door	2, 6, 8, 12, 13, 14, 18

It is important to note that some of the produce (leaf vegetables such as cabbages and lettuce) were watered before transportation to decrease the degradation associated with lower relative humidity. This was analyzed further as sensor measurements varied for this produce. An example of a watered crate of cabbages with sensor 12 (position G1C1L4) is shown in Figure 9.

Figure 9. Crate with cabbages that were watered prior to loading in the truck.

3. Results and Discussion

Data were compiled to better display and compare the results. Averages of different sensors are shown, and the comparison is analyzed below. These averages were made from the readings taken in five-minute periods.

3.1. Average Values of All Sensors

The average of all the sensor readings was the first step to understanding the overall fluctuations in temperature and relative humidity along the route. In Figure 10 the averages for all the readings are presented.

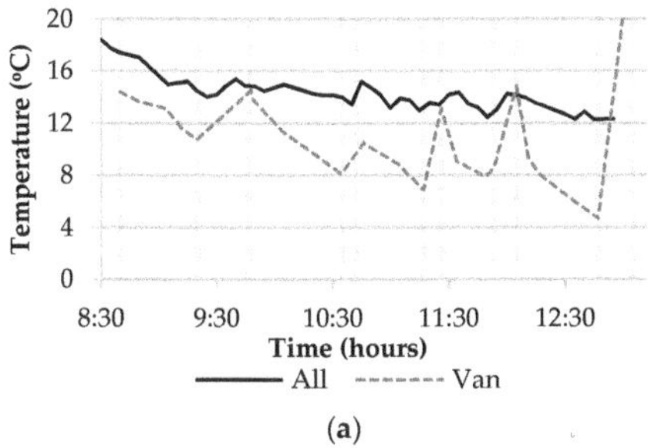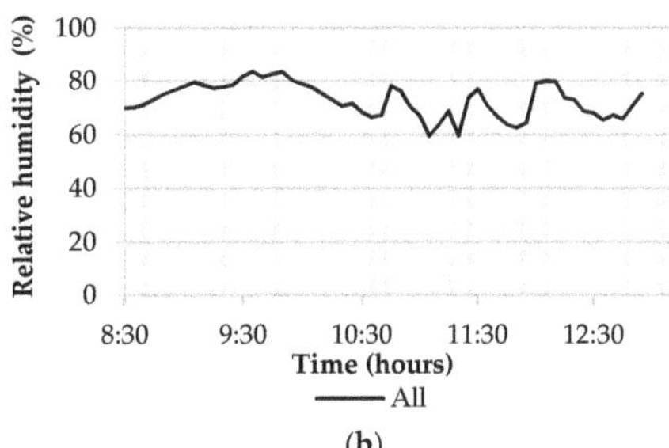

(**a**)

(**b**)

Figure 10. Average measured temperature, and refrigeration system displayed temperature (**a**) and average measured relative humidity (**b**).

Fluctuations were easily observed, especially during stops in the periods that cargo was unloaded. The graph that shows the average temperature of all the sensors also plots the measurements displayed by the refrigeration system installed in the van. This display shows measurements taken right at the outlet of the refrigeration system. It is notable how the temperature readings in the sensors differed from those displayed in the driver's cabinet, only matching when the refrigeration system was turned off. This happened because this sensor measured the outlet of the refrigeration system and not the temperature of the air close to produce. The refrigeration system only measured temperature; therefore, no relative humidity values are presented for comparison. Overall, temperature and relative humidity values decreased during trips, and increased during stops. Although it is known that relative humidity is inversely proportional to temperature, in this case, the refrigeration system while lowering the air temperature also acted as a dehumidifier due to the surface temperature of the heat exchanger being below the dew point temperature. This resulted in measurements showing a decrease in temperature along with a decrease in humidity during trips, while both temperature and relative humidity increased during stops.

3.2. Sensors in the Top Versus Sensors in Lower Crates

Top crates were directly exposed to the refrigerated air during transportation and warm air during unloads, while crates in lower layers should have been more protected against fluctuations. In Figure 11 the average temperatures and relative humidity for sensors in the top (13, 14, 15, 16, 17) and inner layers (3, 6, 8, 7, 5) are shown.

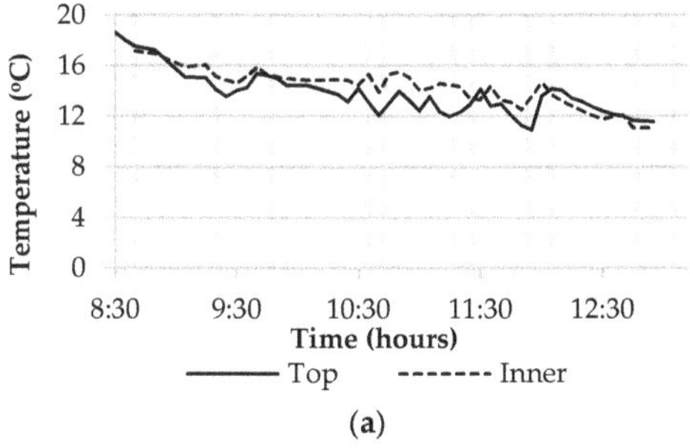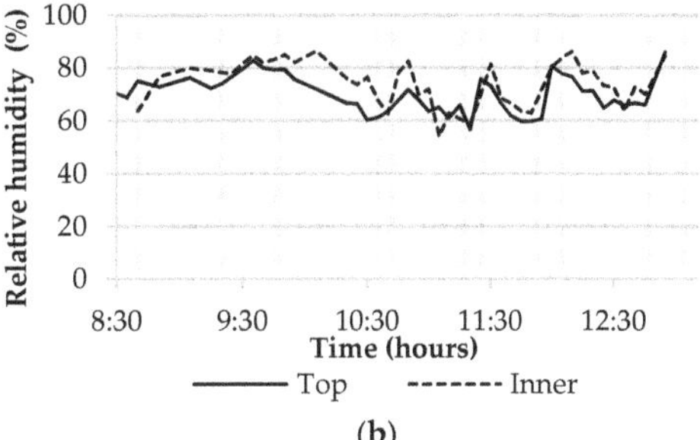

(**a**)

(**b**)

Figure 11. Average of top and inner sensor readings for temperature (**a**) and relative humidity (**b**).

It is possible to observe that as was expected, sensors in the lower layers registered fewer aggressive fluctuations in temperature and relative humidity. The temperature was lower in the top sensors due to them being exposed directly to the outlet of the refrigeration system while traveling, but the values quickly rose to those of the inner sensors while unloading. Relative humidity was also higher for the crates positioned in lower levels, especially between unloads, but the fluctuations were comparable to those in the upper levels, except during unloads.

3.3. Average of All Sensors versus Sensors Placed near the Side Door

It was expected that the sensors placed near the doors would suffer higher fluctuations in temperature and relative humidity when unloading due to the proximity to external environment and larger thermal interaction. In Figure 12, readings from the sensors near the side door (14, 2, 4) are compared to the average. It is worth noting that the side door was only opened from the stop onwards.

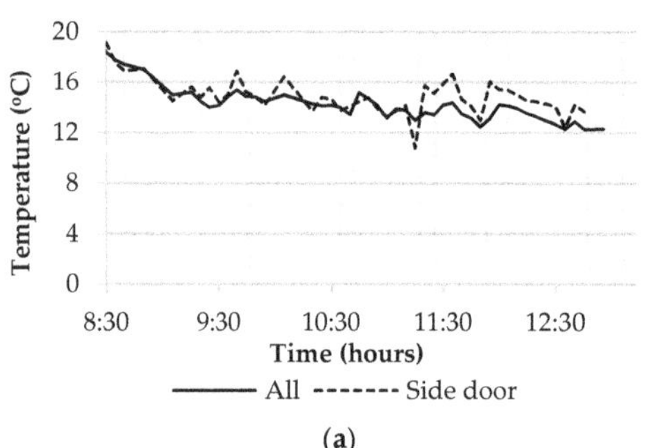

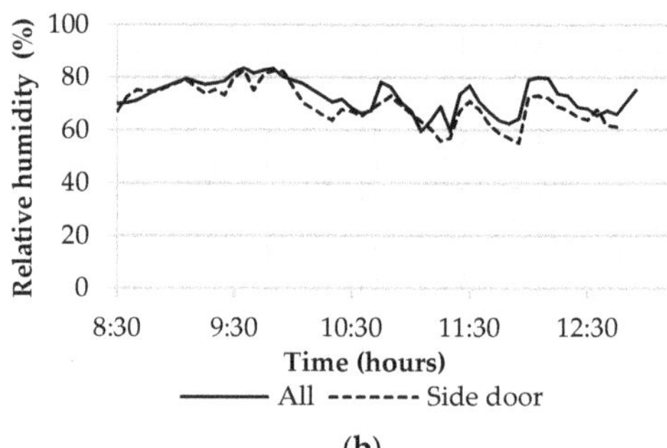

Figure 12. Average of all the sensor readings versus the average of the readings from sensors near the side door for temperature (**a**) and relative humidity (**b**).

As expected, near the unloading door, temperatures were a lot less stable and particularly higher than the average after this door started being used. Furthermore, fluctuations in relative humidity were comparable to the fluctuations in the average relative humidity, but the readings showed consistently lower values after the side door started being used for unloading.

3.4. Sensors Placed in Wet Produce versus Sensors Placed in Dry Produce

To increase the shelf life of some horticultural products, such as cabbage, cauliflower, and lettuce, the procedure established in the warehouse is to hose this produce down with tap water before loading it for distribution. Produce such as potatoes, melons, and onions do not go through this procedure, so the relative humidity measured near them was expected to be lower. In Figure 13, the average of measurements in crates with watered produce (5, 6, 12, 13) is compared to the average of measurements in drier produce (2, 7, 9). It is worth noting that from 11:30 h onwards, all the dry readings shown were from sensor 2, as crates from sensors 7 and 9 were unloaded in stop Ⓒ.

The temperature difference was mostly seen after sensors 7 and 9 are disabled, and the side door started being used. Sensor 2, being placed among potato bags, near the side door measured higher values from stop Ⓒ onward as shown in Section 3.3. For relative humidity, fluctuations were of similar amplitude along the route, but a higher relative humidity was measured near produce watered prior to loading.

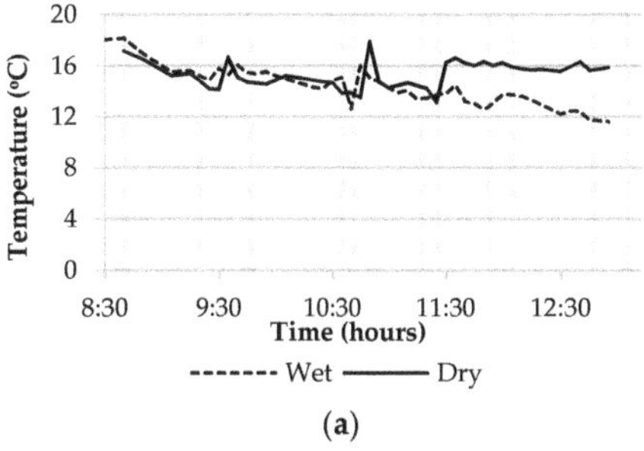
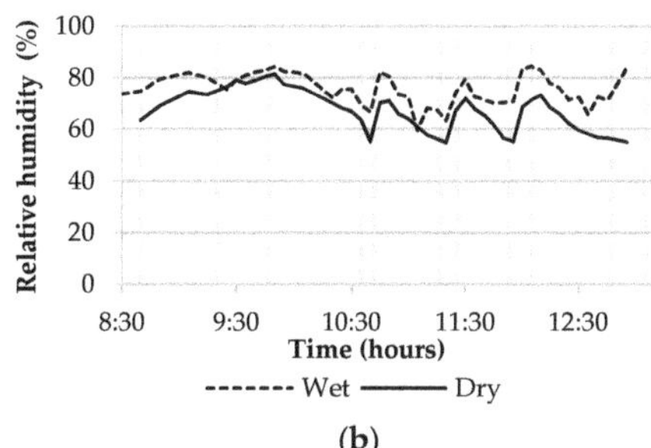

Figure 13. Average of the readings of sensors placed in crates with produce watered previously to being loaded, versus the readings of dry crates for temperature (**a**) and relative humidity (**b**).

3.5. Individual Sensor Measurements with Different Readings

Although the averages help to achieve an overall understanding of how the events during transportation affect the extrinsic conditions along the distribution route, the point of distributing an array of sensors along the cargo volume was to realize how the conditions of individual crates of produce vary independently from each other, and if those variations were relevant enough for individual crate sensitization to be considered. In Figure 14, two different sensors with quite different readings are compared. Sensor 2, in position G1C4L2 was placed among potato bags, while sensor 6, in position G1C1L1 was placed among cabbages. This comparison aimed to display the importance of measuring crates individually, instead of single-point measurements during the transportation of horticultural products.

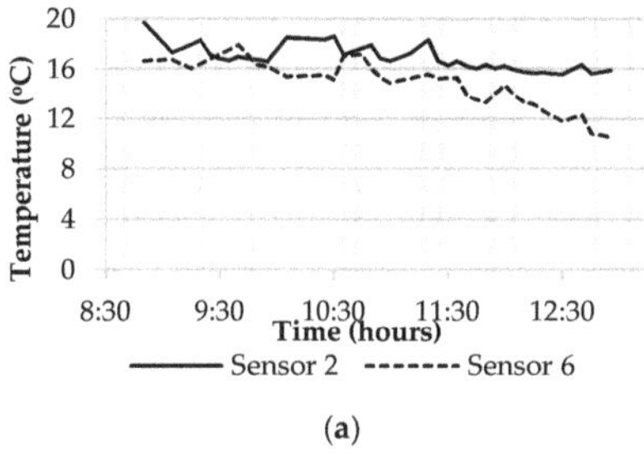
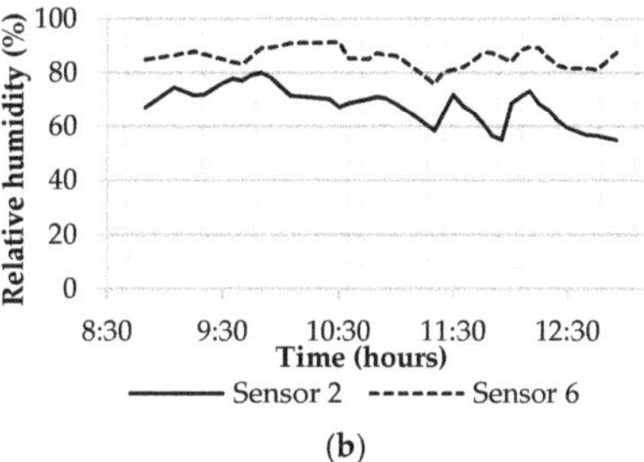

Figure 14. Readings of sensor 2 and sensor 6 for temperature (**a**) and relative humidity (**b**).

The differences between readings were quite surprising, considering the produce spent the night in the same climatized pavilion and were transported in a van adapted for refrigerated transportation of goods. At 12:30 h, for example, the temperature differences were approximately 4 °C, and the relative humidity had a difference of approximately 25%. In this case, the different nature of the horticultural products, their positioning and the procedures undergone before loading were the principal causes for these discrepancies. It is also worth noting that not even the fluctuations were similar.

3.6. Maximum, Minimum and Average Values per Sensor

To better characterize the temperature and relative humidity distribution along the cargo volume, Table 4 shows the absolute maximum shows the highest value measured, the absolute minimum shows the lowest value measured, the average shows the average of all the readings from that sensor, and the range was calculated by subtracting the values of absolute minimum to the values of absolute maximum, to obtain the range within each sensor fluctuates for temperature and relative humidity. The values were measured during the distribution route, while the sensor was active, i.e., before turning off the sensor and unloading the monitored crate. The times when the absolute maximum and absolute minimum values were recorded are also shown.

Table 4. Absolute maximum, absolute minimum, range, average temperature (T) and relative humidity (RH) per sensor, with time of measurement for absolute maximum and absolute minimum values.

Sensor	Absolute Maximum				Absolute Minimum				Range		Average	
	T (°C)	Time (h)	RH (%)	Time (h)	T (°C)	Time (h)	RH (%)	Time (h)	T (°C)	RH (%)	T (°C)	RH (%)
2	18.6	10:30	80.1	9:50	15.6	12:30	55.0	12:55	3.0	25.2	16.6	67.1
3	16.7	9:15	88.3	9:50	12.2	11:50	54.8	11:45	4.4	33.5	14.6	72.9
4	13.4	9:15	91.2	9:50	10.1	10:35	56.0	11:15	3.3	35.3	11.8	78.4
5	16.6	9:15	84.8	9:40	12.0	11:50	50.9	11:20	4.6	34.0	14.2	67.5
6	17.9	9:40	91.3	10:30	10.5	12:55	75.9	11:20	7.4	15.4	14.4	85.3
7	15.7	9:05	77.4	9:50	12.5	11:25	50.1	11:20	3.2	27.3	14.1	65.6
8	16.1	9:45	87.1	9:35	11.4	12:45	59.0	12:45	4.8	28.2	13.5	72.0
9	13.3	9:40	87.0	9:40	10.0	10:35	55.7	11:20	3.3	31.4	11.7	75.5
10	14.8	9:20	86.0	9:30	13.5	9:30	82.6	9:20	1.2	3.4	14.1	84.5
11	15.4	9:15	82.9	9:25	12.9	9:25	71.8	9:15	2.5	11.2	14.4	76.9
12	16.7	9:40	85.5	9:05	12.7	12:55	66.8	12:45	4.1	18.8	14.2	75.2
13	15.4	9:30	85.8	12:55	11.1	11:55	61.6	12:25	4.3	24.3	13.1	71.5
14	16.8	9:40	77.3	9:30	11.3	11:50	55.4	11:20	5.5	21.9	13.7	67.3
15	16.6	9:15	81.3	9:30	10.8	11:55	51.2	11:20	5.8	30.2	13.8	67.7
16	15.2	9:05	84.3	11:25	10.3	11:20	56.4	11:20	5.0	27.9	12.9	74.1
17	15.2	9:40	80.1	9:30	11.2	11:50	57.4	11:20	4.0	22.7	13.5	69.6
18	16.5	10:35	87.1	9:50	11.6	12:55	62.3	12:55	4.9	24.8	14.0	78.6
19	14.0	9:15	86.8	9:45	12.0	9:30	79.8	9:15	1.9	7.0	13.4	82.9

When analyzing Table 4, we can observe that for sensor 2 and sensor 9, the difference between absolute maximum temperatures recorded was 5.3 °C, the difference between absolute minimum temperatures recorded for these same sensors was 5.6 °C and the difference between average recorded temperatures was 4.8 °C. For sensor 10 and sensor 6, the difference between the ranges of recorded temperatures was 6.1 °C.

For relative humidity, sensor 6 and sensor 14 showed a difference between the absolute maximum relative humidity recorded of 14%, while the difference between the absolute minimum relative humidity recorded in sensor 7 and sensor 10 was 32.5%. The difference in average recorded relative humidity for sensor 6 and sensor 7 was 19.7% while the difference in the range of recorded relative humidity for sensor 4 and sensor 10 was 31.9%. These were significant fluctuations among different sensors, which highlights the importance of distributing the measurements between produce crates, instead of single point measurements.

Furthermore, absolute maximum and minimum values for temperature and relative humidity, allowed us to inspect when and where the worse conditions for loss of quality were met, while average values indicated an overall state of preservation, and range highlights fluctuations along the trip. Together, this can be a powerful tool to diagnose faults that otherwise would go undetected.

The best demonstration of the application of this system as a tool for the detection of faults was finding a flaw concerning the procedures taken by the distribution company. After completing the measurements and analyzing the data, the authors were intrigued by the disparities measured in some sensors, namely between sensor 2 and sensor 6, whose values were already compared in Section 3.5. Sensor 2, in position G1C4L2 was placed among potato bags, while sensor 6, in position G1C1L1 was placed among cabbages. What intrigued the authors was how sensor 6, placed in the lower levels, and far from the door, could have a higher range of temperature measurements than sensor 2, placed near the door. Upon close inspection, and as shown in Figure 14a, temperature fluctuations during stops were measured in both sensors, but during the overall trip, the temperature in sensor 6 dropped significantly more than the temperature in sensor 2. In addition, sensor 6 had the second highest value for absolute maximum temperature, which is also unusual, as cabbages were stored at lower temperatures than other products such as potatoes and onions. It was only when looking at the relative humidity which is significantly higher for sensor 6 than for sensor 2 that the following hypothesis arose: cabbages, and other leaved produce, went through the procedure of being watered prior to loading, receiving heat from the tap water, which was aggravated in the warm days of summer the data was collected. This hypothesis would be impossible to achieve if it were not for the array of sensors, together with a diagram of their position and type of horticultural product.

After sensor 6, sensors 15, 14 and 16 registered higher ranges, which was expected, as these were all on the top crates near the side door, and as shown in Figures 11a and 12a the top and door side sensors should have measured higher fluctuations.

4. Conclusions

The impact of inadequate temperature and relative humidity in the preservation of horticultural products is known, but it is not easy to predict if sensors are not placed near the produce. This study presented evidence that reinforces the importance of parameter monitoring within the packages for the precise control of food quality preservation along the supply chain. The disparity between the crate temperature measurements and the readings displayed from the van refrigeration system further reinforces these claims.

The results also showed that the order in which products are stored can have an impact on the temperature and relative humidity fluctuations during distribution. This information can be used to rearrange the order in which products are conditioned to decrease the impact of temperature fluctuations, especially in produce most prone to degradation. For instance, produce that requires higher relative humidity, or is more sensible to temperature and relative humidity fluctuations should be stored in the bottom. This suggestion does not interfere with the duration or unload procedure, as the produce are unloaded in columns, and only the column order is altered, not the position of the column within the cargo area, which is left for the driver to decide how to best accommodate for faster distribution.

The unpredictability of the type and order of transported goods, especially in short-range distribution, does not allow for any assumption of cargo homogeneity, which further reinforces the need for distributed monitoring.

Measurements such as those presented in this work can be used to diagnose and fix flaws along the supply chain. In the case of this study, it was found that a procedure was introducing undesired thermal loads by watering produce with tap water prior to loading it into the distribution vehicle. This may result in lower shelf life for the affected produce due to increased fluctuations in conservation temperature, higher energy consumption associated with refrigeration, to compensate for the thermal load, and energy loss associated with food waste. The negative effects of this procedure would otherwise go undetected if the monitoring of temperature and relative humidity was not spread across the cargo volume. To fix this flaw, chilled water could be sprayed on the produce that require watering, or misting could be applied instead of hosing the produce, to reduce the volume of warm water and associated thermal load. Obviously, the impact of the procedure on the degradation of quality should be studied, to access if changes would result in improvements significant enough to modify it.

The application of the studied sensing system will be specific to every case. The number of variables such as refrigerated or non-refrigerated transportation system, produce variety, season, weather, trip length, number and duration of unloads, produce, and so on will influence results, and that is why this is a good tool to measure what can be almost impossible to predict, helping diagnose flaws that affect the shelf life of produce. This complexity will increase with the number of sensors, as recording positions, and sensed produce is crucial to a complete analysis, which will become a harder task, without a tool specifically designed for the purpose.

The cross analysis of sensor readings is useful to understand the variations during transportation, diagnose flaws and develop improved procedures, although to predict the shelf life of produce in a single crate, data measured close to the specific produce is the most important.

This system has some limitations. The measurement and analysis of temperature and relative humidity fluctuations along an uneven cargo volume was complex enough, given different sides of the van being exposed to the sun, different doors being opened in different phases of the distribution route, and the position of the outlet of the refrigeration system. Added to this was an uneven cargo, with different produce organized not by type, but by client, with different storage temperatures across them and some of them being watered before loading will add to the complexity of the analysis. Furthermore, the fact that some produce is unloaded right at the beginning of the route resulted in a small window of time for these sensors to gather data.

The need for a wireless sensing system means that a battery must be charged from time to time, and although this is a low-power sensor, the logistics associated with charging several dozens of sensors, even if optimized to be a weekly charge, must be addressed and simplified, to make this a viable option.

The robustness of the sensor is also a limitation that must be addressed. The crates of produce endure though mechanical stresses during transportation, being dragged and dropped, or even thrown when empty. If the sensor is attached to the crate, it will need to endure these conditions while remaining undamaged and reliable for the next usage.

A high sample rate is crucial to understanding abrupt changes in temperature and relative humidity, but a low sample rate should be enough to measure slower changes between unloads. The problem with high sample rates is higher power consumption, and therefore lower battery life. This can be solved by implementing a dynamic sample rate that increases sample rate during unloads, by either measuring the state of movement with an accelerometer, or by measuring luminosity within the cargo area, which increases when the doors are open.

Further studies should be made with a larger array of sensors, and a higher frequency of measurements to better understand the fluctuations of temperature and relative humidity along with the horticultural goods during transportation. This will also require a system for data analysis and comparison; as mentioned above, a larger array of sensors will result in a higher complexity of data that in turn will be more difficult to process and understand, find flaws, and monitor quality.

Author Contributions: Conceptualization, P.D.G., M.L.A., L.C.D. and P.D.S.; methodology, P.D.G., M.L.A., L.C.D. and P.D.S.; validation, P.D.G. and P.D.S.; formal analysis, P.D.G. and P.D.S.; investigation, M.L.A., L.C.D., D.M.S.; resources, P.D.G. and M.L.A.; data curation, M.L.A.; writing—original draft preparation, M.L.A.; writing—review and editing, P.D.G., P.D.S. and M.L.A.; supervision, P.D.G. and P.D.S.; project administration, P.D.G.; funding acquisition, P.D.G. All authors have read and agreed to the published version of the manuscript.

Funding: This research was funded within the activities of project "PrunusPós–Optimization of processes for the storage, cold conservation, active and/or intelligent packaging and food quality traceability in post-harvested fruit products", project n. ° PDR2020-101-031695, Partnership n° 87, initiative n. °175, promoted by PDR 2020 and co-funded by FEADER within Portugal 2020. This work was supported in part by the Fundação para a Ciência e Tecnologia (FCT) and C-MAST (Centre for Mechanical and Aerospace Science and Technologies), under project UIDB/00151/2020.

Data Availability Statement: More information on the project PrunusPós and the iTrace platform can be found at https://www.prunuspos.pt/ (accessed on 26 September 2022).

Acknowledgments: The authors are also thankful to the company Albifrutas–Produtos Horticolas, Lda., in the name of José António for allowing us to perform the measurements in a real case scenario, and to the van driver, António, for helping along and being patient with the necessary procedures of the experiment during the route.

Conflicts of Interest: The authors declare no conflict of interest.

References

1. Pina, M.; Gaspar, P.D.; Lima, T.M. Decision support system for dynamic pricing of horticultural products based on the quality decline due to microbial growth. *Appl. Syst. Innov.* **2021**, *4*, 80. [CrossRef]
2. Gaspar, P.D.; Silva, P.D.; Andrade, L.P.; Nunes, J.; Santo, C.E. Technologies for monitoring the safety of perishable food products, Ch, 4. In *Research Anthology on Food Waste Reduction and Alternative Diets for Food and Nutrition Security*; Information Resources Management Association , Ed.; IGI Global: Hershey, PA, USA, 2021; pp. 63–98. ISBN 13. [CrossRef]
3. Rodrigues, C.; Gaspar, P.D.; Simões, M.P.; Silva, P.D.; Andrade, L.P. Review on Techniques and Treatments towards the Mitigation of the Chilling Injury of Peaches. *J. Food Preserv. Process.* **2020**, *2020*, e14358. Available online: http://hdl.handle.net/10400.6/10368 (accessed on 26 September 2022). [CrossRef]
4. Andrade, L.P.; Nunes, J.; Simões, M.P.; Morais, D.; Canavarro, C.; Santo, C.E.; Gaspar, P.D.; Silva, P.D.; Resende, M.; Caseiro, C.; et al. Experimental study of the consequences of controlled atmosphere conservation environment on cherry characteristics. In Proceedings of the 25th IIR International Congress of Refrigeration (ICR 2019), Montreal, QC, Canada, 24–30 August 2019; Available online: http://hdl.handle.net/10400.6/7521 (accessed on 26 September 2022). [CrossRef]
5. Andrade, L.P.; Veloso, A.; Santo, C.E.; Gaspar, P.D.; Silva, P.D.; Resende, M.; Beato, H.; Baptista, C.; Pintado, C.M.; Paulo, L.; et al. Effect of controlled atmospheres and environmental conditions in the physicochemical and sensory characteristics of sweet cherry cultivar Satin. *Agronomy* **2022**, *4*, 188. [CrossRef]
6. Simões, M.P.; Veloso, A.; Gaspar, P.D.; Santo, C.E.; Silva, P.D.; Andrade, L.P. PrunusPÓS—Otimização de processos de armazenamento, conservação em frio, embalamento ativo e/ou inteligente, e rastreabilidade da qualidade alimentar no pós-colheita de produtos frutícolas. In *Grupos Operacionais de Fruticultura 2018–2022*; Simões, M.P., Carmo, M., Eds.; COTHN—Centro Operativo e Tecnológico Hortofrutícola Nacional: Alcobaça, Portugal, 2021; pp. 404–471. ISBN 978-972-8785-18-5.
7. Veloso, A.; Ferreira, D.; Gaspar, P.D.; Andrade, L.P.; Espírito-Santo, C.; Silva, P.D.; Simões, M.P. Influence of storage conditions on fruit quality of 'Royal Time' and 'Royal Summer' peach cultivars. *Rev. Ciênc. Agrár.* **2021**, *44*, 82–90. [CrossRef]
8. Morais, D.; Silva, P.D.; Gaspar, P.D.; Pires, L.C.; Andrade, L.P. Characterization of Refrigeration Systems in the Portuguese Food Processing Industry. In Proceedings of the 25th IIR International Congress of Refrigeration (ICR 2019), Montreal, QC, Canada, 24–30 August 2019; Available online: http://hdl.handle.net/10400.6/7527 (accessed on 26 September 2022). [CrossRef]
9. Nunes, J.; Silva, P.D.; Andrade, L.P.; Gaspar, P.D. Characterization of the specific energy consumption of electricity in the Portuguese sausage industry. *WIT Trans. Ecol. Environ.* **2014**, *186*, 763–774.
10. Silva, P.D.; Gaspar, P.D.; Nunes, J.; Andrade, L.P. Specific electrical energy consumption and CO_2 emissions assessment of agrifood industries in the central region of Portugal. *Appl. Mech. Mater.* **2014**, *675–677*, 1880–1886. [CrossRef]
11. Gaspar, P.D.; Silva, P.D.; Nunes, J.; Andrade, L. Characterization of the Specific Electrical Energy Consumption of Agrifood Industries in the Central Region of Portugal. *Appl. Mech. Mater.* **2014**, *590*, 878–882. [CrossRef]
12. Panoias, P.; Silva, P.D.; Gaspar, P.D.; Pires, L.C.; Nunes, J. Experimental Tests Using Deactivation of Coolant Fluid Circulation to Mitigate Frost Formation on the Heat Exchanger Surface. In Proceedings of the 25th IIR International Congress of Refrigeration (ICR 2019), Montreal, QC, Canada, 24–30 August 2019; Available online: http://hdl.handle.net/10400.6/7518 (accessed on 26 September 2022). [CrossRef]
13. Carneiro, R.; Gaspar, P.D.; Silva, P.D. 3D and transient numerical modeling of door opening and closing processes and its influence on thermal performance of cold rooms. *Appl. Therm. Eng.* **2017**, *113*, 585–600. [CrossRef]
14. Gaspar, P.D.; Gonçalves, L.C.; Pitarma, R.A. Experimental analysis of the thermal entrainment factor of air curtains in vertical open display cabinets for different ambient air conditions. *Appl. Therm. Eng.* **2011**, *31*, 961–969. [CrossRef]
15. Firouz, M.S.; Mohi-Alden, K.; Omid, M. A critical review on intelligent and active packaging in the food industry. *Res. Dev. Food Res. Int.* **2021**, *141*, 110113. [CrossRef] [PubMed]
16. Gaspar, P.D.; Godina, R.; Barrau, R. Influence of orchard cultural practices during the productive process of cherries through Life Cycle Assessment. *Processes* **2021**, *4*, 1065. [CrossRef]
17. Gustavsson, J.; Cederberg, C.; Sonesson, U.; van Otterdijk, R.; Meybeck, A. Global Food Losses and Food Waste—Extent, causes and prevention. Study conducted for the International Congress SAVE FOOD. In Proceedings of the Interpack 2011, Düsseldorf, Germany, 12–18 May 2011; Food and Agriculture Organization of the United Nations (FAO): Rome, Italy, 2011.
18. Gaspar, J.P.; Gaspar, P.D.; Silva, P.D.; Simões, M.P.; Santo, C.E. Energy life-cycle assessment of fruit products—Case study of Beira Interior's peach (Portugal). *Sustainability* **2019**, *10*, 3530. [CrossRef]
19. Aguiar, M.L.; Gaspar, P.D. Refrigerantes naturais: Tendências do mercado, políticas e tecnologias na indústria agroalimentar portuguesa. In Proceedings of the X Iberian Congress and VIII Ibero-American Congress of Cold Sciences and Techniques (CYTEF 2020), Navarra, Spain, 11–13 November 2020; pp. 323–330.
20. Madhan, S.K.; Gaspar, P.D.; Silva, P.D.; Andrade, L.P.; Santo, C.E. A future trends of innovative green packaging of the food products to deplastification. In Proceedings of the 6th IIR International Conference on Sustainability and the Cold Chain (ICCC 2020), Nantes, France, 26–28 August 2020; pp. 436–443, ISBN 978-2-36215-036-47.
21. Leitão, F.; Madhan, S.K.; Silva, P.D.; Gaspar, P.D. Experimental Testing of the Thermal Response of Different Food Alveoli Solutions for Packaging Boxes. Lecture Notes in Engineering and Computer Science. In Proceedings of The World Congress on Engineering 2021 (WCE 2021), London, UK, 7–9 July 2021; pp. 268–272. Available online: http://www.iaeng.org/publication/WCE2021/WCE2021_pp268-272.pdf (accessed on 26 September 2022).

22. Leitão, F.; Madhan, S.K.; Gaspar, P.D.; Silva, P.D. Experimental Study of the Thermal Response of Fruits Alveoli with Different Materials, Structure and Energy Storage. In Proceedings of the 15th International Conference on Heat Transfer, Fluid Mechanics and Thermodynamics (ATE-HEFAT 2021), Online, 25–28 July 2021; Available online: https://hefat2021.org/proceedings/PROCEEDINGS.rar (accessed on 26 September 2022).

23. Madhan, S.K.; Espírito Santo, C.; Andrade, L.P.; Silva, P.D.; Gaspar, P.D. Active and intelligent packaging with phase change materials to promote the shelflife extension of food products. *KnE Eng.* **2020**, *4*, 232–241. [CrossRef]

24. Leitão, F.; Silva, P.D.; Gaspar, P.D.; Pires, L.C. Experimental study on the thermal response of PCM alveoli for fruit packaging boxes. In Proceedings of the III Environmental Innovations: Advances in Engineering, Technology and Management, EIAETM Conference, Online, 27 September–1 October 2021.

25. Curto, J.; Ilangovan, A.; Gaspar, P.D.; Silva, P.D. CFD modelling of PCM alveoli for fruit packaging boxes. III Environmental Innovations: Advances in Engineering, Technology and Management. In Proceedings of the EIAETM Conference, Lisbon, Portugal, 27 September–1 October 2021.

26. Morais, R.; Cunha, J.B.; Cordeiro, M.; Serodio, C.; Salgado, P.; Couto, C. Solar data acquisition wireless network for agricultural applications. In Proceedings of the 19th Convention of Electrical and Electronics Engineers in Israel, Jerusalem, Israel, 5–6 November 1996; pp. 527–530. [CrossRef]

27. Morais, D.; Gaspar, P.D.; Silva, P.D.; Nunes, J.; Andrade, L.P.; Simões, M.P.; Pires, L.C. Current Status and Future Trends of Monitoring Technologies for Food Products Traceability. In Proceedings of the 25th IIR International Congress of Refrigeration (ICR 2019), Montreal, QC, Canada, 24–30 August 2019; Available online: http://hdl.handle.net/10400.6/7526 (accessed on 26 September 2022). [CrossRef]

28. Zöller, S.; Wachtel, M.; Knapp, F.; Steinmetz, R. Going all the way—Detecting and transmitting events with wireless sensor networks in logistics. In Proceedings of the 38th Annual IEEE Conference on Local Computer Networks—Workshops, Sydney, Australia, 21–24 October 2013; pp. 39–47. [CrossRef]

29. Morais, D.; Aguiar, M.L.; Gaspar, P.D.; Silva, P.D. Development of a monitoring device of fruit products along the cold chain. *Procedia Environ. Sci. Eng. Manag.* **2021**, *8*, 195–204.

30. Curto, J.; Gaspar, P.D. SME and quality focused traceability architecture to increase sustainability and profit. In Proceedings of the 6th IIR International Conference on Sustainability and the Cold Chain (ICCC 2020), Nantes, France, 26–28 August 2020; pp. 428–435, ISBN 978-2-36215-036-47.

31. Curto, J.P.; Gaspar, P.D. Traceability in food supply chains: SME focused traceability framework for chain-wide quality and safety—Part 2. *AIMS Agric. Food* **2021**, *4*, 708–736. [CrossRef]

32. Curto, J.P.; Gaspar, P.D. Traceability in food supply chains: Review and SME focused analysis—Part 1. *AIMS Agric. Food* **2021**, *6*, 679–707. [CrossRef]

33. Mendes, A.; Cruz, J.; Saraiva, T.; Lima, T.M.; Gaspar, P.D. Logistics strategy (FIFO, FEFO or LSFO) decision support system for perishable food products. In Proceedings of the 2020 International Conference on Decision Aid Sciences and Applications (DASA'20), Sakheer, Bahrain, 8–9 November 2020; pp. 173–178. [CrossRef]

34. Magalhães, B.; Gaspar, P.D.; Corceiro, A.; João, L.; Bumba, C. Fuzzy Logic Decision Support System to predict peaches marketable period at highest quality. *Climate* **2022**, *4*, 29. [CrossRef]

35. Maciel, V.; Matos, C.; Lima, T.M.; Gaspar, P.D. Decision support system to assign price rebates of fresh horticultural products on the basis of quality decay, Chapter 23. In *Computational Management: Applications of Computational Intelligence in Business Management*; Patnaik, S., Tajeddini, K., Jain, V., Eds.; Springer International Publishing: Cham, Switzerland, 2021; pp. 487–497. ISBN 978-3-030-72928-8. [CrossRef]

36. Ananias, E.; Gaspar, P.D.; Soares, V.N.G.J.; Caldeira, J.M.L.P. Artificial intelligence decision support system based on artificial neural networks to predict the commercialization time by the evolution of peach quality. *Electronics* **2021**, *4*, 2394. [CrossRef]

37. Simões, M.P.; Martins, C. *Grupos Operacionais de Fruticultura no Período 2018–2022*; Centro Operativo e Tecnológico Hortofrutícola Nacional Centro de Competências: Alcobaça, Portugal, 2021; ISBN 978-972-8785-18-5.

38. Morais, D.; Gaspar, P.D.; Silva, P.D. Desenvolvimento de um dispositivo de monitorização de produtos frutícolas na cadeia de frio. In Proceedings of the International Congress on Engineering—Engineering for Evolution (ICEUBI2019), Covilhã, Portugal, 27–29 November 2019.

39. Nanga, R.; Curto, J.; Gaspar, P.D.; Silva, P.D. Numerical parametric study of the influence of the arrangement of fruit packaging boxes in the fluid flow and heat transfer. In Proceedings of the III Environmental Innovations: Advances in Engineering, Technology and Management, EIAETM Conference, Online, Ukraine, 27 September–1 October 2021.

40. Gaspar, P.D.; Gonçalves, L.C.C.; Pitarma, R.A. Three-Dimensional CFD modelling and analysis of the thermal entrainment in open refrigerated display cabinets. In Proceedings of the ASME 2008 Heat Transfer Summer Conference (HT 2008), Jacksonville, FL, USA, 10–14 August 2008; Volume 2, pp. 63–73.

41. Gaspar, P.D.; Gonçalves, L.C.C.; Pitarma, R.A. CFD parametric studies for global performance improvement of open refrigerated display cabinets. *Model. Simul. Eng.* **2012**, *2012*, 54. [CrossRef]

42. IPMA—Portuguese Institute for Sea and Atmosphere. *Boletim Climático Portugal Continental. Junho 2022*; IPMA: Lisboa, Portugal, 2022.

Article

Evaluating Indoor Carbon Dioxide Concentration and Ventilation Rate of Research Student Offices in Chinese Universities: A Case Study

Guangtao Fan, Haoran Chang, Chenkai Sang, Yibo Chen, Baisong Ning and Changhai Liu *

School of Civil Engineering, Zhengzhou University, Zhengzhou 450001, China; guangtaofan@zzu.edu.cn (G.F.); chrzzu@163.com (H.C.); sck51577@163.com (C.S.); yb_chen77@zzu.edu.cn (Y.C.); bsning@foxmail.com (B.N.)
* Correspondence: liuchanghai@zzu.edu.cn

Abstract: This work provides a case study on the indoor environment and ventilation rate of naturally ventilated research student rooms in Chinese universities. In the measured room, air temperature, relative humidity and carbon dioxide (CO_2) concentration were monitored during the heating period for 4 weeks. The number of indoor occupants, occupied time of the room and window/door-opening cases were simultaneously recorded. Results showed the research student room was occupied for an average of 12.0 h each day. Due to a large indoor and outdoor temperature difference during the heating season, and occupants' adaption to indoor environment, indoor occupants seldom open windows/doors for ventilation. Air exchange of the room only by air infiltration cannot meet the ventilation requirement. As a result, an average of 77.6% of measured CO_2 data each day exceeded 1000 ppm during occupied time. In fact, according to CO_2 data, it was observed that window/door opening could effectively decrease indoor CO_2 concentration. Therefore, intermittent window/door opening or CO_2-based demand-controlled ventilation facilities were suggested for improving indoor air quality of such rooms. Additionally, special attention should be paid to other possible outdoor pollution.

Keywords: indoor air quality; high occupancy rate; natural ventilation; university building

Citation: Fan, G.; Chang, H.; Sang, C.; Chen, Y.; Ning, B.; Liu, C. Evaluating Indoor Carbon Dioxide Concentration and Ventilation Rate of Research Student Offices in Chinese Universities: A Case Study. *Processes* **2022**, *10*, 1434. https://doi.org/10.3390/pr10081434

Academic Editor: Cherng-Yuan Lin

Received: 27 June 2022
Accepted: 19 July 2022
Published: 22 July 2022

Publisher's Note: MDPI stays neutral with regard to jurisdictional claims in published maps and institutional affiliations.

1. Introduction

Since entering the new century, China's higher education has achieved a leapfrog in development. According to China's statistics yearbook 2021 released by the National Bureau of Statistics of China [1], in 2020 there were about 36 million students in colleges or universities, and the student population was the largest in the world. For these students, almost all of their time was spent in university buildings, including classrooms, dormitories and other possible rooms (e.g., laboratory, library and lecture hall). Indoor air quality (IAQ) inside these rooms is one of the key issues for a pleasant stay for students. However, these rooms usually present a high occupancy rate; as a result, ventilation of these rooms is generally insufficient. A few published studies have confirmed this point. Regarding classrooms, Sarbu et al. measured indoor carbon dioxide (CO_2) levels in two air conditioned classrooms at a university in the west of Romania during both cooling and heating seasons [2]. Results indicated that indoor CO_2 concentration could reach a value of 2400 ppm in the case of inadequate ventilation. In addition, the CO_2 concentration could decrease significantly to 1500 ppm by manually opening the windows. Chang et al. evaluated IAQ in two computer classrooms and one general classroom in a southern Taiwan college, and indicated that low ventilation rates were likely responsible for the very high indoor CO_2 concentration [3]. By measuring and comparing IAQ in 15 classrooms in Brazilian universities, Jurado et al. found that the CO_2 level in the air-conditioned rooms was significantly higher than that in naturally ventilated rooms [4]. Asif et al. investigated the IAQ and thermal comfort in classrooms of four buildings of an educational institute and found that indoor CO_2 concentration in naturally ventilated classrooms exceeded safe levels at higher frequency [5]. Argunhan and Avci measured indoor temperature, relative humidity (RH) and IAQ of the classrooms in universities in Turkey and found that average indoor CO_2 level is higher than the ASHRAE standards, due to closed doors and windows in winters [6]. With regard to college student dormitories, Li et al. tested indoor CO_2 concentration in naturally ventilated student dormitories in a college in Beijing, China and showed that indoor CO_2 concentration most of the time exceeded the referenced guideline of 1000 ppm provided by IAQ standard [7], due to the low ventilation rate of the dormitories by air infiltration. However, through single-sided natural ventilation, sufficient outdoor air could be provided to dilute the indoor CO_2. Zhang et al. monitored indoor CO_2 concentration in three selected rooms of university dormitory buildings in winter in Shanghai, China and found that the occupants tended to close the windows to maintain an indoor thermal environment in winter, thereby resulting in elevated CO_2 concentrations during sleeping hours [8]. The above literature review indicated that ventilation in these university rooms was insufficient, even in developed countries. Furthermore, IAQ of other rooms such as libraries [9], lecture rooms [10,11], and research laboratories [12,13] in university buildings was also studied by some researchers.

As a matter of fact, in university buildings, there is also another type of room, namely the research student room. A research student room is often the place where postgraduate students carry out scientific research and administrative work in universities. Figure 1 shows indoor views of several typical research student offices in some universities of China. In recent years, due to a sharp increase in the number of postgraduate students in China, these rooms present a much higher occupancy rate than other offices in general office buildings. Additionally, these rooms usually have no fresh air unit in Chinese universities. For these rooms, the most effective ways to provide necessary outdoor fresh air are usually natural ventilation, which refers to air change through the intentional openings of building envelopes (windows/doors); and air infiltration, which refers to the air change through unintentional leakage areas of building envelopes when the doors and windows are closed [14]. In fact, in China most research student rooms in universities are not originally designed for high occupancy rates. Instead, with a rapidly increasing population of postgraduates in universities, some unoccupied rooms in universities have to be used as research student rooms. This inconsistency between design and utilization of the room is bound to bring about poor IAQ.

Figure 1. Indoor views of typical research student offices in some universities of China.

To date, however, indoor environment, ventilation rate and occupancy characteristics of such rooms in Chinese universities are rarely investigated. How about the indoor environment in such rooms? Can natural ventilation meet the requirements of postgraduate students? How to improve the ventilation rate of such rooms?

With these questions, a case study on naturally ventilated research student rooms in Chinese university is provided in the present paper. We selected a representative research student room with a high occupancy density in a university of China as the measured room and monitored indoor environmental parameters (including temperature, relative humidity and CO_2 concentration) for four weeks during the heating period. At the same time, occupied and unoccupied profiles of the room and open/closed cases of the windows/doors were recorded. Based on CO_2 measurement data and occupancy profile of the room, indoor environment and ventilation rate of the room were assessed. The results would be instructive to improve Chinese research students' studying environment and to improve the ventilation rate of such rooms.

2. Materials and Methods

2.1. Description of the Measured Room

In this study, a representative research student room in a university of Beijing (latitude: 39°54′ N, longitude: 116°23′ E) was selected as the measured room. The room is on the second floor of a five–story university building, as shown in Figure 2. The five-story university building was built in the 1980s and is close to the urban main road. The layout of the measured room is given in Figure 3. The room has a height of 2.8 m and a floor area of 23 m^2. In addition, the room has a door, which is set into the corridor. The fully opened door has an area of 1.8 m^2. There are two in-swinging casement windows, which are located at a height of 0.9 m above the floor. Each window comprises two window sashes. Each fully opened window sash has an area of 0.55 m^2, as shown in Figure 4. There is no air supply system in the room. Beijing is located in the cold region; in winter, there are three months during which the monthly mean temperature is below 0 °C. Due to low outdoor air temperature, building heating is imperative for occupants' thermal comfort in winter. Accordingly, central heating system is used in this room in winter. In general, seven postgraduate students work in this room. During monitoring period, however, one of the postgraduates was not staying in the room. Thus, the room was generally occupied with six postgraduate students during measuring period. The details of the occupancy characteristics of the room are described in the Results, Section 3.3.

Figure 2. Aerial view (**a**) and outdoor view (**b**) of the measured room.

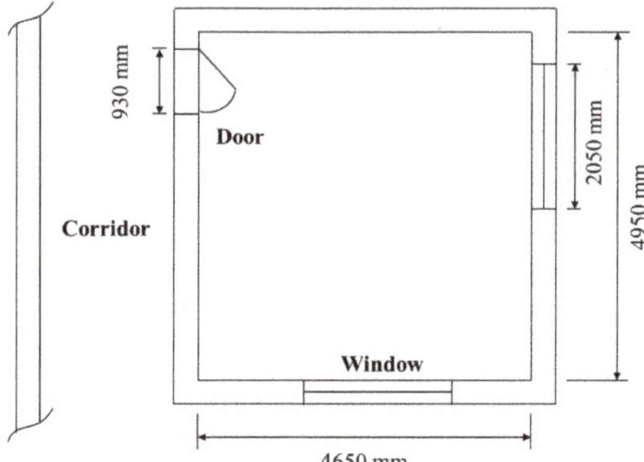

Figure 3. Layout of the measured room.

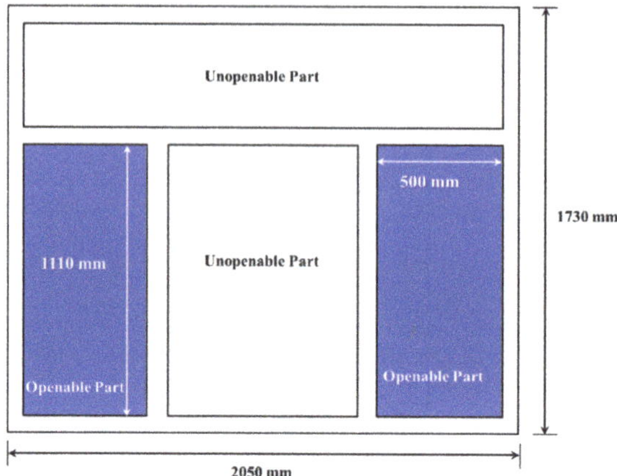

Figure 4. Layout of the window.

2.2. Measuring Process

The measurement was conducted in the research student room for 4 weeks during the typical heating period. Measured parameters included air temperature, RH and CO_2 concentration. Considering probably inadequate full mixing of indoor air, two sets of equipment were placed in the room. The data loggers (Testo 175H1, shown in Figure 5) with temperature and humidity sensors were used to monitor indoor air temperature and RH. Outdoor meteorological data were recorded by a small-sized automatic weather station. Indoor and outdoor CO_2 concentration were monitored using data loggers with CO_2 sensor (MCH–383SD, shown in Figure 5). The CO_2 data loggers were installed on the convenient position to avoid the influence of the occupants' exhalation. The locations of all used instruments were at the height of 1.3 m (the breathing zone of a seated postgraduate) above the floor level, away from the windows and door. These instruments recorded data at an interval of 5 min. Table 1 summarizes the test range and accuracy of the instruments used in the current study. Furthermore, the postgraduates were asked to complete a Room Occupation Log Sheet so that the number of occupants, occupied and unoccupied periods of the room and open/closed cases of the windows/door were clarified.

Figure 5. Recording instruments for temperature and RH (Testo 175H1) and CO_2 (MCH-383SD).

Table 1. Summary of instrument range and accuracy.

Parameters	Range	Accuracy
CO_2 concentration	0–4000 ppm	±40 ppm (below 1000 ppm) ±5% rdg (1001~3000 ppm) ±250 ppm (above 3000 ppm)
Indoor temperature	−20–55 °C	±0.4 °C
Indoor RH	0–100%	±2%

2.3. Calculation of Ventilation Rate

There are many studies using exhaled CO_2 as a tracer gas to evaluate the ventilation rate [15,16]. In this study, a 24-hour day is split into occupied period when postgraduate is present during daylight and unoccupied period when postgraduate is absent during nighttime. Postgraduates can generate CO_2 into the room when they are present. During unoccupied period, there is no indoor source of CO_2, so air exchange rate (AER) of the room by air infiltration can be evaluated using CO_2 decay method. The tracer-gas method is based on the principle of mass balance in a designated room and the assumption of uniform distribution of indoor air. The tracer-gas mass-balance equation can be expressed as [7]:

$$V\frac{d(C_{in}(t))}{dt} = Q(C_{out}(t) - C_{in}(t)) \tag{1}$$

where V is room volume (m^3), t is the time (h), $C_{in}(t)$ is indoor CO_2 concentration at time t (m^3/m^3), Q is volumetric airflow rate into (and out of) the space (m^3/h), $C_{out}(t)$ is outdoor CO_2 concentration (m^3/m^3).

If Q is assumed, and $C_{out}(t)$ is constant, Equation (1) can be integrated to obtain the following equation for $C_{in}(t)$:

$$C_{in}(t) = C_{out}(t) + [C_{in}(0) - C_{out}(t)]e^{-Nt} \tag{2}$$

where $C_{in}(0)$ is the indoor CO_2 concentration at time $t = 0$, $N=Q/V$ is AER.

If air exchange rate (i.e., N) is given a value (adjustable variable parameter) and outdoor CO_2 concentration (i.e., $C_{out}(t)$) is given a constant parameter, a series of theoretical concentrations $C_{in}(t)$, can be calculated at the end of each time interval by Equation (2). Then, the difference between the measured and theoretical values is calculated by the least-squares method:

$$Error = \sum_{1}^{n}[C_{in,t}(n) - C_{in,m}(n)]^2 \tag{3}$$

where n is the number of measured concentrations, $C_{in,t}(n)$ and $C_{in,m}(n)$ is the computed concentration and measured concentration at the end of the nth time interval, respectively. Finally, air exchange rate is fitted out by obtaining theoretical concentration close to the measured concentration.

2.4. Model for Human CO_2 Generation Rate

To analyze the association among CO_2 concentration, occupancy rate and air change rate, CO_2 generation rate from human respiration needs to be calculated. According to ASHRAE Handbook Fundamentals [17], an empirical formula to calculate CO_2 generation rate can be expressed as follows:

$$V_{CO_2} = RQ \cdot \frac{0.55887W^{0.425}H^{0.725}M}{0.23RQ + 0.77} \tag{4}$$

where RQ is the respiratory quotient, volumetric ratio of CO_2 produced to oxygen (O_2) consumed, dimensionless; a good estimate for the average adult is $RQ = 0.83$ for light or sedentary activities ($M < 1.5$ met), 1met = 58.1 W/m^2. M is the metabolic rate, W/m^2, depending on the level of physical activity, M=1.2 for occupant working in research student room; W is the body weight (kg); H is the body height (m).

3. Results

3.1. Long-Term Monitoring CO_2 Concentration

We firstly compared the CO_2 data from indoor two monitoring points and found that the maximum nonuniformity coefficient of indoor CO_2 distribution is 5.68%, less than 10% required by ASTM E741. Thus, this study considered indoor air to be uniformly distributed and reported an average CO_2 value of two measuring points as indoor CO_2 concentration. Figure 6 shows indoor and outdoor CO_2 concentrations during the monitoring period of four weeks. During the period, indoor CO_2 concentration in the room ranged from 442 to 3491 ppm, with an average of 1161 ppm, while outdoor CO_2 value varied very little, with an average of 458 ppm. Obviously, indoor CO_2 concentration variations appeared with peaks and valleys. The reason is that the room was generally occupied in the daytime and unoccupied in the evening.

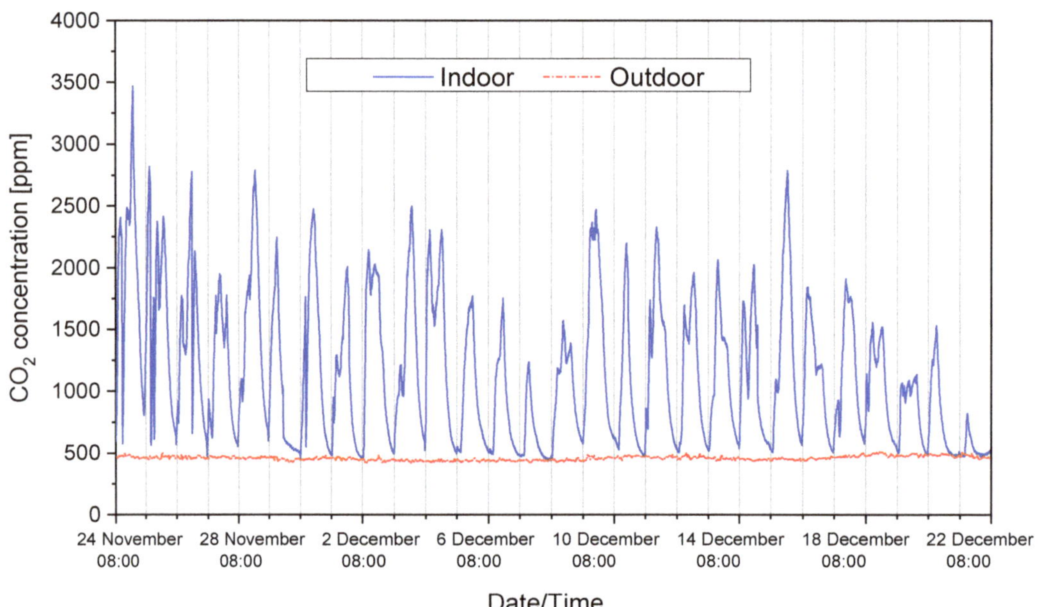

Figure 6. Indoor and outdoor CO_2 concentration during monitoring period.

According to ASHRAE Standard 62.1 [18] and the Chinese indoor air quality standard GB/T 18883 [19], indoor CO_2 concentration should be kept below the commonly referenced guideline of 1000 ppm (8–h average). However, based on the measured data, there was a large chunk of time when indoor CO_2 concentration exceeded the guideline each day. Figure 7 shows the cumulative frequency of indoor CO_2 concentration during the measuring period. Over 50% of the measured data were more than 1000 ppm. A total of 27.8% of the measured data were more than 1500 ppm and 1.7% of the measured data exceeded 2500 ppm.

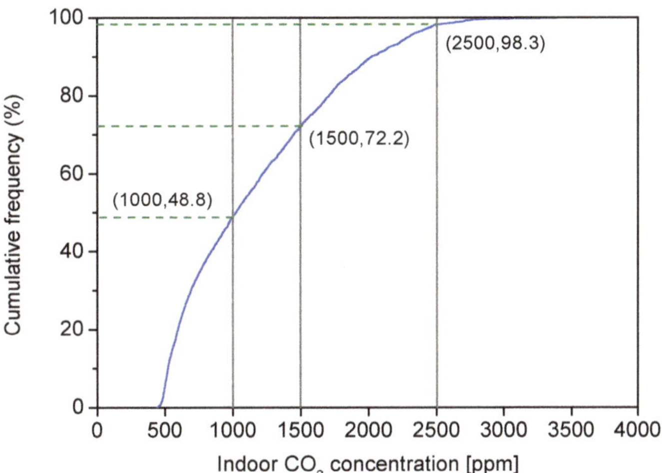

Figure 7. Cumulative frequency of indoor CO_2 concentration during monitoring period.

Figure 8 summarizes the time of indoor CO_2 level over 1000 ppm each day. As a whole, indoor CO_2 concentration exceed 1000 ppm during nearly all monitoring days. During the occupied time each day, the period when indoor CO_2 concentration exceeded 1000 ppm ranged from 2.0 to 18.8 h, with an average value of 10.2 h. The percentage of occupied times with a CO_2 concentration higher than 1000 ppm ranged from 58.3% to 96.2%, with an average value of 77.6% each day. These results indicate that indoor air quality in the research student room was poor during heating season.

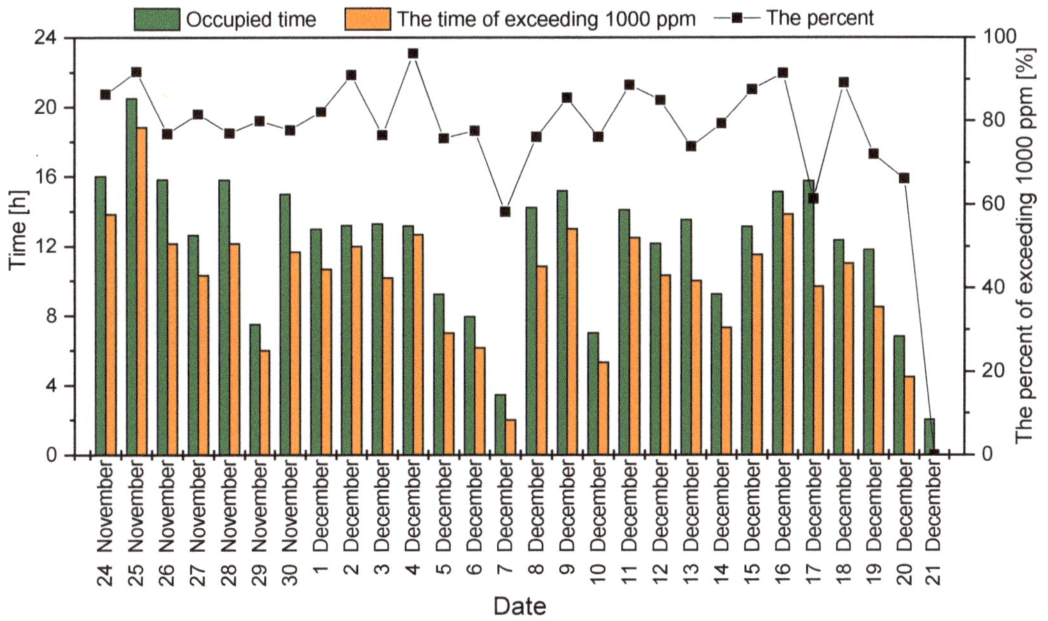

Figure 8. Summary of duration for indoor CO_2 level over 1000 ppm each day during monitoring period.

3.2. Air Exchange Rate of the Room by Infiltration

Air exchange rate (AER) of the room by air infiltration can be evaluated using the CO_2 decay method. As a general requirement, during the unoccupied period, the CO_2 steady decay period when the initial CO_2 level exceeds 1000 ppm is selected to calculate the air exchange rate of the room. Outdoor CO_2 concentration is given as a constant 458 ppm (average value during monitoring period). Table 2 summarizes the AERs of the room calculated using the CO_2 decay method during the unoccupied period each day (Adj. $R^2 > 0.99$). There were two days when indoor CO_2 concentration could not meet the above requirement. During the decay periods for calculation, the difference between indoor and outdoor temperature ranged from 18.7 to 26.3 °C, with an average of 22.3 ± 3.6 °C. Differences between indoor and outdoor temperature can cause different air infiltration, thereby generating different AERs. From Table 2, calculated AERs have a range of 0.247–0.409 h^{-1}, and an average value of 0.313 h^{-1}; that is, 0.9 L/ (s·person) when all postgraduate students are in the room, which is significantly lower than the minimum ventilation rates of 2.5 L/ (s·person) in breathing zone for office space.

Table 2. Summary of AER calculated by using CO_2 decay method during unoccupied period.

Date	Decay Period	Initial CO_2 Level [ppm]	Final CO_2 Level [ppm]	AER [h^{-1}]	Adj. R^2
24 November	-	-	-	-	
25 November	03:10–06:20	1500	828	0.332	0.9987
26 November	00:00–07:00	1777	643	0.303	0.9975
27 November	00:00–06:20	1506	615	0.291	0.9986
27 November	23:00–06:00	1474	586	0.317	0.9968
28 November	23:10–06:30	2135	702	0.271	0.9980
29 November	15:00–18:00	1771	983	0.316	0.9928
30 November	23:00–08:00	1373	484	0.373	0.9991
1 December	22:10–07:30	1171	475	0.409	0.9989
2 December	23:30–06:30	1253	521	0.343	0.9975
3 December	23:10–06:00	1945	642	0.317	0.9996
4 December	21:30–06:00	1924	538	0.351	0.9996
5 December	20:30–05:00	1544	584	0.285	0.9961
6 December	20:00–03:00	1360	559	0.320	0.9974
7 December	15:30–22:30	1132	550	0.293	0.9929
8 December	23:50–07:00	1150	597	0.247	0.9905
9 December	23:30–07:00	1494	635	0.280	0.9911
10 December	19:30–02:00	1361	586	0.306	0.9968
11 December	23:10–05:30	1427	607	0.296	0.9991
12 December	23:00–07:00	1459	550	0.301	0.9995
13 December	23:00–06:00	1282	598	0.255	0.9954
14 December	-	-	-	-	
15 December	23:00–06:00	1898	607	0.325	0.9995
16 December	23:10–06:10	1148	527	0.341	0.9969
17 December	23:20–06:20	1449	605	0.301	0.9946
18 December	21:40–05:00	1248	596	0.277	0.9914
19 December	23:20–06:00	1067	506	0.376	0.9990
20 December	15:00–23:50	1278	507	0.310	0.9992

3.3. Indoor Occupancy Characteristics

Indoor occupants are an important component of the built environment, since both occupancy rate and occupant activities have an effect on the requirement of ventilation. Figure 9 shows indoor occupancy cases of the research student room in a representative week. From Figure 9, in a day, the first occupant normally entered the room at 7:00–8:00, and the last occupant left the room at 23:00–24:00, while at mealtimes the room was sometimes unoccupied. The peak occupancy rate of the room each day mainly occurred at 10:00 in the morning and 16:00 in the afternoon. In addition, sometimes there were visitors entering the room, so the peak occupancy rate of the room was more than six. When the room was occupied with six persons, the per-capita area was 3.8 m^2/p. There is no specific definition of high-occupancy density in offices in any building design standard of China. According to the Building Area Index for Regular Institutions of Higher Education [20] released by Ministry of Education of the People's Republic of China, the subsidy building area per person for graduate student learning was 6 m^2/p for Master's degree candidates and 8 m^2/p for PhD degree candidates. Obviously, the measured research student room was crowded. In terms of occupied time, the occupied time of the research student room each day overall varied from 3.4 to 20.5 h, with an average value of 12.0 h. Compared with other rooms in university buildings, the occupied time of the research student room was longer. In general, both the occupied time and occupancy rate of the research student room on weekdays were more than that on weekends, because most postgraduate students tended to have a rest on weekends. In this measurement, the average occupied time on weekdays and weekends was 13.5 h and 8.2 h, respectively. Postgraduate students in the room were commonly at the activity level of typing or reading.

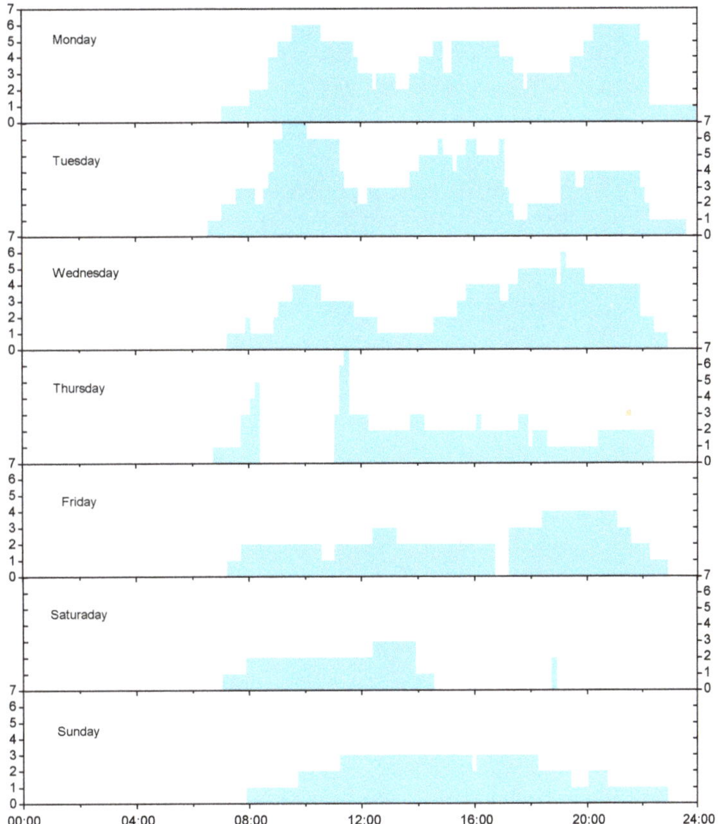

Figure 9. Indoor occupancy case of the research student room in a representative week.

3.4. Window/Door-Opening Behaviors of Postgraduate Students

Table 3 summarizes the window/door-opening behaviors of postgraduate students. During the monitoring days, there were only five days when postgraduate students actively opened windows/doors during the occupied period. We also find that, when window/door-opening behavior occurred, indoor CO_2 concentrations were always at a high level. In addition, when both the window and door were open, the duration was relatively short (12 min and 8 min). The reason might be that when both the window and door are open, cross-ventilation is produced through the room. Cross-ventilation for a long time might reduce indoor thermal comfort due to a great indoor and outdoor temperature difference. When only the door was open, because the temperature in the corridor was not as low as outdoors, the duration was relatively long (16 min, 22 min and 31 min, respectively). Interestingly, it was observed that window/door-opening behaviors generally occurred when postgraduate students just went from outside into the room. A possible explanation for this case was that when occupants just went from outside into the room, they had completely adapted to outdoor fresh air, so that they could not bear indoor high CO_2 concentration and bioeffluents. Consequently, they opened the window/door of the room for ventilation and let outdoor fresh air in. This indicated that occupants tended to adapt to high CO_2 concentrations and bioeffluents, and had less control over window/door opening for ventilation requirement, when staying in the environment for a long time. In addition, during four weeks of continuous monitoring, as a general rule, the first occupant going into the room in the morning customarily opened the window/door to have proper ventilation for 5–10 min, although indoor CO_2 concentration was not high. This might be greatly ascribed by the inadaption induced by going into the room from outside. The standard of ASTM D6245 [21] also pointed out that for adapted persons (i.e., indoor occupants), the ventilation rate per person to provide the same acceptance was approximately one-third of the value for unadapted persons (i.e., visitors), and the corresponding CO_2 concentrations that outdoors-adapted occupants could accept were three times higher than those for unadapted persons. Although adapted occupants were able to accept high CO_2 concentrations caused by such a reduction in the ventilation rate, the effect of exposure to high CO_2 concentration on occupants' health and productivity was not negligible. In addition, a reduction in the ventilation rate might give rise to high concentration of other contaminants. Thus, the method where the occupants control window/door opening for ventilation in such rooms by perceived air quality was inadvisable.

Table 3. Summary of opening window/door operation during occupied time.

Days	Windows	Door	Window/Door-Opening Duration (min)	Indoor CO_2 Level When Occupants Open Windows/Door (ppm)
1	x	o	16	2254
2	o	o	12	2694
3	x	o	31	1759
4	x	o	22	2776
5	o	o	8	1746

Notes: x is closed; o is opened.

4. Discussion

4.1. Occupancy Rate Affects Indoor CO_2 Level

Figure 10 shows indoor CO_2 concentration and occupancy rate at typical weekdays and weekends, when the door/windows are closed. It was observed that the variation in indoor CO_2 concentration was sensitive to the indoor occupancy rate. Indoor CO_2 concentration usually decreased to outdoor levels during the unoccupied period (i.e., nighttime). In general, indoor CO_2 concentration increased with occupancy time. The variation in CO_2 concentration began to grow relatively fast, and then gradually became slower. However, when CO_2 concentration reached a certain level, indoor CO_2 concentration also decreased with time, although the room was still occupied. The reason was that indoor CO_2 generation was not enough to offset the CO_2 decrement induced by indoor and outdoor air exchange. Moreover, when indoor CO_2 generation was equal to the CO_2 decrement induced by indoor and outdoor air exchange, indoor CO_2 generation tended to stabilize. On the whole, due to high occupancy rates and long occupancy time on weekdays, the duration times of indoor CO_2 concentration above the reference level (dashed line in Figure 10) were longer than that on weekends. Therefore, in such kinds of room, it was recommended to reduce the number of students to ensure a low carbon dioxide concentration in the room.

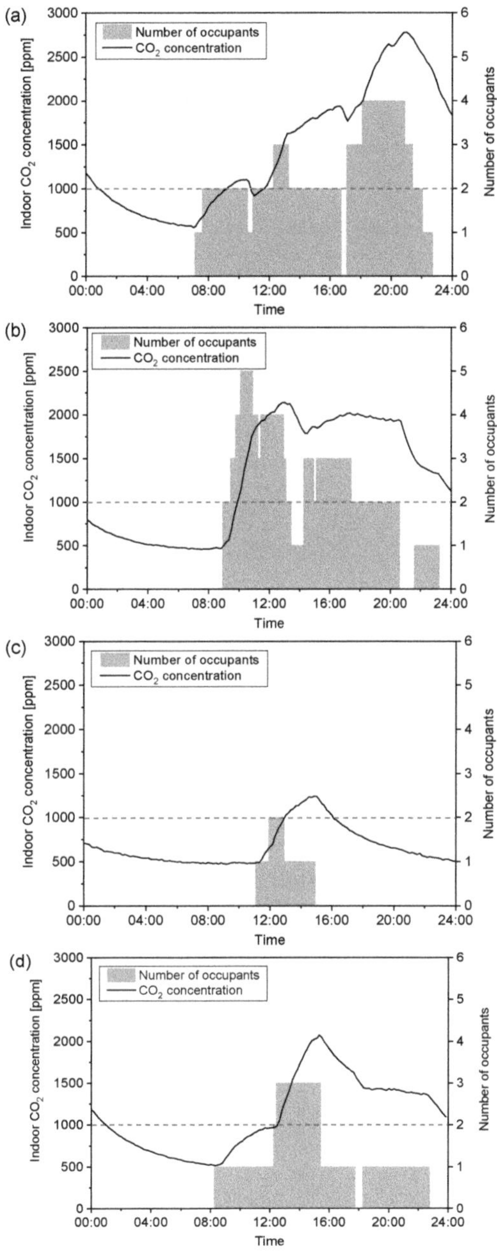

Figure 10. Indoor CO_2 concentration and occupancy rate (door/window closed): (**a**,**b**) at weekday; (**c**,**d**) at weekend.

4.2. Opening Door/Windows Decreases Indoor CO_2 Concentration

As aforementioned in the Section Result, the average air exchange rate of the measured room by air infiltration was $0.313\ \mathrm{h}^{-1}$. In fact, this air infiltration rate could not meet indoor ventilation requirements, due to a high occupancy rate. This point was also verified by the monitoring data of indoor CO_2 concentration. From Figure 11, it is seen that opening windows/doors could give rise to a considerable AER, a few dozen times higher than the air infiltration rate. Thereby, the shorter duration of opening windows/doors could significantly decrease indoor CO_2 concentration close to outdoor level. Therefore, for such research student rooms with high occupancy rates, intermittent window/door opening might be an effective way to avoid continuous indoor high CO_2 concentration. However, due to the great indoor and outdoor temperature difference during the heating season, postgraduate students tended to close windows/doors for a long time. When staying in the environment for a long time, postgraduate students had adapted to high CO_2 concentrations and had less control over opening windows/doors for indoor ventilation.

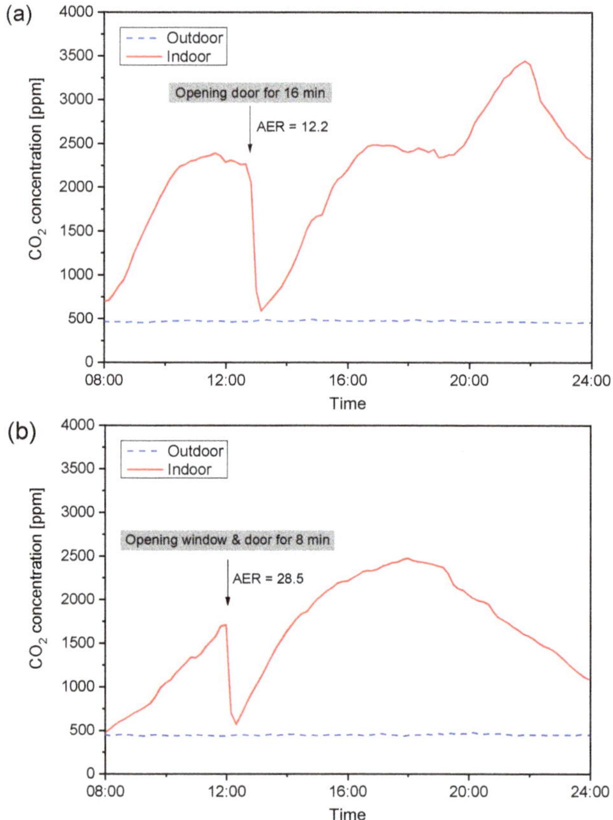

Figure 11. Indoor CO_2 concentration and door/window opening: (**a**) 24 November; (**b**) 30 November.

4.3. Suggestion for Ventilation in Such Rooms

Based on the aforementioned discussion, the high occupancy rate and low air exchange rate of the room are the principal determinants of high indoor CO_2 concentration of the naturally ventilated room. If any of these two parameters are adjusted, indoor air quality can be improved. However, with a rapidly increasing population of graduate students in universities, it seems to be difficult to decrease the occupancy rate of such rooms; instead, the occupancy rate may even increase. For this situation, adjusting ventilation rate of the room may be an effective way to maintain good IAQ. A basic way to enlarge the ventilation rate is for indoor occupants (graduate students) to open windows/doors. In particular, indoor occupants may need to be reminded to implement the action of opening windows/doors. For example, we can install a CO_2 monitor with a warning device in such rooms. When indoor CO_2 level exceeds a preset level, the device can remind indoor occupants to open window or doors.

In fact, indoor CO_2 concentration can be roughly estimated based on indoor occupancy characteristics and ventilation rate. For a single-zone room, when the only indoor source of CO_2 comes from occupants in the room, the indoor CO_2 concentration mass balance equation can be expressed as:

$$V\frac{d(C_{in}(t))}{dt} = Q(C_{out}(t) - C_{in}(t)) + G(t) \tag{5}$$

where V is room volume (m^3); t is the time (h); $C_{in}(t)$ is indoor CO_2 concentration at time t (m^3/m^3); Q is volumetric airflow rate into (and out of) the space (m^3/h); $C_{out}(t)$ is outdoor CO_2 concentration (m^3/m^3), and the value varied very little; $G(t)$ is CO_2 generation rate in the room at time t (m^3/h).

If Q is assumed, and $C_{out}(t)$ and $G(t)$ are constant, Equation (5) can be integrated to obtain the following equation for $C_{in}(t)$:

$$C_{in}(t) = C_{out}(t) + \frac{G(t)}{Q} + \left[C_{in}(0) - C_{out}(t) - \frac{G(t)}{Q}\right]e^{-Nt} \tag{6}$$

where $C_{in}(0)$ is the indoor CO_2 concentration at time $t = 0$; N is AER, Q/V.

For Equation (6), if outdoor CO_2 concentration—$C_{out}(t)$, indoor initial CO_2 concentration—$C_{in}(0)$, and room volume—V are known, there will be close associations among indoor CO_2 concentration—$C_{in}(t)$, AER—N, and indoor CO_2 generation rate—$G(t)$. Once two parameters of them are known, the remaining one will be obtained. Based on the aforesaid model for human CO_2 generation rate in the "Materials and Methods" section, The CO_2 generation rate corresponding to an average-sized adult (BSA = 1.8 m^2) engaged in office work (1.2 m) is about 18720 mL/h. The number of indoor occupants in the room hypothetically is m; Equation (6) can be converted into the following equation:

$$C_{in}(t) = C_{out} + \frac{m \cdot 18720}{N \cdot V} + \left[C_{in}(0) - C_{out} - \frac{m \cdot 18720}{N \cdot V}\right]e^{-N \cdot t} \tag{7}$$

The duration of CO_2 concentration from indoor initial level to the referenced guideline (1000 ppm) is obtained as:

$$t = \frac{1}{N}\{\ln[m \cdot 18720 - (C_{in}(0) - C_{out}) \cdot N \cdot V] - \ln[m \cdot 18720 - (1000 - C_{out}) \cdot N \cdot V]\} \tag{8}$$

If indoor initial CO_2 concentration $C_{in}(0)$ is equal to outdoor CO_2 concentration, Equation (8) can be converted into the following equation:

$$t = \frac{1}{N}\{\ln(m \cdot 18720) - \ln[m \cdot 18720 - (1000 - C_{out}) \cdot N \cdot V]\} \tag{9}$$

Generally speaking, outdoor CO_2 concentration C_{out} varies very little in an area. Therefore, from Equation (9), the main factors of the duration of indoor CO_2 concentration increasing from the outdoor level to 1000 ppm are the number of indoor occupants, room volume and AER. If the three factors are known, the duration will be clear. Taking a room with four persons, a volume of 100 m^3 and an AER of 0.5 h^{-1} as an example, the duration of indoor CO_2 concentration increasing from the initial level to 1000 ppm in the room can be calculated as 0.9 h. This duration can also be considered as the reference of the interval of opening windows/doors for ventilation. Of course, for some economically viable universities, CO_2-based demand-controlled ventilation facilities were also proposed to be installed in such rooms. An appropriate way to control the ventilation facilities is to directly measure the indoor CO_2 level and activate the facilities if the level exceeds a preset level. Considering the serious outdoor air pollution due to particulate matter in winter in some regions of China, such as the Beijing–Tianjin–Hebei region, the simple idea to provide more outdoor air to dilute indoor CO_2 concentration may even worsen indoor air quality. In this case, ventilation facilities with air-filter apparatus should be required. This is also the requirement for mechanical ventilation equipment stipulated by the current building ventilation standards in China. At the same time, it is recommended that monitoring devices for environmental pollution parameters (e.g., $PM_{2.5}$, PM_{10}, O_3, NO_2, SO_2, CO) should be installed indoors to achieve real-time evaluation of indoor air quality [22,23]. When the concentration of indoor pollutants exceeds the threshold, indoor air-purification devices are also recommended to ensure good indoor air quality and the health of indoor occupants.

4.4. Limitations

There are several limitations for this study. A first obvious limitation was that this study only took one room as a case study, which might affect the representativeness and universality of the results. Another limitation is the accuracy of the measuring device. Compared with some instruments with high precision, the error of the instrument used in this study may affect the accuracy of measuring data, but it does not seem to make much influence on our conclusion.

5. Conclusions

In Chinese universities, postgraduate students usually carry out scientific research in a research student room in university buildings. This paper selected a naturally ventilated research student room in Chinese universities to conduct continuous field measurements of indoor CO_2 level, air temperature and RH during the heating period, and then evaluated indoor environment and ventilation rate of the measured room based on measured CO_2 concentration. Results showed that the research student room was crowed, and occupied time of the research student room each day varied from 3.4 to 17.5 h, with an average of 12.0 h, which was longer than general offices. During the occupied time, 58.3% to 96.2% (average 77.6%) of measured CO_2 data each day exceeded 1000 ppm. Indoor high CO_2 concentration was attributed to the high occupancy rate and low air infiltration rate of the room. Therefore, in such kind of rooms, it was recommended to reduce the number of students. Furthermore, it was found that opening windows/doors for several minutes could significantly decrease the indoor CO_2 level close to the outdoor level. However, due to the great indoor and outdoor temperature difference during the heating season, and occupants' adaption to indoor environment, indoor occupants (postgraduate students) generally have less control over window/door opening for ventilation requirements in such rooms. Therefore, intermittent window/door opening reminded by warning devices or estimated by indoor occupancy and air infiltration volume were suggested in such rooms. For economically viable universities, CO_2-based demand-controlled ventilation facilities were also advocated for improving indoor air quality of such rooms. These results would be helpful to improve Chinese research students' studying environment.

Author Contributions: Conceptualization, G.F.; methodology, C.L.; formal analysis, G.F.; investigation, G.F.; data curation, G.F.; writing—original draft preparation, G.F. and H.C.; writing—review and editing, H.C. and C.S.; supervision, C.L.; funding acquisition, G.F., Y.C. and B.N. All authors have read and agreed to the published version of the manuscript.

Funding: The authors would like to express their gratitude to the China Postdoctoral Science Foundation (Grant No. 2020M672279), National Natural Science Foundation of China (No. 51808505 and 52108100) for their financial supports.

Institutional Review Board Statement: Not applicable.

Informed Consent Statement: Not applicable.

Data Availability Statement: Not applicable.

Acknowledgments: The authors would like to sincerely thank the postgraduate students involved in this study for their cooperation.

Conflicts of Interest: The authors declare no conflict of interest.

Abbreviations

IAQ	Indoor Air Quality
CO_2	Carbon Dioxide
RH	Relative Humidity
RQ	Respiratory Quotient
ASHRAE	American Society of Heating, Refrigerating and Air-Conditioning Engineers

References

1. China's Statistics Yearbook 2021. Available online: http://www.stats.gov.cn/tjsj/ndsj/2021/indexch.htm (accessed on 1 May 2022).
2. Sarbu, I.; Pacurar, C. Experimental and numerical research to assess indoor environment quality and schoolwork performance in university classrooms. *Build. Environ.* **2015**, *93*, 141–154. [CrossRef]
3. Chang, F.H.; Li, Y.Y.; Tsai, C.Y.; Yang, C.R. Specific indoor environmental quality parameters in college computer classrooms. *Int. J. Environ. Res.* **2009**, *3*, 517–524. [CrossRef]
4. Jurado, S.R.; Bankoff, A.D.; Sanchez, A. Indoor air quality in Brazilian universities. *Int. J. Environ. Res. Public Health* **2014**, *11*, 7081–7093. [CrossRef] [PubMed]
5. Argunhan, Z.; Avci, A.S. Statistical evaluation of indoor air quality parameters in classrooms of a university. *Adv. Meteorol.* **2018**, *2018*, 4391579. [CrossRef]
6. Asif, A.; Zeeshan, M.; Jahanzaib, M. Indoor temperature, relative humidity and CO_2 levels assessment in academic buildings with different heating, ventilation and air-conditioning systems. *Build. Environ.* **2018**, *133*, 83–90. [CrossRef]
7. Li, H.R.; Li, X.F.; Qi, M.W. Field testing of natural ventilation in college student dormitories (Beijing, China). *Build. Environ.* **2014**, *78*, 36–43. [CrossRef]
8. Zhang, N.B.; Kang, Y.M.; Zhong, K.; Liu, J.P. Indoor environmental quality of high occupancy dormitory buildings in winter in Shanghai, China. *Indoor Built Environ.* **2016**, *25*, 712–722. [CrossRef]
9. Righi, E.; Aggazzotti, G.; Fantuzzi, G.; Ciccarese, V.; Predieri, G. Air quality and well-being perception in subjects attending university libraries in Modena (Italy). *Sci. Total Environ.* **2002**, *286*, 41–50. [CrossRef]
10. Sulaiman, S.A.; Isa, N.; Raskan, N.I.; Harun, N.F.C. Study of indoor air quality in academic buildings of a university. *Appl. Mech. Mater.* **2013**, *315*, 389–393. [CrossRef]
11. Harun, H.A.; Buyamin, N.; Othman, M.A.; Sulaiman, S.A. A case study on indoor comfort of lecture rooms in university buildings. *Appl. Mech. Mater.* **2013**, *393*, 821–826. [CrossRef]
12. Giulio, M.D.; Grande, R.; Campli, E.D.; Bartolomeo, S.D.; Cellini, L. Indoor air quality in university environments. *Environ. Monit. Assess.* **2010**, *170*, 509–517. [CrossRef] [PubMed]
13. Ugranli, T.; Toprak, M.; Gursoy, G.; Cimrin, A.H.; Sofuoglu, S.C. Indoor environmental quality in chemistry and chemical engineering laboratories at Izmir Institute of Technology. *Atmos. Pollut. Res.* **2015**, *6*, 147–153. [CrossRef]
14. Shi, S.S.; Chen, C.; Zhao, B. Air infiltration rate distributions of residences in Beijing. *Build. Environ.* **2015**, *92*, 528–537. [CrossRef]
15. Fujikawa, M.; Yoshino, H.; Takaki, R.; Okuyama, H.; Hayashi, M.; Sugawara, M. Experimental study of the multi-zonal airflow measurement method using human expiration. *J. Environ. Eng.* **2010**, *75*, 499–508. [CrossRef]
16. Lawrence, T.M.; Braun, J.E. A methodology for estimating occupant CO_2 source generation rates from measurements in small commercial buildings. *Build. Environ.* **2007**, *42*, 623–639. [CrossRef]
17. *ASHRAE Handbook Fundamentals*; American Society of Heating, Air Conditioning and Refrigeration Engineers: Atlanta, GA, USA, 2013.
18. *ANSI/ASHRAE Standard 62.1*; Ventilation for Acceptable Indoor Air Quality. American Society of Heating, Refrigerating and Air-Conditioning Engineers, Inc.: Atlanta GA, USA, 2013.
19. *GB/T 18883*; Indoor Air Quality Standard. Ministry of Environmental Protection of the People's Republic of China: Beijing, China, 2002.

20. *Building Area Index for Regular Institutions of Higher Education*; Ministry of Education and National Development and Reform Commission of the People's Republic of China: Beijing, China, 2018.
21. *ASTM D6245*; Standard Guide for Using Indoor Carbon Dioxide Concentrations to Evaluate Indoor Air Quality and Ventilation. American Society for Testing and Materials: West Conshohocken, PA, USA, 2012.
22. Sarkheil, H.; Rahbari, S. Development of case historical logical air quality indices via fuzzy mathematics (Mamdani and Takagi–Sugeno systems), a case study for Shahre Rey Town. *Environ. Earth Sci.* **2016**, *75*, 1319. [CrossRef]
23. Dionovaa, B.W.; Mohammed, M.N.; Al-Zubaidi, S.; Yusuf, E. Environment indoor air quality assessment using fuzzy inference system. *ICT Express* **2020**, *6*, 185–194. [CrossRef]

MDPI

St. Alban-Anlage 66

4052 Basel

Switzerland

www.mdpi.com

Processes Editorial Office

E-mail: processes@mdpi.com

www.mdpi.com/journal/processes

www.ingramcontent.com/pod-product-compliance
Lightning Source LLC
LaVergne TN
LVHW070144170726
843515LV00007B/1859